Maker Innovations Series

Jump start your path to discovery with the Apress Maker Innovations series! From the basics of electricity and components through to the most advanced options in robotics and Machine Learning, you'll forge a path to building ingenious hardware and controlling it with cutting-edge software. All while gaining new skills and experience with common toolsets you can take to new projects or even into a whole new career.

The Apress Maker Innovations series offers projects-based learning, while keeping theory and best processes front and center. So you get hands-on experience while also learning the terms of the trade and how entrepreneurs, inventors, and engineers think through creating and executing hardware projects. You can learn to design circuits, program AI, create IoT systems for your home or even city, and so much more!

Whether you're a beginning hobbyist or a seasoned entrepreneur working out of your basement or garage, you'll scale up your skillset to become a hardware design and engineering pro. And often using low-cost and open-source software such as the Raspberry Pi, Arduino, PIC microcontroller, and Robot Operating System (ROS). Programmers and software engineers have great opportunities to learn, too, as many projects and control environments are based in popular languages and operating systems, such as Python and Linux.

If you want to build a robot, set up a smart home, tackle assembling a weather-ready meteorology system, or create a brand-new circuit using breadboards and circuit design software, this series has all that and more! Written by creative and seasoned Makers, every book in the series tackles both tested and leading-edge approaches and technologies for bringing your visions and projects to life.

More information about this series at https://link.springer.com/bookseries/17311

Industrial Robotics Control

Mathematical Models, Software Architecture, and Electronics Design

Fabrizio Frigeni

Apress®

Industrial Robotics Control: Mathematical Models, Software Architecture, and Electronics Design

Fabrizio Frigeni
Shanghai, China

ISBN-13 (pbk): 978-1-4842-8988-4   ISBN-13 (electronic): 978-1-4842-8989-1
https://doi.org/10.1007/978-1-4842-8989-1

Managing Director, Apress Media LLC: Welmoed Spahr
Acquisitions Editor: Susan McDermott
Development Editor: James Markham
Coordinating Editor: Jessica Vakili

Distributed to the book trade worldwide by Springer Science+Business Media New York, 1 NY Plaza, New York, NY 10004. Phone 1-800-SPRINGER, fax (201) 348-4505, e-mail orders-ny@springer-sbm.com, or visit www.springeronline.com. Apress Media, LLC is a California LLC and the sole member (owner) is Springer Science + Business Media Finance Inc (SSBM Finance Inc). SSBM Finance Inc is a **Delaware** corporation.

For information on translations, please e-mail booktranslations@springernature.com; for reprint, paperback, or audio rights, please e-mail bookpermissions@springernature.com.

Apress titles may be purchased in bulk for academic, corporate, or promotional use. eBook versions and licenses are also available for most titles. For more information, reference our Print and eBook Bulk Sales web page at http://www.apress.com/bulk-sales.

Any source code or other supplementary material referenced by the author in this book is available to readers on the Github repository: https://github.com/Apress/Industrial-Robotics-Control. For more detailed information, please visit http://www.apress.com/source-code.

Printed on acid-free paper

A Klara e Alice, con tanto affetto dal vostro papà.

Table of Contents

TABLE OF CONTENTS

About the Author

Fabrizio Frigeni is a multicultural engineer, hobbyist, teacher, and entrepreneur. Hailing from Italy, he obtained a degree in electrical engineering from the Polytechnic University of Milan; then worked on laser physics for his master's degree at the University of Illinois at Chicago, USA; and later completed a doctoral degree in microelectronics at the Technical University of Dresden, Germany.

He has gained extensive experience in the fields of automation, robotics, computer vision, and machine learning while working for several years on assignments in Austria, Brazil, India, and China. He has been based in Shanghai since 2009, where he has first managed local engineering teams, and then started his own company to teach, consult, and design control systems for automated machines and robots.

About the Technical Reviewer

Zbynek Uher is a Czech developer, graduated in mathematics and computer science. Since 2005, his professional career is connected with industrial robot software development in Austria and the Czech Republic. His areas of expertise include robotic transformations, robotics control system, and its software architecture.

Preface

Robotics is a very exciting field to be working in, and there is no better way for me to share that excitement with you than collecting and presenting all the knowledge and experience I have gained through many years of hard work with several kinds of robots and all sorts of automated machines. Also, I believe there is no better way of understanding how a device works than actually building it entirely from scratch. Therefore, the goal of this book is to provide a general overview of how a robotics control system works and how to design and build one by yourself.

Modern industrial manufacturing plans are highly automated, and a large amount of work is performed by robotic arms. They are efficient, accurate, and fast. They are often employed in tasks that are too repetitive or dangerous for humans. They increase the productivity and reliability of a production chain. However, their control is far from being easy. They require a good deal of understanding of mathematics, software algorithms, and electronic hardware.

You will be guided through the details of the robots' geometrical models, their movements in space, their driving algorithms, and their electronic control circuits. By the end of the book, you should be able to build a fully functional control system capable of driving any industrial robot.

The original version of this book was revised. A correction to this book is available at https://doi.org/10.1007/978-1-4842-8989-1_20

The author working on a large welding robot back in 2011

Who Is This Book For?

This book is thought for students, hobbyists, and engineers who would like to understand the theory behind a robotics control system and then put that knowledge into practice to build their own working solution.

It is not meant to be an academic textbook, and the theory is presented in a rather informal way to make it a bit lighter and friendlier. Nevertheless, university students can still find useful information to complement their more formal academic volumes.

Practicing engineers can find the solutions to several kinematic models that are rarely included in detail in robotics textbooks. Also, they will find plenty of practical suggestions on problems that are less of theoretical interest but very often faced in daily robotics activities (workspace monitoring, calibration, servo loop tuning). I tried to bring in as much information as possible from the practical side of things: tips and observations from several years of hands-on work in the industry.

Hobbyists will probably focus more on the second half of the book, where I describe how to build a complete control system for robots: from the software architecture to the electronic hardware design. Here too, I show details from real-life examples and products I built during my engineering career.

Anyone interested in technology with a basic engineering background will be able to follow along. There is indeed quite a bit of mathematics to go through, which is essential for understanding how a complex robotic arm moves in space. But as long as you are familiar with basic trigonometry and linear algebra, you will be fine. Many formulas are presented directly without formal derivation to avoid lengthy pages filled with equations.

There are also different control algorithms to digest and general software architecture to think over. However, no specific programming skill is required. I will leave the actual code implementation for you to work on, according to your background and preferences.

In general, I liked to focus on the reasoning behind certain functions more than on their direct implementation. The book is meant as a source of inspiration for creative people, not as a source of scripts to copy-paste in your system.

As for the hardware implementation, some electronics knowledge is required to put together a working circuit board. However, you could also skip that part and use third-party ready-made components. Alternatively, you could even work with a simulation environment to test out your software without using a real robot at all.

Structure of the Book

The core of the book can be split into two distinct sections: the first half (Parts I and II) presents the mathematical models needed to describe robots' structures and their movements in space; the second half (Parts III and IV) presents the software and hardware components needed to build a complete control system for the robots.

The first chapter provides an introduction to industrial robots: what different kinds there are and what applications they are used for. We also go through the various nomenclatures commonly used in the industry to make sure we all understand what we are talking about. Then we introduce a generic structure of a robotics control system to show what components you will study over the rest of the book. Finally, we briefly present a simulation environment you could use to test your controller while you build it, in case you do not have access to a real robot. Most of the images in this book are captured from that simulation environment.

Parts I and II (Robot Geometry and Robot Movements) provide the theoretical foundations for robot control. These two sections are a bit heavy on the math, but if your goal is to write a fully functional control software, you need to be patient and work through the details. We first define a generic geometrical framework for robotic arms (Chapter 2) and use it to solve the kinematic chain of a standard six-axes industrial manipulator: the forward kinematics in Chapter 3 and the inverse kinematics in Chapter 4. In other words, we learn how to define the position of a robot in space. The next step is to learn how to move the robot between different positions: we talk about path-planning (Chapter 5); we make sure the path is valid and safe using workspace monitoring functions (Chapter 6); we then describe the motion equations to generate and execute the trajectory (Chapter 7). We also give a brief overview of some advanced control techniques using static and dynamic models (Chapter 8).

Part III (Robot Software) analyzes the functions and characteristics of a typical control software package for robots. We start with the core firmware (Chapter 9) showing its internal structure and how to make it accessible via a simple interface. We describe in detail the motion control and motor commutation algorithms typically found in industrial servo drives. We also present common procedures, tips, and guidelines to calibrate the robot according to its specific application (Chapter 10) and to commission it safely to the customer's site (Chapter 11). We then show how to create a virtual world using the game engine Unity (Chapter 12), so that you can directly test all the features of your code and receive immediate visual feedback of the resulting actions performed by the robot. This approach is useful in case you have no access to a real industrial robot or also to test your code in simulation before running it on a real machine to avoid costly and dangerous surprises. Finally, we also describe the basics of machine vision algorithms (Chapter 13), in order to augment the functionalities and range of applications of your robot.

Part IV (Robot Hardware) is all about the hardware you need to build a complete robotic arm. We start with the actuators, by describing different kinds of electric motors (Chapter 14) and positional encoders (Chapter 15). Then we delve into the electronic components and learn how to design our own inverter drives (Chapter 16), power management systems (Chapter 17), and logic controller (Chapter 18). Lastly, we take a quick look at the actual fabrication steps (Chapter 19), both for PCB implementations and for mechanical parts.

An appendix at the end of the book provides the solutions for kinematic chains other than the standard six-axes manipulator. We show how to solve a COBOT, a SCARA arm, a Palletizer manipulator, and a Delta (Tripod) parallel kinematic.

CHAPTER 1

Industrial Robots

The word robot is very general and normally identifies all sorts of automatically moving machines. This book focuses on robotic arms, also called manipulators, typically employed in factories for manufacturing tasks (see example shown in Figure 1-1).

Figure 1-1. *Industrial robots at work in a production line (with permission from EFORT)*

These industrial robots come in many shapes and sizes, depending on the applications they are going to work on. The smallest robots can only lift loads of about a kilogram or two. The largest ones can easily move loads exceeding a couple of tons.

© Fabrizio Frigeni 2023
F. Frigeni, *Industrial Robotics Control*, Maker Innovations Series,
https://doi.org/10.1007/978-1-4842-8989-1_1

Their mechanical structures are usually very rigid in order to guarantee high accuracy and simplify the control algorithms, especially at high speed. Their movements are fast and precise, but they usually lack any intelligence. In fact, the majority of their applications are focused on simple, repetitive tasks: for instance, cutting, welding, and painting metal pieces in the automotive industry; packaging or palletizing in food or medical industries; and assembling in consumer electronics production lines.

Adding sensors and cameras to the robots makes them able to solve increasingly complex tasks and makes them more aware of their surroundings. The additional information is input in smarter algorithms and allows the control system to adapt its behavior to different situations. Typical examples are robots autonomously sorting packages in warehouses or collaborative robots (cobots) able to detect and soften collisions with human operators.

Nomenclature

Before we start learning more details about the world of industrial robots, we need to introduce a few technical terms to make sure we all understand what their parts and features are called.

The mechanical structure of a robotic arm is essentially a chain of **links**. These links are connected together with **joints**. The joints are usually actuated by motors, so that the links can move (see Figure 1-2).

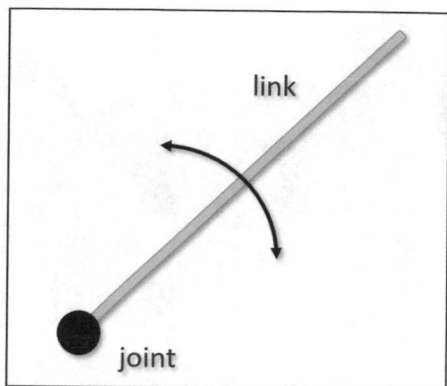

Figure 1-2. *Rotational joint and link*

The field of physics that describes the position of linked objects in space is called **kinematics**. This description only considers the geometrical position of the links and does not account for the forces acting on them. In Part 1 of this book, we will study how to write and solve the kinematic equations for a robotic arm.

Using a kinematic model is the most common approach of controlling a robot in the industry, by using position-based commands. However, in some cases, taking into account the forces acting on the individual links is also required: either to improve the quality of the motion control or to solve force control applications. For example, screw tightening and bottle capping are performed with force commands, not position control. The field of physics that describes the movement of objects under the action of forces is called **dynamics**. We will briefly introduce the dynamic model of industrial robots in Chapter 8.

We mentioned earlier that robotic arms can be built with different mechanical configurations. We can now make a clear distinction based on the form of their kinematic chain: serial and parallel kinematics (see Figure 1-3).

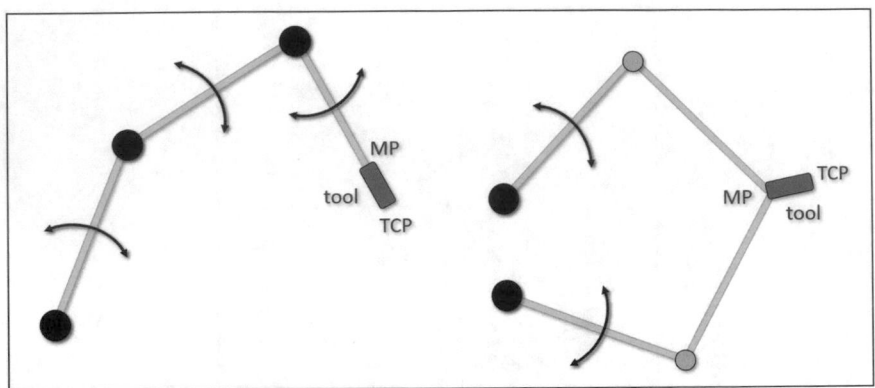

Figure 1-3. *Serial (left) vs. parallel (right) kinematics*

Serial kinematics consist of an open chain of links, similar to a human arm: from the shoulder, to the elbow, down to the wrist and fingers.

Parallel kinematics, on the other hand, are composed of links positioned parallel to each other, somewhat similar to a group of legs or to a tripod.

No matter what kind of kinematic chain we are dealing with, we call the point at the end of the structure the mounting point (**MP**) that is where a tool is normally mounted to perform certain tasks (a gripper, a drill, a welding gun). The point at the very end of the tool is called the tool center point (**TCP**). That is a fundamentally important definition because the TCP is the point where the robot interacts with the environment. *Learning how to control the position, orientation, speed, and force of the TCP is the ultimate goal of this entire book.*

The range of all the possible positions in space that can be reached by the TCP is called the **workspace** of the robot. Serial kinematics normally offer larger workspaces when compared to parallel kinematics, which means they can reach objects farther away from their base. However, larger workspaces also increase the risk of collisions with other robots or other objects in the surrounding environment. Workspace monitoring and **collision detection** are important topics that will be covered in Chapter 6.

The TCP, even though called point, is defined not only by its position but also by its orientation, naturally derived from the tool body. As an analogy, think of how you can reach for an object with your hand from different angles. The target position is the same, but the orientation is not. In the field of robotics, the combination of position and orientation of a target point is often called a **pose**.

The ability to position and orient the TCP along and around different axes is measured by the number of **degrees of freedom** (DoFs) of the robot.

A cart travelling on a linear rail can only position its load with one degree of freedom: the position is measured with a single scalar value along the only axis of motion. However, a generic object in space can be placed with up to six degrees of freedom (see Figure 1-4): three degrees of positions (*along* the X-Y-Z axes) and three degrees of orientation (*around* the X-Y-Z axes).

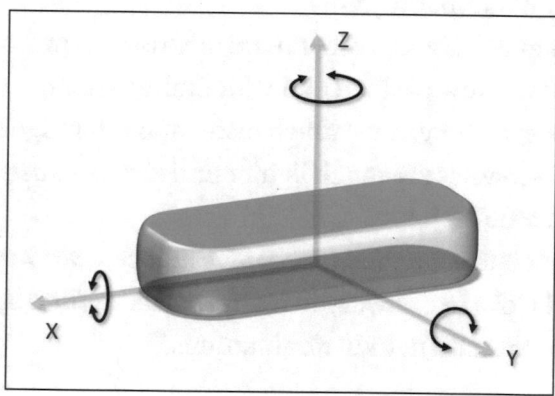

Figure 1-4. *A generic object in space has six DoFs*

In practice, not all industrial robots offer six DoFs because the applications they are designed for do not require them. Think, for example, of a typical use case for industrial robots: picking cookies off a running conveyor and placing them in a box for market distribution. If the cookies are round, then they can fit in the box at any angle. The robot is required to reach the correct target position in space, but the orientation of the cookie is irrelevant. In this case, three degrees of freedom are enough. However, if the cookies have a specific asymmetric shape, say rectangular, then they need to be oriented at a specific angle with respect to the box in order to fit inside. In that case, an additional degree of freedom is required: a rotational movement around the vertical axis.

Adding degrees of freedom to a robot means adding links, motors, weight, and control complexity to the system. You could ideally build a robot with as many joints as you wanted, but it would get harder and harder to control. Keeping the mechanics light and small usually results in faster and more accurate movements.

Serial kinematics often offer more flexibility, thanks to their high number of joints, but they lack speed and accuracy. That is because each link adds some weight and some errors to the chain, so that the total error builds up with the number of joints.

Parallel kinematics, on the other hand, are used in industrial settings for applications with low payload and with limited number of degrees of freedom. However, they move much faster and offer higher accuracy, because their mechanics are much lighter and also because errors do not add up between parallel joints.

There are, of course, parallel mechanics that can carry very high loads and have six full DoFs (e.g., hexapod platforms for motion simulators), but these are not as common in industrial settings.

Mechanical Configurations

Now that we know how to differentiate robotic structures, we can look at some examples of kinematical configurations typically found on the market.

SCARA: The name stands for *Selective Compliance Assembly Robot Arm*. It is a serial kinematic that comes in two, three, or four DoFs versions, with four being the most common (see Figure 1-5). It is a compact, cheap, and fast robot, with a low payload and a small workspace. It finds application mostly in pick-and-place or assembly tasks in the electronics manufacturing lines.

This is by far the easiest and cheapest robot to manufacture, and I strongly suggest starting from here if you want to build your own working robot at home.

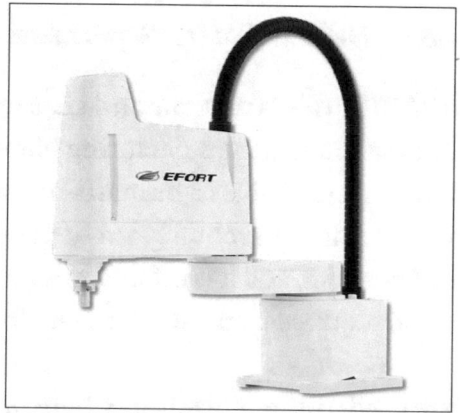

Figure 1-5. *A four-axes SCARA Robot (with permission from EFORT)*

DELTA: Also known as tripod robot. It is a parallel kinematic that, just like the SCARA, is sold in two, three, or four DoFs versions, with four-axes being the most common configuration (see Figure 1-6). In the three-axes and four-axes versions, its parallel arms are positioned at the vertex of an equilateral triangle, hence the name delta.

It is a lightweight robot that supports small payloads but can reach incredible speeds. It is typically used to speed up production lines where the size of the product is small: food, electronics, and medical industry.

Figure 1-6. *A four-axes Delta Robot (with permission from CODIAN)*

ANTHROPOMORPHIC: This is the standard six-axes robot and is by far the most common robot in manufacturing plans. It is called anthropomorphic because it resembles a human arm (see Figure 1-7), but most people refer to it simply as robotic arm or manipulator. It has six axes and six degrees of freedom, and it is offered in a large range of sizes according to the payload required by the application: from a few kilograms up to a full ton.

Six-axes robots are used in all applications where high dexterity is needed, the largest field being automotive industry: cutting, welding, polishing, painting, etc.

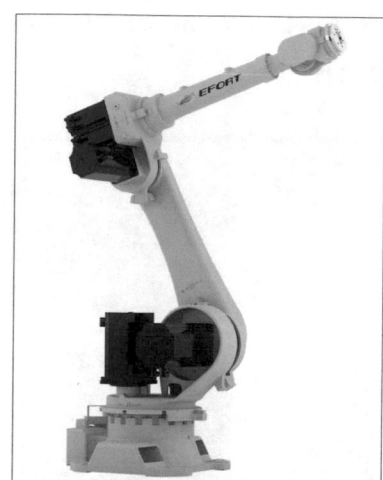

Figure 1-7. *A six-axes Anthropomorphic Robot (with permission from EFORT)*

This is the most generic type of robot, and we will study it intensively through the book. However, most of the concepts are general enough and can be quickly applied to all other kinds of robots without modifications. The main exception is the solution of the kinematic model, which is unique to each robot's structure. The equations and solutions for this robot are presented in detail in Chapter 3 and Chapter 4, while all the other models are solved in the Appendix at the end of the book.

PALLETIZER: This kind of robot looks similar to the anthropomorphic version, but it only has four axes and therefore two fewer degrees of freedom (see Figure 1-8). As the name suggests, it is used in palletizing applications, loading and unloading large and heavy objects, usually by means of a mechanical gripper or a vacuum suction system.

The robot picks the objects from the top and places them on a flat surface. This operation requires only four DoFs: three for the target position and one for the orientation angle around the vertical axis. No orientation around the two horizontal axes is possible nor needed. Giving up two degrees of freedom results in the advantage of removing two motors and a lot of complex mechanics, thereby saving costs and weight.

Palletizers are usually the largest industrial robots, with the furthest reach and the heaviest payload.

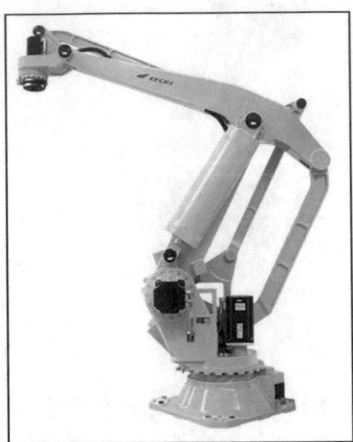

Figure 1-8. *A four-axes Palletizer Robot (with permission from EFORT)*

COBOT: This is a six-axes lightweight robot (see Figure 1-9), where lightweight means that its mechanical structure and its control system are engineered to make collisions soft. The main purpose of such a design is that these robots can safely work alongside human operators without causing injuries, unlike all other kinds of industrial robots, which must be protected and isolated behind metal cages. That is where the name cobot comes from (short for *collaborative robot*).

They find application in several fields where they can directly assist humans, instead of completely replacing them: assembly lines, domestic environments, and medical settings.

Figure 1-9. *A six-axes cobot (with permission from JAKA)*

Another reason we put this robot in a different class is that its kinematic chain is slightly different than the standard six-axes manipulator. The solution of its kinematic model is presented in the Appendix.

Structure of a Robot Control System

After learning what industrial robots look like, we need to understand how we can make them move to perform different useful tasks. In this section we analyze the general structure of the control system (shown in Figure 1-10), which by the end of the book you should be able to design and build.

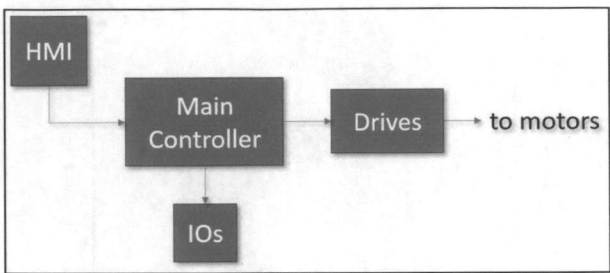

Figure 1-10. *General structure of a robot's control system*

A robot, just like any other automated machine, needs a brain to work. The brain, technically called the **main controller**, performs three fundamental tasks:

1. It reads information from user input, external sensors, motor encoders, cameras, etc.

2. It internally processes that information based on specific rules and algorithms.

3. It generates output signals to control the actuators: motors, pumps, lights, etc.

The controller is in practice a processor of some kind (CPU, MCU, FPGA) with various interfaces to the input and output signals. In our specific case, we will work with a microcontroller unit (MCU) because of its low cost, simplicity to program, and large number of readily available interfaces.

It is our task to program the controller with functions to plan and execute robot's movements. The quality and complexity of our algorithms will be reflected in the performance of the robot. It is often said that a machine will only be as smart as its programmer. For instance, the robot controller will not think about safety issues unless you specifically programmed it to do so.

There are also situations where artificial intelligence techniques can be used to improve existing algorithms and outperform human performances. Machine vision analysis and motion control optimization are typical applications.

Once the controller is programmed, it needs to be interfaced to the external world in order to receive and execute commands. The first basic connection is with a human operator and is technically called **HMI** (human-machine interface). That is what the operator will use to access the robot, make it move, check its status, and change some of its configuration parameters. A critical property of this interface is that it has to be simple and intuitive to use: the operator is not a programmer and is not required to understand the internal details of the algorithms controlling the robot. An important selling point of any technology is its ability of being user-friendly.

The second connection that the controller needs with the external world is a series of input and output signal lines (**IOs**). These signals are used for a multitude of reasons and are often synchronized with the internal motion calculations: communication and coordination with other robots or machines, monitoring of safety areas, control of application-specific tools (e.g., a welding gun), feedback from a sensor or a camera to signal the presence of a new workpiece, turning on a warning light or alarm siren, and many others.

Finally, the absolutely essential interface is that between the controller and the motion actuators. The robot can physically move because of the action of its motors, and the hardware components responsible for the motor control are called **motor drives**. The controller cyclically sends commands to the drives to tell them how and where to position the motors, so that the entire robot will accurately follow the correct path and reach the final position as desired by the operator.

In industrial robots, the motor drives are usually separate hardware components with their own processor units and communicate with the robot's main controller via high-speed data lines known as **fieldbuses**. Alternatively, in smaller and simpler robots, the motor driving function can be directly incorporated in the main controller to save space and costs.

Combining the controller and the drives into a single product results in what is known as *smart motion controller* and represents a very efficient solution for low-power applications. We will present this approach when designing our robot control system (see Section on Integrated Solution in Chapter 18). However, from a conceptual point of view, you should still think of the two components as separate entities, each responsible for its own task: the main controller is responsible for the overall control of the robot, whereas each motor drive is only responsible for the control of its own mechanical axis and knows nothing about the rest of the robot.

In this book, we describe how to design both the software and the hardware for the main robot controller and the motor drives. Technically, once you learn how to build your own motor drive, you can use it in wide range of motion applications: from industrial robots and machines to household appliances and electric vehicles.

Digital Twin

The process of building a product as complex as a robot controller requires a lot of testing in order to find mistakes and correct them. However, testing a new software function directly on a real robot can lead to unpredictable movements and dangerous outcomes.

Simulation environments are always used by robotics engineers to show the action of a robot in response to a certain function and make sure that the behavior is correct before carrying on the same test on the actual physical machine.

A three-dimensional model that represents our robot and responds according to motion commands in the same way the real robot does is called a **digital twin**. Different simulation options exist: from basic kinematic modeling to more advanced workspace monitoring and dynamic modeling functionalities.

In the course of this book, we will use a simple kinematic model of a six-axes anthropomorphic robot (see Figure 1-11) to show the movements of the robot in response to our calculations.

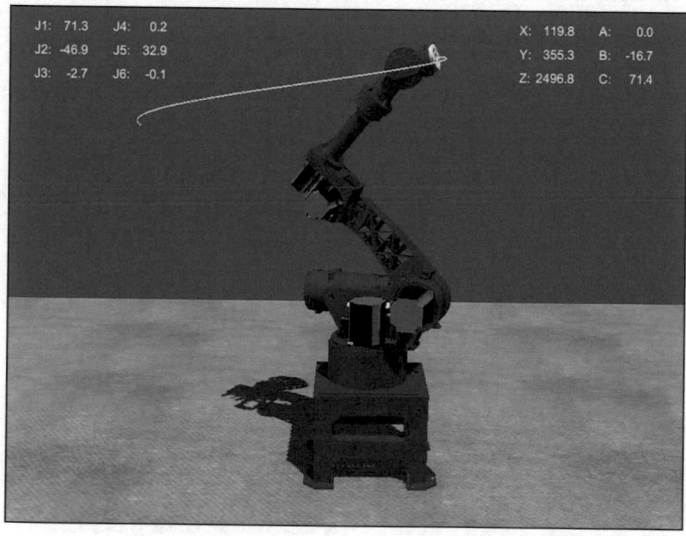

Figure 1-11. *The digital twin we will use in the rest of this book*

The digital twin approach is not limited to the simulation of an individual robot but can be easily extended to an entire production line. Besides testing new functions, a typical and useful application is to test different configuration of production parameters in order to optimize a manufacturing process. The savings in terms of time and costs can be significant.

Normally, software packages like MapleSim and Matlab/Simulink are standard choices when the need comes to simulate algorithms in control engineering. However, when simulating the movement of a robot, we are not only interested in results of numeric calculations. We want to do more:

- Visualize the robot moving in space and interacting with the external environment.

- Observe the robot picking objects off a conveyor and placing them in a box in the right position and with the correct orientation.

- Make sure that several welding robots working on a same metal structure do not collide with each other.

- Train a reinforcement learning algorithm to optimize motion parameters based on a large number of random situations.

- Take our potential customers on a virtual tour of a manufacturing floor to show them the control capabilities of our software libraries.

- Remotely connect from our office to the real robot located in a factory across the country, and perform monitoring or maintenance activities.

These and many other requirements necessitate a realistic environment with which the controller can interact. Two examples of software specifically designed for that purpose are Gazebo and industrialPhysics.

Another creative approach consists of using a generic game engine like Unity and take advantage of its powerful physics engine and machine learning toolkit to create a virtual world in which you can place models of your robots and let them move, interact, and crash, while you test and optimize your motion control algorithms. The possibilities are endless and the actual cost is zero.

More details on how to build a simple simulator in Unity and connect it to the hardware controller can be found in Chapter 12.

Summary

This first chapter served as a basic introduction to the topic of industrial robotics. At this point, you should be able to recognize a typical industrial manipulator and identify its main components: the passive links, the actuated joints, the TCP, and its control hardware. You should also be able to tell apart a serial kinematics from a parallel one and have a decent idea of what workspace the robot is going to cover.

We are now ready to start modeling the robot in a more formal mathematical way, so that we can later write software code to accurately control its movements.

PART I

Robot Geometry

In order to control a robot, we need to understand its geometry. This part of the book introduces a geometrical framework that can be used to solve all kinds of kinematic chains. We will apply it directly to derive the forward and inverse kinematics of a six-axes robot. The details are a bit heavy on the math but necessary.

CHAPTER 2

Geometrical Framework

We present here the concept of frames and their operations. Frames are the geometrical foundation of robotic arms, and understanding how to handle them is an important step toward solving kinematic models. The framework we present has general validity, regardless of the robotic structure we need to solve.

Reference Frames

A **frame** is essentially a coordinate system with a specific position and orientation (see Figure 2-1). When describing the position of a generic point in space, we always need to do that relative to a specific frame. In other words, a point has different coordinates according to what frame we are using as reference.

© Fabrizio Frigeni 2023
F. Frigeni, *Industrial Robotics Control*, Maker Innovations Series,
https://doi.org/10.1007/978-1-4842-8989-1_2

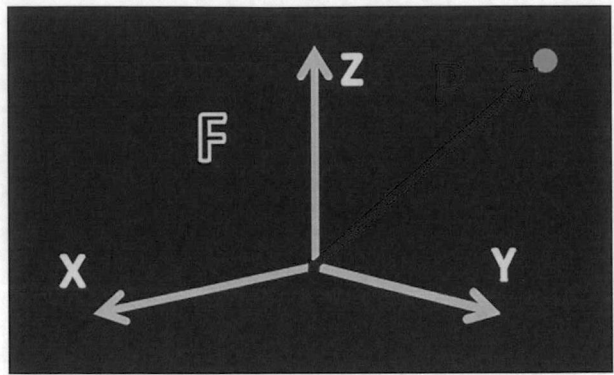

Figure 2-1. *Generic frame F and point P*

When working with robots, we can immediately identify a few fundamental reference frames.

The **global coordinate system** (or world coordinate system) defines the origin and orientation of our world. When your space includes several machines and robots, it is always important to define a unique global reference system, so that everyone understands the same global coordinates.

Each robot or machine will then have its own **local coordinate system** (or machine coordinate system). That is convenient when programming individual robots: other machines will not see this system and will not understand coordinates specified in this frame. Of course, if you work with only one robot, then you can let the global and local coordinate systems coincide.

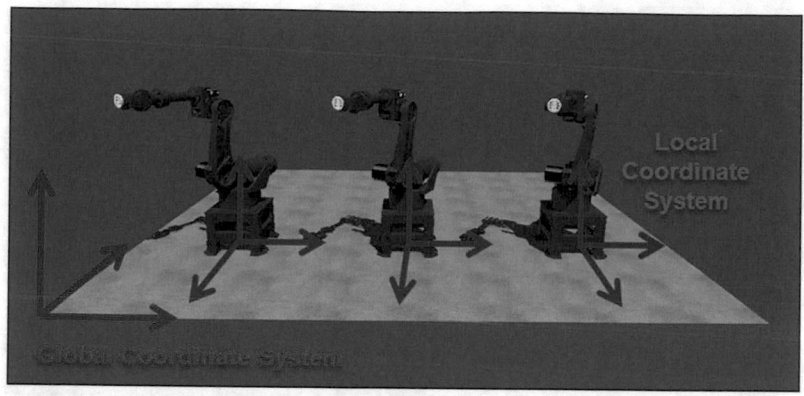

Figure 2-2. *Global (red) vs. local (purple) reference frames*

For example, the three robots in Figure 2-2 are all placed at the origin $O = [0 \ 0 \ 0]$ of their local (purple) reference frames. But their positions in the global (red) frame are $O_1 = \begin{bmatrix} 2000 & 2000 & 0 \end{bmatrix}$, $O_2 = \begin{bmatrix} 4000 & 2000 & 0 \end{bmatrix}$, and $O_3 = \begin{bmatrix} 6000 & 2000 & 0 \end{bmatrix}$, meaning that they are all spaced two meters apart from each other along the global X axis. We will explain later how to move from one frame to another.

The next important frame to consider is the **tool coordinate system**, whose origin is located at the TCP of the robot (shown in yellow in Figure 2-3). In this case, we have no tool attached to the robot, so the tool coordinate system is based at the mounting point on the sixth axis.

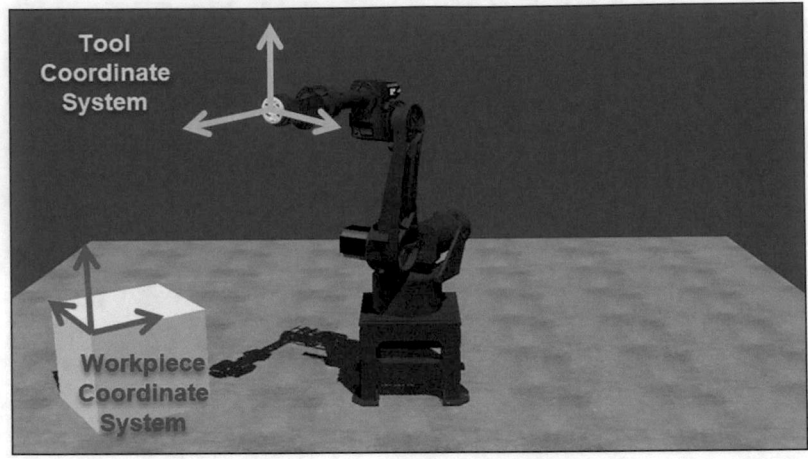

Figure 2-3. *Tool (yellow) and workpiece (blue) reference frames*

Finally, the most meaningful frame for the operator is the **workpiece coordinate system** (shown in blue in Figure 2-3). This frame is what the robot's operator sees and considers as the origin of its workspace: all the programmed points and movements refer to this system. It is often also called product coordinate system, and there are actually several of them in a normal working cell, because the robot performs tasks on a multitude of products, each of them having its own unique origin and orientation.

We now need to understand how to move from one frame to another one and how to transform coordinate points across different frames.

Frame Operations

Imagine we have two different observers, one based in frame F_1, the other based in frame F_2, and they are both looking at the same point in space. The first observer will measure the point with coordinates P_1 and the second with coordinates P_2 (see Figure 2-4).

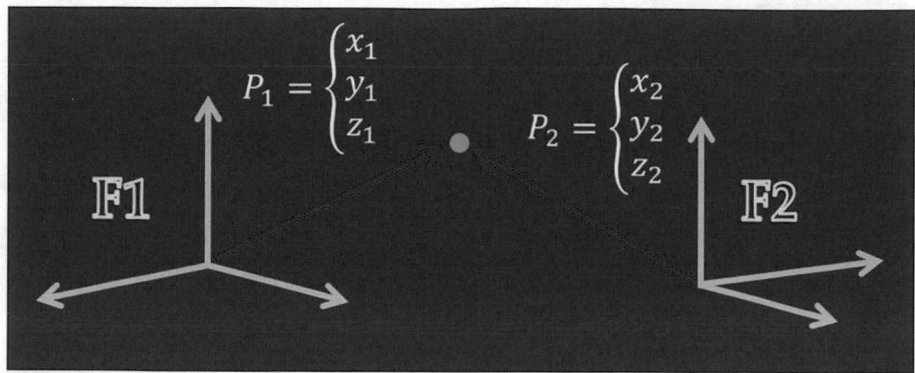

Figure 2-4. *The same point seen from different frames*

In the previous section, we learned that robotic applications express their target positions in several different coordinate systems. Therefore, a typical requirement is how do we find the position of that point from the perspective of frame F_2 if we already know its position in frame F_1?

In order to find a way to convert from P_1 to P_2, we need to know how the two original frames F_1 and F_2 are related to each other. This is where frame operations come into play.

We can translate a frame and rotate it, and we can also rotate and translate at the same time, as shown in Figure 2-5.

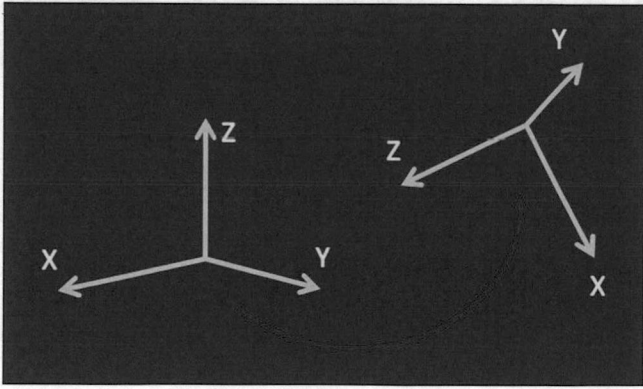

Figure 2-5. *Generic roto-translation of a frame*

Frame Translations

Let's start with translations, considering the example in Figure 2-6. We are based in frame F_1 and observe the blue point. Its position from this perspective is $P_1 = \begin{bmatrix} x_1 & y_1 & z_1 \end{bmatrix}$. Let's use some numeric values to make the understanding easier. Assume $P_1 = \begin{bmatrix} 1 & 1 & 0 \end{bmatrix}$.

Then we move our base to a new frame F_2, which is translated by an **offset** $\Delta = \begin{bmatrix} \delta_X & \delta_Y & \delta_Z \end{bmatrix}$. The same blue point seen from this perspective has a different position, which we call $P_2 = \begin{bmatrix} x_2 & y_2 & z_2 \end{bmatrix}$. Make sure you understand the meaning of this operation here: the physical point in space (the blue point) is absolutely the same, regardless of the observation frame, and has not moved. However, our observation base has moved, so the point's coordinates in our new frame (P_2) are different than those in the old frame (P_1).

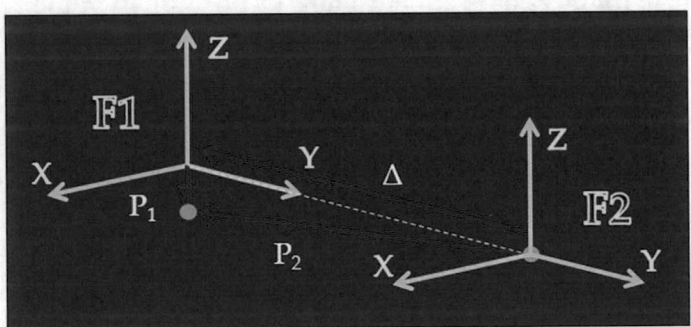

Figure 2-6. *Frame translation*

Translation is a linear operation and the formula is very simple:

$P_2 = \Delta + P_1$.

Keep in mind that we are now basing our observation point in frame F_2, so *the value of the offset Δ is also measured from the F_2 frame's perspective*. In other words, the offset Δ expresses how the old frame F_1 is seen from the new frame F_2. To continue our numerical example, let's use $\Delta = \begin{bmatrix} 0 & -10 & 0 \end{bmatrix}$, i.e., a negative offset along the Y axis.

Intuitively, you can think as starting from the origin of F_2, first moving back to the origin of F_1, and then from there to P_1.

Breaking up for each coordinate gives the system of linear equations:

$$\begin{cases} x_2 = x_1 + \delta_X \\ y_2 = y_1 + \delta_Y \\ z_2 = z_1 + \delta_Z \end{cases} \tag{2-1}$$

Numerically, if the original point was $P_1 = \begin{bmatrix} 1 & 1 & 0 \end{bmatrix}$ and the translation offset is $\Delta = \begin{bmatrix} 0 & -10 & 0 \end{bmatrix}$, then we can calculate the new point coordinates as $P_2 = \begin{bmatrix} 1 & -9 & 0 \end{bmatrix}$, which can be easily verified in Figure 2-6.

Frame Rotations

Rotations are a bit more complicated than translations: they are not linear operations.

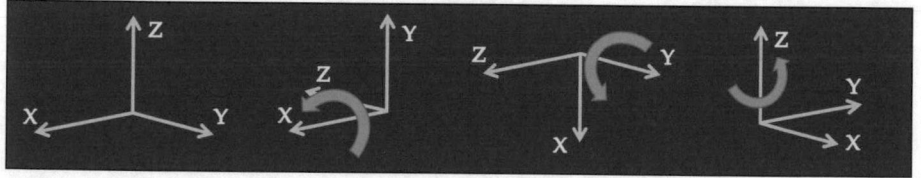

Figure 2-7. *Rotating the original frame around its X, Y, and Z axes*

In order to calculate the coordinates of a point after a rotation of the observation frame, we need to pre-multiply the original coordinates by a so-called rotation matrix R, which is built according to the axis around which the frame rotates, and the angle of rotation θ:

$$P_2 = R(\theta)P_1 \tag{2-2}$$

Let's start with some simple examples shown in Figure 2-7. If we rotate around the X axis by an angle θ, the rotation matrix assumes the following form:

$$R_X(\theta) = \begin{bmatrix} 1 & 0 & 0 \\ 0 & \cos\theta & -\sin\theta \\ 0 & \sin\theta & \cos\theta \end{bmatrix} \qquad (2\text{-}3)$$

Note that we have cosines on the diagonal and sines outside of it. That is not a coincidence, because the rotation matrix reduces to the identity matrix when the angle of rotation is zero: the cosines on the diagonal are equal to 1, and the rest is 0. In other words, no rotation occurs, and clearly the final position is the same as where we started.

Similar matrices can be derived for rotations around the Y and Z axes:

$$R_Y(\theta) = \begin{bmatrix} \cos\theta & 0 & \sin\theta \\ 0 & 1 & 0 \\ -\sin\theta & 0 & \cos\theta \end{bmatrix} \qquad (2\text{-}4)$$

$$R_Z(\theta) = \begin{bmatrix} \cos\theta & -\sin\theta & 0 \\ \sin\theta & \cos\theta & 0 \\ 0 & 0 & 1 \end{bmatrix} \qquad (2\text{-}5)$$

In all cases, a rotation of zero degrees is equivalent to the identity matrix:

$$R(\theta = 0) = \begin{bmatrix} 1 & 0 & 0 \\ 0 & 1 & 0 \\ 0 & 0 & 1 \end{bmatrix} \equiv I \qquad (2\text{-}6)$$

Another property we can immediately notice is that by transposing a rotation matrix, we find the same matrix but with an opposite rotation angle. Geometrically, that is equivalent to a rotation in the opposite direction.

$$R(-\theta) \equiv R^T \tag{2-7}$$

The direction of the rotation angle can be positive or negative. In this book, we follow the **right-hand rule**: if we align our right thumb along the axis of rotation, the movement of the other fingers closing on the hand's palm shows the positive direction of rotation (see Figure 2-8).

Figure 2-8. *We use the right-hand rule to define a positive rotation angle*

Let's look at the numerical example shown in Figure 2-9 to clarify. We start with a point P_1 with coordinates $P_1 = \begin{bmatrix} 1 & 1 & 0 \end{bmatrix}$ in frame F_1. We then rotate the frame around the vertical Z axis by an angle of 90 degrees.

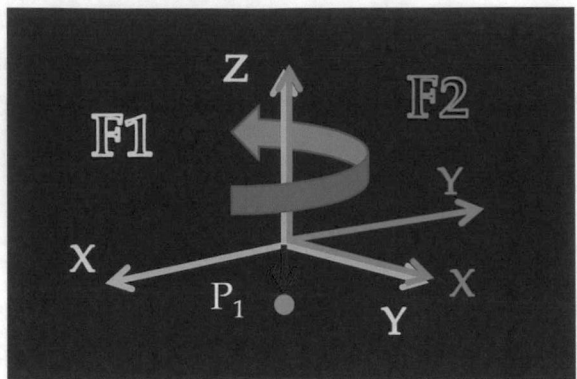

Figure 2-9. *Frame rotation of 90 degrees around Z from F1 to F2*

Note that the rotation angle expresses how the old frame is seen from the new frame. Just like we defined the offset in case of translations, *we express the rotation angle in the new frame coordinates.* Therefore, in this case, we actually have $\theta = -90$.

The rotation matrix for rotations around the Z axis was shown in Equation (2-5). If we plug the -90 degrees ($-\dfrac{\pi}{2}$ radians) rotation angle in that matrix and multiply it by the old position P_1, we find the new P_2:

$$P_2 = R_z\left(-\frac{\pi}{2}\right)P_1 = \begin{bmatrix} \cos-\dfrac{\pi}{2} & -\sin-\dfrac{\pi}{2} & 0 \\ \sin-\dfrac{\pi}{2} & \cos-\dfrac{\pi}{2} & 0 \\ 0 & 0 & 1 \end{bmatrix}\begin{bmatrix} 1 \\ 1 \\ 0 \end{bmatrix} = \begin{bmatrix} 0 & 1 & 0 \\ -1 & 0 & 0 \\ 0 & 0 & 1 \end{bmatrix}\begin{bmatrix} 1 \\ 1 \\ 0 \end{bmatrix} = \begin{bmatrix} 1 \\ -1 \\ 0 \end{bmatrix} \quad (2\text{-}8)$$

If you look at the blue point from the perspective of the green rotated frame in Figure 2-9, you can verify that the calculated values are correct. For example, the position along the Y axis is now negative.

Properties of a Rotation Matrix

A generic rotation matrix has some nice properties that are worth exploring here before we move on with frame operations.

Most and foremost the rotation matrix is an **orthogonal matrix**. We do not need to understand all the details of orthogonal matrices, but we can list here some important results:

- Given an orthogonal matrix R, its transpose is equal to its inverse:

$$R^T = R^{-1} \qquad\qquad (2\text{-}9)$$

    We will see that this property simplifies some of our calculations tremendously, because transposing a matrix is an immediate operation, while inverting it can be a lot of work.

- The determinant of an orthogonal matrix is always ±1.

$$|R| = \pm 1 \qquad\qquad (2\text{-}10)$$

    In particular, a rotation matrix is an orthogonal matrix with determinant equal to +1. This property makes convenient to check whether the matrix we are working with is a correct rotation matrix. If the determinant is not 1, then the matrix is not a rotational matrix and cannot be used to describe rotations. However, the opposite is not true: there are many other matrices with determinant equal to 1 that are also not rotational.

- The product of two rotation matrices is still a rotation matrix. This property is very helpful when dealing with multiple rotations as we will see in the next section. For example, if we want to rotate around the X and Y axes, we simply multiply the two matrices R_X and R_Y to find the matrix for the entire rotation:

$$R_Y R_X = R_{YX} \tag{2-11}$$

- Products between rotation matrices are *associative*, so you can simplify your calculations grouping them as it fits best:

$$R_{ZYX} = R_Z R_Y R_X = \left(R_Z R_Y \right) R_X = R_Z \left(R_Y R_X \right) \tag{2-12}$$

- However, the product is *not commutative*! This is a very important property, and it physically means that rotating around Z first and then X is not the same as rotating around X first and then Z:

$$R_{ZX} \neq R_{XZ} \tag{2-13}$$

The rule is simple: adding a new rotation requires *pre-multiplying* the existing matrix by the new rotation matrix.

Composing Rotations: Euler Angles

So far, we have looked at rotations around a single axis. If we want to describe a more general rotation in space, we need to combine individual rotations into a multi-axis one.

Mathematically, the combination of two consecutive rotations can simply be expressed by the product of the two rotation matrices. In the most general case, a frame can be rotated in space around three axes, so we use the product between three individual matrices.

We pick the base X-Y-Z axes as references and generate a rotation matrix R as the product of the three individual rotations around the X-Y-Z axes by the angles A-B-C, respectively:

$$R = R_Z(C) R_Y(B) R_X(A) \tag{2-14}$$

That means, we first rotate of an angle A around the X axis, then by an angle B around Y, and finally by an angle C around Z. The angles [A, B, C] are called **Euler angles**.

If you think about it, we could have selected a different order or even different combinations of rotation axes: *X-Z-Y*, or *Z-X-Z*, or many others. Those are all different valid ways of expressing a generic rotation in space. They all represent the same rotation but using different values because of the different reference system.

The particular choice we made is probably the simplest to understand, especially for operators in factories who have to program the target poses for the robot: they see the fixed base coordinate system X-Y-Z and understand A-B-C rotations to be around those axes.

The angles A-B-C in the particular notation we chose (around the base *X-Y-Z* axes) are called improper Euler angles, also known as Cardan or RPY angles (from the roll-pitch-yaw angles used in aerospace).

What is important to understand is that no matter what notation we choose, *Euler angles can always represent any rotation in space with only three values*. That is a major advantage in terms of convenience.

Equivalently, given any rotation matrix, we can always decompose it in three Euler angles.

From a computational point of view, composing a generic rotation matrix given three Euler angles is a very straightforward multiplication as we saw in Equation (2-14). However, decomposing the matrix into individual Euler angles is not always so easy. We will see in the next section how to do that.

Decomposing a Rotation Matrix

Imagine we have a rotation matrix, which describes a specific rotation in space. We now want to find out the corresponding Euler angles required to achieve that rotation. This operation is calling decomposition of the matrix and is used often in the control software of a robot.

A typical application is finding the wrist axes of a six-axes robot when solving its inverse kinematic function.

Keep in mind that the result of the decomposition depends on the specific notation of angles we choose. In this paragraph, we continue with the choice of A-B-C Euler angles around the base X-Y-Z axes. The wrist joints of a six-axes robot rotate around a different set of axes (X-Y-X), so the choice of notation will be different in that case. The underlying concept is identical in all cases.

Given our chosen notation, a generic rotation in space is given by the product of the three generic rotation matrices around the base axes. We start with $R_X(A)$ (a rotation of an angle A around the X axis), then premultiply by $R_Y(B)$, and finally by $R_Z(C)$:

$$R = R_Z\left(C\right)R_Y\left(B\right)R_X\left(A\right) = \begin{bmatrix} c_z c_y & c_z s_y s_x - s_z c_x & s_z s_x + c_z s_y c_x \\ s_z c_y & c_z c_x + s_z s_y s_x & s_z s_y c_x - c_z s_x \\ -s_y & c_y s_x & c_y c_x \end{bmatrix} \tag{2-15}$$

The matrix looks complicated, and to simplify the notation, we expressed individual sines and cosines elements with more compact symbols. For example, s_y is the sine of the rotation angle around the Y axis: that is, the sine of B, in our case.

To directly address each element of the matrix, we use the symbol R_{ij}, where i and j are the row and column of the element.

Extracting the A-B-C angles from the rotation matrix works by small individual steps.

First, we look at the element in the third row and first column: $R_{31} = -s_y$. That is, the sine of B is already given, so we can quickly derive B. From a programming point of view, instead of taking the arcsine of that value, it is numerically more stable to use the arctangent function, which works well even for values of B close to zero.

$$B = atan2\left(-R_{31}, \pm\sqrt{1-\left(R_{31}\right)^2}\right) \tag{2-16}$$

Note that we can choose two different values for the cosine of B, either positive or negative. This choice will give two possible results for the final tuple A-B-C. In other words, we can reach the same global rotation with different individual rotations around the base axes. Euler angles are not unique!

Once we know the value of B, we can find the rotation A around the X axis from the elements on the third row of the matrix:

$$A = atan2\left(\pm R_{32}, \pm R_{33}\right) \tag{2-17}$$

Similarly, we can use the elements on the first column to find the angle C of rotation around the Z axis:

$$C = atan2\left(\pm R_{21}, \pm R_{11}\right) \tag{2-18}$$

Notice again how the final values of both A and C are affected by the choice of sign we initially made for B.

You might have noticed already the presence of a particular case: if B is ±90 degrees, its cosine becomes 0 and the matrix simplifies:

$$R = \begin{bmatrix} 0 & \pm c_z s_x - s_z c_x & s_z s_x \pm c_z c_x \\ 0 & c_z c_x \pm s_z s_x & \pm s_z c_x - c_z s_x \\ \mp 1 & 0 & 0 \end{bmatrix} \tag{2-19}$$

This special case actually complicates calculations, because A and C are not linearly independent any more. We are facing a so-called singularity, a particular configuration, for which an infinite number of solutions exist. We can only compute the sum or difference of A and C, but not their individual values.

$$\begin{cases} A - C = atan2\left(R_{12}, R_{22}\right) & \text{for } s_y = 1 \\ A + C = atan2\left(R_{12}, - R_{22}\right) & \text{for } s_y = -1 \end{cases} \qquad (2\text{-}20)$$

The practical way to solve this system is fixing an arbitrary value for one of the two angles (e.g., A = 0) and then derive the other one.

The physical meaning of a singular rotation matrix is that we can select from an infinite number of possible rotations around the individual axes to reach the final desired orientation. The best solution to select is usually determined by the movement's actual conditions, as we will see in Chapter 5 when studying path-planning.

Note that we decomposed a rotation matrix into the A-B-C angles of rotations around the X-Y-Z axes of the fixed base frame. In case you want to decompose the same matrix into other Euler angles notations, you proceed in a very similar way, but use a different initial composition of rotation matrices.

Column Vectors

Understanding the target rotation given a tuple of Euler angles can usually be done intuitively by visualizing the combination of the three individual rotations around the selected reference axes. On the other hand, a rotation matrix looks too abstract to visualize in practice.

That is actually not the case after we present an interesting property here: *the column vectors of a rotation matrix represent the axes of the old frame (before the rotation) as seen from the new frame (after the rotation).*

In order to demonstrate this property, we recall that the rotation between two frames can be expressed by pre-multiplying the old frame by the rotation matrix.

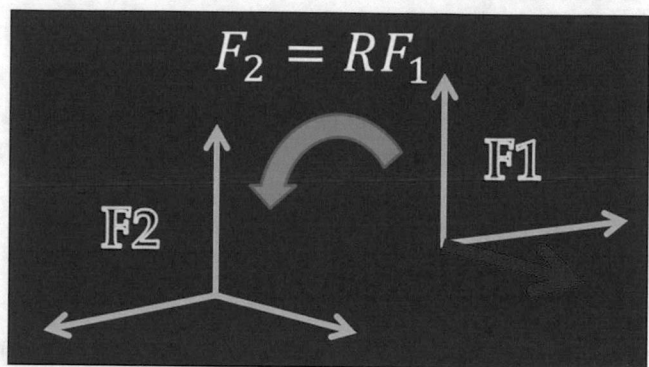

Figure 2-10. *Rotation from the old frame F1 to the new frame F2. The X axis of F1 is highlighted in red*

Let's observe Figure 2-10 as reference. The X axis of the old frame F_1 is highlighted in red. Its coordinates in the F_1 frame are $X = \begin{bmatrix} 1 & 0 & 0 \end{bmatrix}$. Now, if we want to find out the coordinates of that same vector as seen from the new frame F_2, we need to pre-multiply that vector by the rotation matrix R. It is easy to see that the result is exactly the first column vector of the rotation matrix itself.

$$RX = \begin{bmatrix} R_{11} & R_{12} & R_{13} \\ R_{21} & R_{22} & R_{23} \\ R_{31} & R_{32} & R_{33} \end{bmatrix} \begin{bmatrix} 1 \\ 0 \\ 0 \end{bmatrix} = \begin{bmatrix} R_{11} \\ R_{21} \\ R_{31} \end{bmatrix} \tag{2-21}$$

Similarly, we can proceed for the Y and Z axis and find the second and third columns of the matrix.

This property provides a very simple way to build a rotation matrix in practice, if we know the vector axes of a frame with respect to another one. A typical example is when a robot's operator wants to find out what rotation a workpiece has with respect to the base of the robot: that is, the rotation between the workpiece coordinate system and the local coordinate system, as described in the Reference Frames Section. The operator can position the TCP of the robot on two points along each axis of the workpiece in order to let the system identify the axes vectors and from there immediately build the rotation matrix without any further calculation. The details are explained in Cell Calibration Section in Chapter 10 where we talk about calibrating frames in a working cell.

Expressing Rotations

So far, we have seen two alternative ways of expressing a general rotation in space: a 3x3 rotation matrix and a three-tuple of Euler angles. We have also seen how to convert back and forth from one representation to the other.

Both choices have their strengths and weaknesses:

- *Simplicity*: Euler angles are very simple to understand because they can represent a generic rotation with only three angles. Rotation matrices are large and complicated: they need nine elements to represent a single rotation, and they are not at all intuitive. Not to mention that they are computationally much more expensive.

- *Uniqueness*: Euler angles are not uniquely defined as there are 12 different kinds of possible notations. Also, they do not provide unique solutions: many combinations of Euler angles can represent the same rotation in space. In the worst case, an infinite number of solutions are possible, as we have seen when decomposing a rotation matrix around a singularity. All that can be confusing. Rotation matrices, on the other hand, are unique.

- *Interpolation*: The operation of smoothly moving from one starting point to a final target point is called interpolation. We will describe the details in Chapter 5, but we anticipate here that Euler angles do not interpolate linearly. If our robot starts from a rotation and wants to reach another target rotation, we cannot simply take the Euler angles and linearly move them from start to end, like we would do for positions. The resulting movement could look totally random and wrong to an external observer. Rotation matrices are even worse: there is no practical way to interpolate between two matrices directly.

Actually, there is also other way to express rotations that we have not introduced yet: quaternions. We will present them in Chapter 5. Quaternions are somewhat in between Euler angles and rotation matrices, and they are actually the only choice when it comes to interpolations.

So, the question is which representation do we pick when developing our control software? Well, we need all of them really. Since Euler angles are simpler to understand, we normally use them for the interface to the operator (e.g., for visualization purposes). Internally, we use quaternions for interpolations (e.g., for path-planning operations) and rotation matrices for all frame operations (e.g., for kinematic transformations).

Combining Translations and Rotations

We have first studied translations of frames and then rotations, and now we need to put them together into one individual operation to derive the most generic transformation of a frame into another.

For that purpose, we introduce a tool called **homogeneous matrix**: a matrix T, which combines a rotation matrix R and a translation offset Δ into a single transformation. The resulting combination is a 4x4 matrix, filled with zeros under the rotation matrix and with a 1 under the translational offset.

$$T = \begin{bmatrix} R & \Delta \\ 0 & 1 \end{bmatrix} = \begin{bmatrix} R_{11} & R_{12} & R_{13} & \delta_X \\ R_{21} & R_{22} & R_{23} & \delta_Y \\ R_{31} & R_{32} & R_{33} & \delta_Z \\ 0 & 0 & 0 & 1 \end{bmatrix} \tag{2-22}$$

Joining the two operations together means that with a single matrix multiplication, we can perform both translation and rotation of a frame at the same time.

For example, if we want to rotate around the Z axis by θ and then translate along the Y axis by δ_Y, we can use the following matrix:

$$T = \begin{bmatrix} cos\theta & -sin\theta & 0 & 0 \\ sin\theta & cos\theta & 0 & \delta_Y \\ 0 & 0 & 1 & 0 \\ 0 & 0 & 0 & 1 \end{bmatrix} \tag{2-23}$$

It is critical to remember that the order is not commutative: when moving from the old frame to a new one, we first rotate and then translate. If we would first translate and then rotate, the result would be totally different (see Figure 2-11).

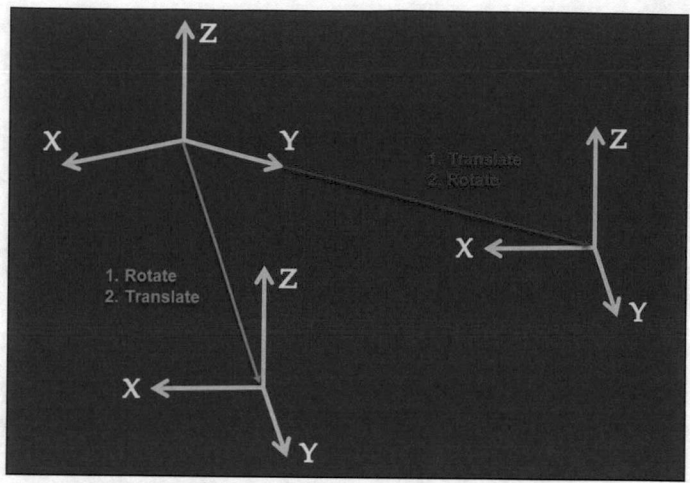

Figure 2-11. *Roto-translations are not commutative!*

Also critical is to remember that the values of the offset and rotation angles are to be expressed as seen from the new frame coordinates.

The homogeneous matrix is a powerful tool to express generic transformation between frames. Actually, as far as industrial robots are concerned, this tool is all we need to derive geometrical models of mechanical structures.

Using it in practice is very simple. The typical application is shown in Figure 2-12: we have a unique point in space described by the coordinates P_1 in an old frame F_1 and P_2 in a new frame F_2; we know the value of P_1, and we know the transformation matrix T_{21} between F_1 and F_2; the question is how to find P_2.

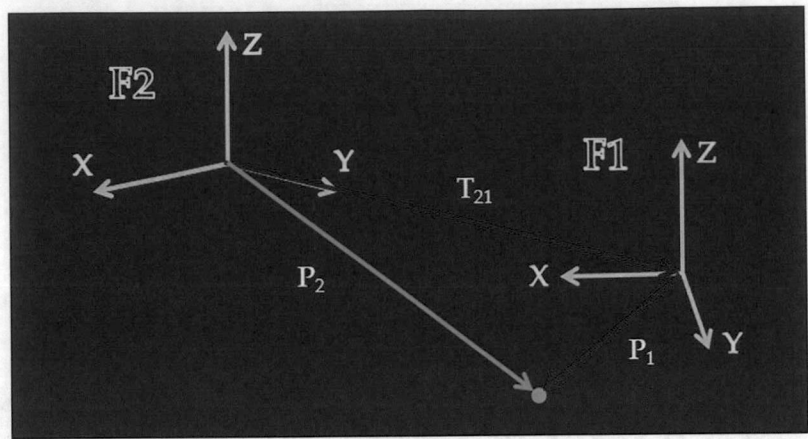

Figure 2-12. *Applying the homogeneous transformation T_{21}*

The answer is $P_2 = T_{21} P_1$. Just like we used to pre-multiply by a rotation matrix to account for a frame rotation, now we pre-multiply by the entire homogeneous matrix to account for a frame rotation and translation combined.

Moving from the old F_1 to the new F_2, we first rotate and then translate. The angle and offset values are measured as seen by the new frame F_2.

Note that because the homogeneous matrix is a 4x4 matrix, while a Cartesian point only has three coordinates, we also need to express the point in homogeneous coordinates, by padding its vector with a 1 at the end.

$$P_2 = T_{21} P_1 = \begin{bmatrix} R_{11} & R_{12} & R_{13} & \delta_X \\ R_{21} & R_{22} & R_{23} & \delta_Y \\ R_{31} & R_{32} & R_{33} & \delta_Z \\ 0 & 0 & 0 & 1 \end{bmatrix} \begin{bmatrix} x_1 \\ y_1 \\ z_1 \\ 1 \end{bmatrix} \qquad (2\text{-}24)$$

Solving the product gives the following solution for P_2:

$$P_2 = \begin{cases} x_2 = R_{11}x_1 + R_{12}y_1 + R_{13}z_1 + \delta_X \\ y_2 = R_{21}x_1 + R_{22}y_1 + R_{23}z_1 + \delta_Y \\ z_2 = R_{31}x_1 + R_{32}y_1 + R_{33}z_1 + \delta_Z \end{cases} \tag{2-25}$$

It is easy to verify that if the two frames only have a translation and no rotation in between, the elements along the diagonal of the rotation part will all be 1s and the elements outside the diagonal will be 0s. Then we fall back to a simple linear offset addition for the translational effects.

Similarly, if the translation offset vector is 0, then we only have the rotation elements left.

Just like rotation matrices can be chained multiplied together to express a series of rotations, also homogeneous matrices can build up a chain of multiplication to describe a series of generic frame transformations. We will use this property when solving the direct kinematics of robots. As usual, the resulting product is associative but not commutative.

Example

Let's look at a practical example to make things a bit clearer.

Imagine we have a very simple mechanical structure with one single rotating axis, as shown in Figure 2-13.

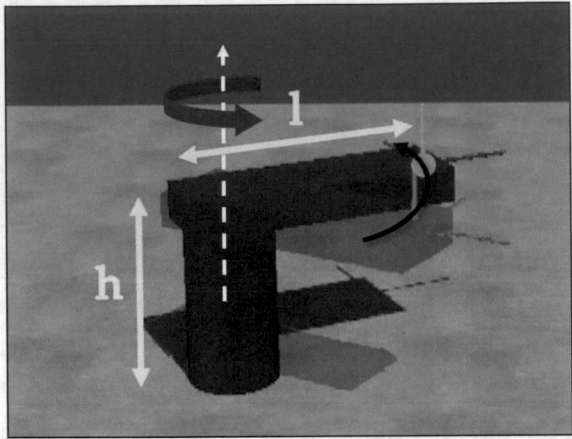

Figure 2-13. *Simple mechanical structure with one rotating joint*

The position of the TCP is indicated by the blue dot, and it moves along the black curve when the rotating axis is turning. We now want to derive the actual position of the TCP referenced to the fixed robot base frame as a function of the rotating joint angle.

We start by providing some fixed geometrical parameters: the length of the arm l and its height h.

Then we identify two frames: the mobile frame F_1 and the fixed base frame F_0 (see Figure 2-14). The position of the TCP as seen from the mobile frame F_1 (the robot's moving link) is actually constant. We call it P_1. The position of the TCP as seen from the fixed frame F_0 (the robot's base) is moving. We call it P_0.

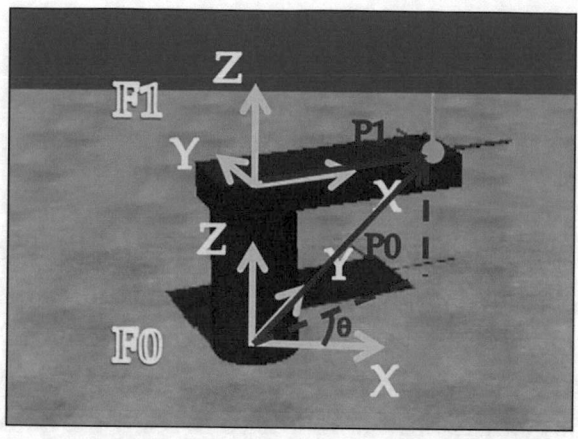

Figure 2-14. *Identifying the mobile frame F1 and the fixed frame F0*

In order to find P_0, we need to know P_1 and the transformation matrix T_{01} that modifies the mobile frame F_1 into the fixed frame F_0. The solution will then be $P_0 = T_{01} P_1$.

P_1 is simple: the TCP position in frame F_1 is fixed at $[l, 0, 0]$.

As for the homogeneous transformations, we first take into account a rotation by an angle θ around the Z axis and then a translation along the Z axis by an offset h. Note that both θ and h are positive as seen from the fixed frame F_0. However, while h is always constant, the angle θ can change all the time because it is driven by the rotating joint of the robot.

$$T_{01} = \begin{bmatrix} cos\theta & -sin\theta & 0 & 0 \\ sin\theta & cos\theta & 0 & 0 \\ 0 & 0 & 1 & h \\ 0 & 0 & 0 & 1 \end{bmatrix} \tag{2-26}$$

We can now pre-multiply P_1 by the matrix T_{01} to find P_0:

$$P_0 = T_{01} \, P_1 = \begin{bmatrix} cos\theta & -sin\theta & 0 & 0 \\ sin\theta & cos\theta & 0 & 0 \\ 0 & 0 & 1 & h \\ 0 & 0 & 0 & 1 \end{bmatrix} \begin{bmatrix} l \\ 0 \\ 0 \\ 1 \end{bmatrix} = \begin{bmatrix} lcos\theta \\ lsin\theta \\ h \\ 1 \end{bmatrix} \tag{2-27}$$

The solution is $P_0 = \begin{bmatrix} lcos\theta & lsin\theta & h \end{bmatrix}$. These are the coordinates of the TCP as seen by an observer fixed at the base of the robot.

In such a simple example, we could have easily derived the correct answer by mere geometrical considerations. However, solving a full six-axes robot is going to be much harder, and using the homogeneous transformations is certainly more convenient.

If you have understood this example, you are a good half-way through solving the kinematic model for the six-axes robot.

Inverted Transformation

Once you are able to transform points from one frame to another, you can also quickly learn how to perform the inverse operation. Essentially, instead of finding $P_1 = T_{10} \, P_0$, we want to find $P_0 = T_{01}P_1 = T_{10}^{-1} \, P_1$. We know $T_{10} = \begin{bmatrix} R & \Delta \\ 0 & 1 \end{bmatrix}$ but we do not know how to compose T_{10}^{-1}.

Inverting an individual translation is as easy as simply taking the same offset in the negative direction: $\Delta \rightarrow -\Delta$.

Inverting an individual rotation is also easy by recalling one of the properties of rotation matrices that we presented earlier in this chapter: inverting a rotation matrix is the equivalent of taking its transpose, which is an elementary operation: $R \rightarrow R^{-1} = R^T$.

However, when translations and rotations are combined together into a homogeneous matrix, we need to take an extra step: the reason being that the order of rotation and translation is not commutative in a frame transformation, as already stressed earlier.

Intuitively, we can proceed with the following reasoning: A rotation angle is not modified by a translation, but the direction of a translation is modified by a rotation of the frame. Therefore, when doing the operation in reverse, we can safely inverse the rotation matrix, but we cannot simply take the same offset in the negative direction. The correct solution is to take a translation offset along the rotated frame, so $-\Delta$ actually becomes $-R^T\Delta$.

For a simple proof, consider the following example. We first transform a point P_1 in P_0 by a rotation R and a translation Δ:

$$P_0 = RP_1 + \Delta \tag{2-28}$$

Then we derive P_1 as a function of P_0 and observe the resulting inverted values for rotation and translation (in **bold**):

$$RP_1 = P_0 - \Delta$$

$$P_1 = R^{-1}(P_0 - \Delta) = R^{-1}P_0 - \boldsymbol{R^{-1}}\Delta \tag{2-29}$$

The combined expression for the inverted transformation matrix is then as follows:

$$T_{10}^{-1} = \begin{bmatrix} R^T & -R^T\Delta \\ 0 & 1 \end{bmatrix} \tag{2-30}$$

Summary

This was a heavy but fundamental chapter to lay the foundations for robotics control. You have now learned the geometrical tools needed to tackle all kinematic models, including the most generic six-axes manipulator that we are going to study in the next chapters. Make sure you understand well how to combine the translations and rotations of frames into a homogeneous matrix before moving forward.

After all, the whole process of solving robotics kinematics relies on finding the position of a point (the TCP) in different frames created by the mechanical structure of joints and links, where joints introduce angular rotations, while links add offset translations.

CHAPTER 3

Forward Kinematics

There are two fundamental functions in a robot's kinematic model:
the **forward transformations** and the **inverse transformations**. The
forward operation, also often called direct transformations, is the process
of calculating the position and orientation of the TCP given the current
values of the joint axes. The inverse operation, as the word suggests, is the
opposite process: calculate the values of the joint axes given the current
TCP pose.

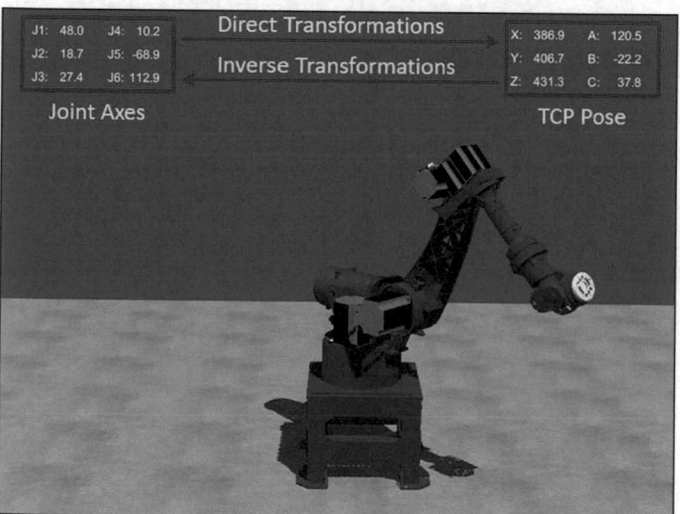

Figure 3-1. *Direct and inverse transformations between the joints
and the TCP of a six-axes robot*

© Fabrizio Frigeni 2023
F. Frigeni, *Industrial Robotics Control*, Maker Innovations Series,
https://doi.org/10.1007/978-1-4842-8989-1_3

We are now going to derive in detail both the forward and inverse transformations for a standard six-axes anthropomorphic robot (see Figure 3-1), the kind already described in Mechanical Configurations Section in Chapter 1. We will study the forward pass in this chapter and the backward pass in the next one.

Mechanical Structure

A kinematic model depends on the mechanical structure of the robot. Let's look at our robot and analyze its structural details.

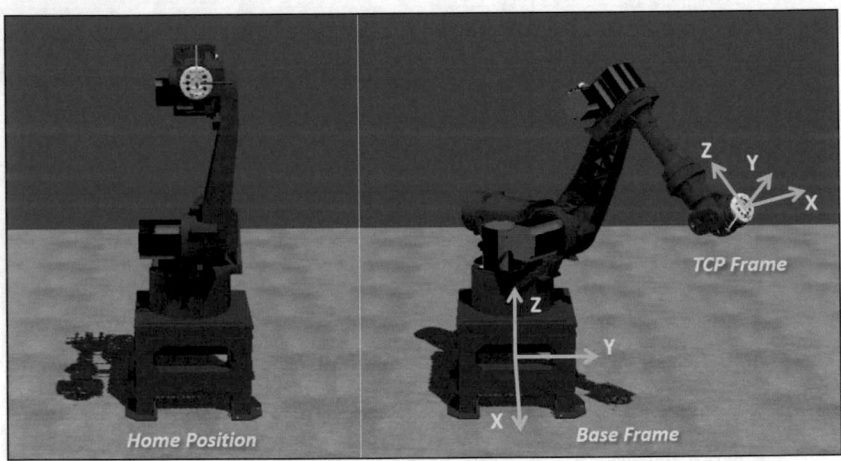

Figure 3-2. *Standard six-axes robot in its home position (left) and in a generic position (right)*

We first identify the fixed base frame and the moving TCP frame. The position and orientation of the TCP frame with respect to the base frame is determined by the values of the joint axes and the length of the links. When all the joint axes have zero value, the robot is said to be in its **home position**. The robot's arm points along the base X axis as shown in Figure 3-2.

The movement of each joint affects the pose of the entire robot by modifying the position and orientation of all the frames of the joints following along in the kinematic chain. For example, the first joint introduces a rotation around the Z axis and displaces all the other frames. Figure 3-3 shows how the TCP frame and the J_2 frame are modified by a rotation of the first joint.

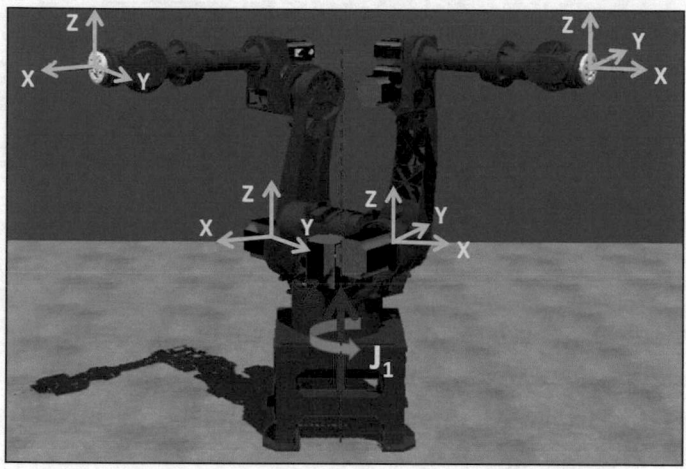

Figure 3-3. *Joint 1 introduces a rotation around its local Z axis*

The second and third joints both introduce rotations around the Y axis of their local frames (see Figure 3-4).

Figure 3-4. *Joints 2 and 3 introduce rotations around their local Y axis*

The first three joints constitute the first half of the robot's body, similar to what the shoulder and the elbow represent in our human arm. On the other hand, the last three joints affect the second half at the end of the robot, which is called the **wrist**.

Joints 4 and 6 rotate around their X axes, while Joint 5 rotates around its Y axis (see Figure 3-5).

Figure 3-5. *Joint 4, 5, and 6, respectively, rotate the wrist around the X, Y, and X axes of their local frames*

The six joints of the robot are connected with each other in a chain by means of the mechanical links. Unlike the joints, whose angles are variable, the links of this robotic arm always have a fixed length. They determine the overall size of the robot and of its workspace.

There exist, of course, other robotic structures with variable link lengths. Those are called translational axes, as opposed to the rotational axes of the joints. The way to solve those kinematic models is absolutely the same as the one we are solving here.

We name the links according to the joint index they point to and the direction axis they move along, as shown in Figure 3-6.

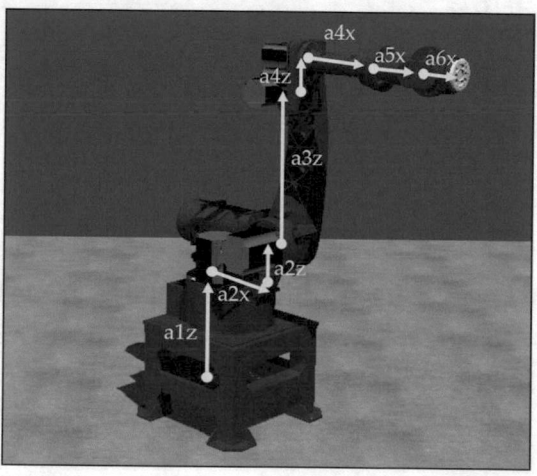

Figure 3-6. *Mechanical parameters of a six-axes robot*

Geometrically speaking, joints and links introduce a chain of rotations and translations, starting from the base frame all the way up to the TCP frame. As we already know from the frame operations, such a system can be mathematically described by homogeneous transformations between frames, where the rotation angles are given by the values of the joint axes and the translational offsets are given by the lengths of the links. The joint angles are normally read in real time from the encoders mounted on the electric motors driving the joints. The mechanical size of the arms can be found in the datasheet provided by the robot's manufacturer.

The purpose of the forward transformations is to find the coordinates of the TCP in the base frame. We start from the TCP frame and move backward along the kinematic chain to express the coordinates of the TCP in each local frame of the robot's joints, until we reach the base. Every step introduces a rotation (because of the joint's orientation) and a translation (because of the link offset), which are combined together into a homogeneous matrix to be pre-multiplied by the point's coordinates. By the end of the chain, we will have generated the final global transformation matrix:

$$TCP(X,Y,Z) = T_{01}T_{12}T_{23}T_{34}T_{45}T_{56}O = T_{06}\begin{bmatrix} 0 \\ 0 \\ 0 \\ 1 \end{bmatrix} \quad (3\text{-}1)$$

This solution is very general and can be applied to any kind of serial kinematic chain, no matter how many axes it has and no matter what kind of axes they are (rotational or translational).

Step-by-Step Solution

We now begin solving the FKs (forward kinematics) by putting together all the homogeneous transformations introduced by each joint of the kinematic chain, starting from the TCP moving all the way down to the base of the robot.

The first transformation is between F_6 and F_5 (see Figure 3-7) and involves two steps: first, the sixth joint modifies the TCP by adding a rotation around its local X axis; then, we add a translation by $a6z$ along the X axis of the F_5 frame. With this operation, we reach the middle of the wrist: this is a very important point, called **wrist center point**, and will play a critical role when solving the inverse transformations.

Figure 3-7. *FK Step #1: from Frame 6 to Frame 5*

We call the homogeneous transformation for this step T_{56}, and we build it from the rotation matrix R_{56} around the X axis as given in Equation (2-3) and the offset $a6z$:

$$T_{56} = \begin{bmatrix} 1 & 0 & 0 & a6x \\ 0 & cosJ_6 & -sinJ_6 & 0 \\ 0 & sinJ_6 & cosJ_6 & 0 \\ 0 & 0 & 0 & 1 \end{bmatrix} \tag{3-2}$$

Remember that by applying the transformation T_{56} we are actually moving backward from the frame F_6 (centered in J_6) to the frame F_5 (centered in J_5). From that perspective, the rotation comes before the translation, which is the correct way to do it, according to the rule we followed when defining a homogeneous transformation matrix (see Section on Combining Translations and Rotations in Chapter 2). The values of offset and rotation angle are all expressed from the perspective of the new target frame F_5. This observation is valid for all the next steps and will not be repeated each time.

Also, note that we technically started from the mounting point, which only corresponds to the TCP when no tool is mounted on the flange. In the generic case of a nonzero tool, an additional (fixed) frame needs to be added: the solution is shown later on in the Tool Frame Section.

Moving on with the next step as shown in Figure 3-8, we see a rotation around the local Y axis of F_5 introduced by J_5, plus a translation along the X axis of F_4:

$$T_{45} = \begin{bmatrix} cosJ_5 & 0 & sinJ_5 & a5x \\ 0 & 1 & 0 & 0 \\ -sinJ_5 & 0 & cosJ_5 & 0 \\ 0 & 0 & 0 & 1 \end{bmatrix} \tag{3-3}$$

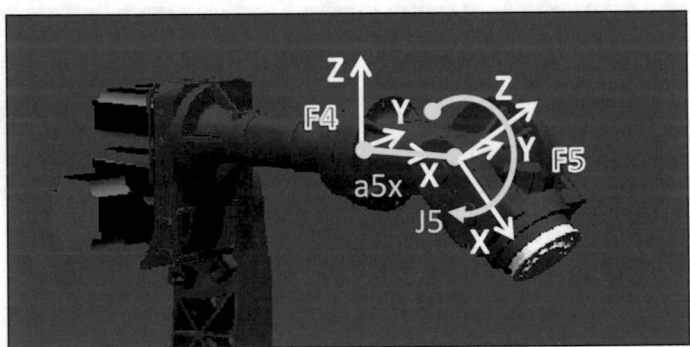

Figure 3-8. *FK Step #2: from Frame 5 to Frame 4*

The next step goes from the fourth to the third joint (Figure 3-9). The actual mechanical configuration can change a little bit here, according to the way the robot is built: in our specific case, we have two offsets: one along the Z axis and another one along the X axis. In some mechanical structures, there are also offsets along the Y axis at this point.

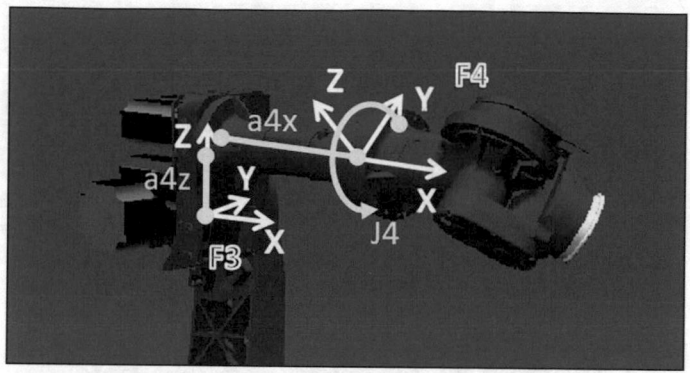

Figure 3-9. *FK Step #3: from Frame 4 to Frame 3*

The rotation of the fourth joint happens around the X axis.

$$
T_{34} = \begin{bmatrix} 1 & 0 & 0 & a4x \\ 0 & cosJ_4 & -sinJ_4 & 0 \\ 0 & sinJ_4 & cosJ_4 & a4z \\ 0 & 0 & 0 & 1 \end{bmatrix}
$$

(3-4)

Next, the link from Joint 3 to Joint 2 introduces a vertical translation along the Z axis of F_2 by $a3z$ (see Figure 3-10). This is usually the longest and heaviest arm of the robot, and the motor driving Joint 2 needs to be quite large. Often, an additional external mechanical spring is added in parallel to the second joint to help the motor by supporting some of the arm's weight.

Figure 3-10. *FK Step #4: from Frame 3 to Frame 2*

Mathematically, we build the homogeneous transformation matrix T_{23} using a rotation around the Y axis and a translation along the Z axis:

$$T_{23} = \begin{bmatrix} cosJ_3 & 0 & sinJ_3 & 0 \\ 0 & 1 & 0 & 0 \\ -sinJ_3 & 0 & cosJ_3 & a3z \\ 0 & 0 & 0 & 1 \end{bmatrix} \qquad (3\text{-}5)$$

The next step goes from the second to the first frame (Figure 3-11). Here we have the rotation of the second joint around its local Y axis. Additionally, we have two translations: one along the X axis and one along the Z axis of F_1.

Figure 3-11. *FK Step #5: from Frame 2 to Frame 1*

We build the homogeneous transformation matrix T_{12} accordingly:

$$T_{12} = \begin{bmatrix} cosJ_2 & 0 & sinJ_2 & a2x \\ 0 & 1 & 0 & 0 \\ -sinJ_2 & 0 & cosJ_2 & a2z \\ 0 & 0 & 0 & 1 \end{bmatrix} \tag{3-6}$$

Finally, we reach the last step of the chain: from the first joint down to the base frame F_0 (see Figure 3-12). We observe a rotation around the Z axis and a translation along the Z axis by $a1z$.

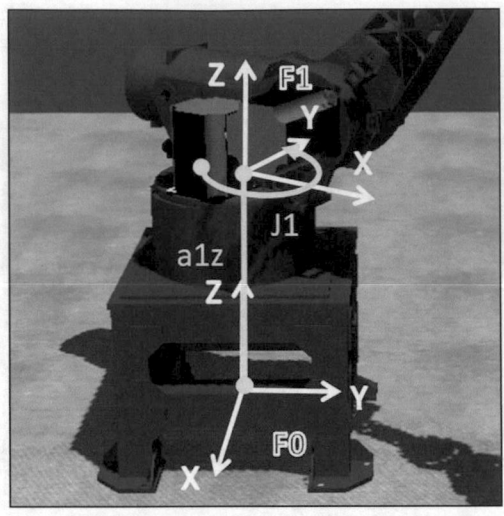

Figure 3-12. *FK Step #6: from Frame 1 to the base frame*

We call the homogeneous transformation for this first step T_{01}, and we build it from the rotation matrix R_{01} around the Z axis as given in Equation (2-5) and the offset $a1z$:

$$T_{01} = \begin{bmatrix} cosJ_1 & -sinJ_1 & 0 & 0 \\ sinJ_1 & cosJ_1 & 0 & 0 \\ 0 & 0 & 1 & a1z \\ 0 & 0 & 0 & 1 \end{bmatrix} \qquad (3\text{-}7)$$

This is actually the same matrix we used to solve the example in Chapter 2: a rotation around Z plus a vertical offset.

Combined Transformation Matrix

It is now time to put all those steps together and derive the general formula for the direct transformation, from the base to the TCP (see Figure 3-13).

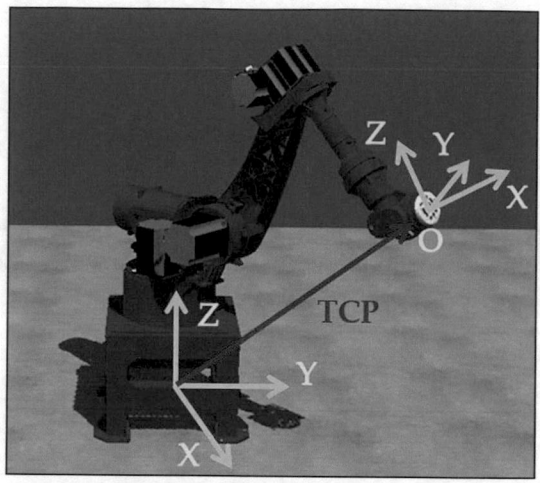

Figure 3-13. *Direct transformation from base to TCP*

The TCP is the origin of the TCP frame, so its coordinates in that frame are $\begin{bmatrix} 0 & 0 & 0 \end{bmatrix}$. The goal is to express that position in the base frame. The way to achieve that is pre-multiplying the origin of the TCP frame by all the homogeneous transformations we obtained for each frame in the previous six steps.

$$TCP(X,Y,Z) = T_{01}T_{12}T_{23}T_{34}T_{45}T_{56}O = T_{06}\begin{bmatrix} 0 \\ 0 \\ 0 \\ 1 \end{bmatrix} \qquad (3\text{-}8)$$

Multiplying all those 4x4 matrices looks like a lot of work, but since many elements are 0, it can be done very quickly numerically. Once again, we stress that this procedure is very generic and valid for any open kinematic chain.

To find the rotation of the TCP frame with respect to the base frame, it is enough to consider the rotation part of those homogeneous matrices:

$$TCP(A,B,C) = R_{01}R_{12}R_{23}R_{34}R_{45}R_{56} = R_{06} \qquad (3\text{-}9)$$

The resulting rotation matrix can be decomposed in Euler angles [A, B, C], as we learned in Chapter 2. Also, recall that the columns of the rotation matrix represent the coordinates of the TCP frame axes as seen from the base frame.

The product of rotation matrices is associative, so we can split the calculations in two for convenience:

$$R_{06} = (R_{01}R_{12}R_{23})(R_{34}R_{45}R_{56}) = R_{arm}R_{wrist} \qquad (3\text{-}10)$$

The first three joints determine the rotation of the arm, from the base to the wrist center point. The last three joints determine the rotation of the wrist from its center point to the TCP. This observation will turn out to be helpful when solving the inverse kinematics.

Numerical Test

In case you are implementing the calculations in a program and want to make sure that your results are correct, we provide here an example for you to replicate. The mechanical dimensions for the example robot are reported in Table 3-1.

Table 3-1. *Values of mechanical parameters for the example robot*

a1z	a2x	a2z	a3z	a4x	a4z	a5x	a6x
650	400	680	1100	766	230	345	244

The values or the mechanical parameters are in mm, so the calculated position of the TCP will be in mm as well.

We start from the home position of the robot, where all the joint axes are 0 (see Figure 3-14).

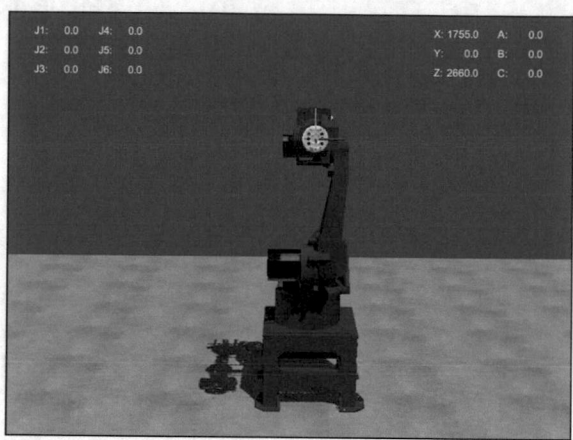

Figure 3-14. *Test Position #1*

The resulting coordinates of the TCP are shown in Table 3-2:

Table 3-2. *Test Position #1 (home position)*

Joints		TCP	
J_1	0	X	1755
J_2	0	Y	0
J_3	0	Z	2660
J_4	0	A	0
J_5	0	B	0
J_6	0	C	0

We notice that the Y position is 0, which was to be expected, because the robot's arm in its home position is aligned along the base X axis. Also, the orientation angles are all 0, because the TCP frame is perfectly aligned with the base frame.

We now move the robot's first axis by an angle of 90 degrees. The arm will point toward the base Y axis, as shown in Figure 3-15.

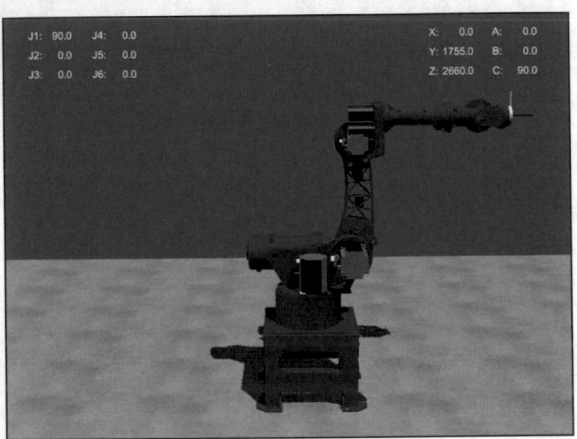

Figure 3-15. *Test Position #2*

The TCP Euler C angle shows a rotation around the base Z axis (see Table 3-3).

Table 3-3. *Test Position #2*

Joints		TCP	
J_1	90	X	0
J_2	0	Y	1755
J_3	0	Z	2660
J_4	0	A	0
J_5	0	B	0
J_6	0	C	90

We now move the robot to two random positions in space (shown in Figure 3-16) so that you can verify the results of your calculations.

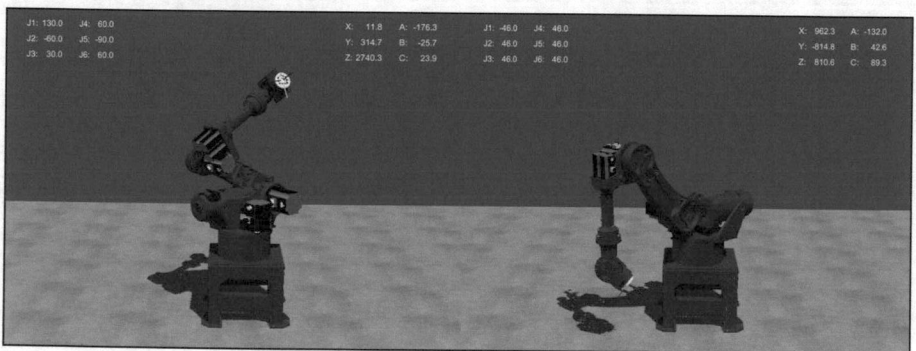

Figure 3-16. *Test Positions #3 and #4*

The resulting coordinates of the TCP are shown in Table 3-4 and Table 3-5.

Table 3-4. *Test Position #3*

Joints		TCP	
J_1	130	X	11.8
J_2	-60	Y	314.7
J_3	30	Z	2740.3
J_4	60	A	-176.3
J_5	-90	B	-25.7
J_6	60	C	23.9

Table 3-5. *Test Position #4*

Joints		TCP	
J_1	-46	X	962.3
J_2	46	Y	-814.8
J_3	46	Z	810.6
J_4	46	A	-132
J_5	46	B	42.6
J_6	46	C	89.3

Zero Frame

So far, we have calculated the position and orientation of the TCP as
seen from the local robot's frame, sometimes also referred to as machine
coordinate system (MCS). That is the most natural choice of global system
if you are working with one individual robot.

However, you might happen to commission a system with a number of robots working together on the same part. Imagine, for example, three different robotic arms: one cutting, one welding, and one painting a single metal piece. Their base positions are different, but their target workpiece location is the same. In order for them to understand the same target values input by the operator, we need to adopt a unique global coordinate system (GCS) for all robots.

The additional offset and rotation between the global coordinate system and the base of each robot is sometimes called base frame or zero frame. Note that the zero frame does not only introduce a translational offset, but often also entire rotations along all their degrees of freedom. For instance, robots can have their bases mounted on the walls or on the ceilings, as shown in Figure 3-17.

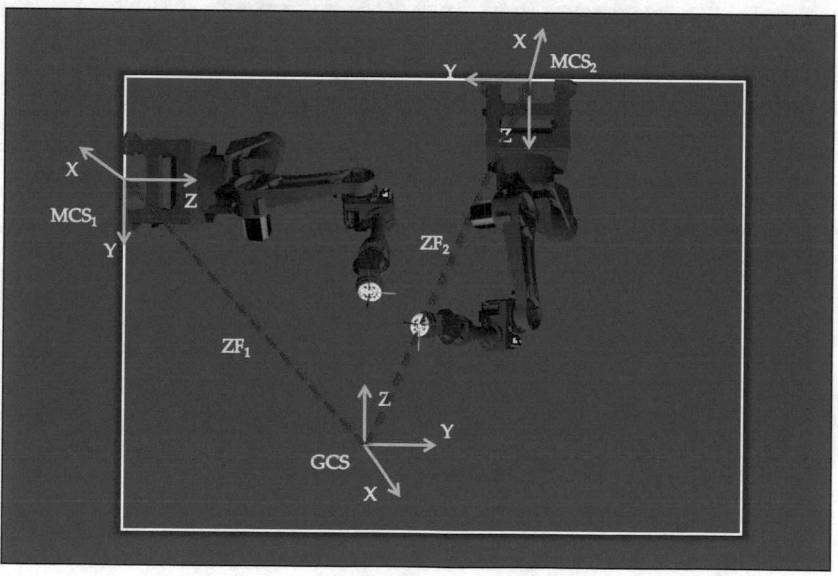

Figure 3-17. *Introducing an additional zero frame (ZF) for each robot*

When solving the forward kinematics, we need to express the resulting TCP position and orientation in the unique GCS. Mathematically, that corresponds to adding an additional frame transformation at the base of the robot (see Figure 3-18).

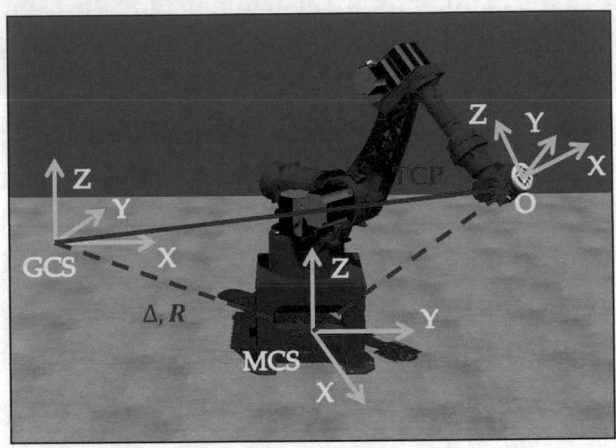

Figure 3-18. *Adding a zero frame to the transformations*

The frame transformation between GCS and MCS is described by a homogeneous matrix built using the translational offset Δ and rotation matrix R between the base of the robot and the global system:

$$T_{Base} = \begin{bmatrix} R & \Delta \\ 0 & 1 \end{bmatrix} \tag{3-11}$$

Transforming the coordinates of the TCP from MCS to GCS requires a frame pre-multiplication. We use the whole homogeneous matrix to find the position:

$$TCP_{XYZ} = T_{Base}\, T_{06} \begin{bmatrix} 0 \\ 0 \\ 0 \\ 1 \end{bmatrix} \tag{3-12}$$

To find the orientation, we only need the rotation part:

$$TCP_{ABC} = R_{Base}\, R_{06} \tag{3-13}$$

Remember that a quick way to build the rotation matrix between two frames is using the column vectors, which can be identified by simple measurements.

Tool Frame

Another additional frame operation that is commonly needed in practice is shifting from the mounting point to the actual TCP, in case the robot carries a working tool (see Figure 3-19).

Every tool has a finite size, which translates and possibly rotates the end-tip of the robot. By knowing the tool dimensions, it is possible to build the homogeneous matrix for the additional frame transformation.

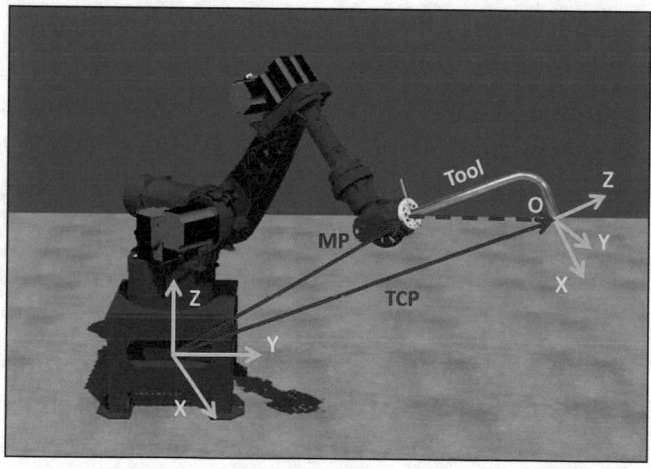

Figure 3-19. *Adding a frame for the tool*

Since the tool frame comes after the last joint, we need to add the homogeneous matrix and the rotation matrix after the multiplication chain:

$$TCP_{XYZ} = T_{06}\, T_{Tool} \begin{bmatrix} 0 \\ 0 \\ 0 \\ 1 \end{bmatrix} \tag{3-14}$$

$$TCP_{ABC} = R_{06}\, R_{Tool} \tag{3-15}$$

In case the exact size of the attached tool is unknown, it is possible to use an empirical calibration method to identify it. We will present it in the Tool Calibration Section in Chapter 10.

Mechanical Coupling

Before we conclude the forward kinematics section, we present one more related topic: **mechanical coupling**.

This is something that common robotics textbooks do not normally mention, because it is not of theoretical interest. However, it is a very useful practical note, because arguably all industrial anthropomorphic robots (but also four-axes palletizers and SCARA robots) have this characteristic, and it is helpful for you to know how to handle it in your control program.

A coupling between two axes of a robot means that the movement of one axis is internally (mechanically) linked to that of another axis. For example, a movement of the fifth axis J_5 causes a movement of the sixth axis J_6, even if the motor of J_6 is not moving!

The resulting behavior can appear puzzling to an external observer and, more importantly, causes the TCP to drift away from its programmed path. We clearly need to compensate for this effect in our calculations of the forward and inverse kinematics.

Let's consider a simple example for a better understanding. We build a robot with two rotational joint axes J_1 and J_2 (see Figure 3-20 left) and observe the different behavior of the TCP when *only driving the motor in* J_1: in case the axes are independent (middle) and mechanically coupled (right).

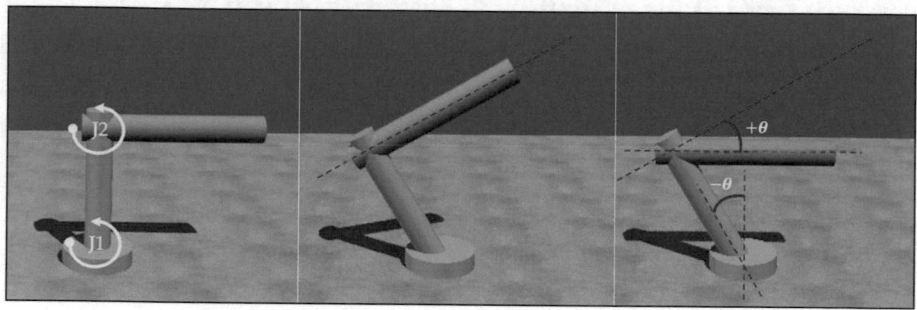

Figure 3-20. *Difference between uncoupled and coupled axes*

There is a clear difference between the actual positions reached by the arm with and without coupling. The difference is a function of the **coupling coefficient** between J_1 and J_2. The coefficient used in this example is -1: a movement of the first axis by an angle θ forces a movement of the second axis by an angle $-\theta$. Incidentally, that is the typical behavior of the palletizer robot shown in Figure 1-8 and solved in the Appendix.

Notice that the rotation of the second axis is entirely caused by the mechanical coupling with the first axis. The motor driving J_2 has not moved. However, if we had turned the motor on J_2 by an angle of $-\theta$, the final position of the arm would have been coincident with that of the uncoupled robot. That is exactly the extra compensation we need to add in our calculations, in order to remove the effect of the mechanical coupling.

Let's see how to introduce coupling coefficients in the forward kinematic model so that the actual position of the TCP is calculated correctly.

We normally solve the direct transformations between joints and TCP with the forward kinematic function. The inputs to the function are the joint positions that we read from the encoders, which detect the actual position of the motors. However, those values are not necessarily equal to the actual position of the joint angles, in case an additional mechanical coupling effect is present. That would lead to a wrong TCP position calculation.

The joint values need to be adjusted by the coupling coefficient before being processed by the forward kinematics, as described with the diagram in Figure 3-21.

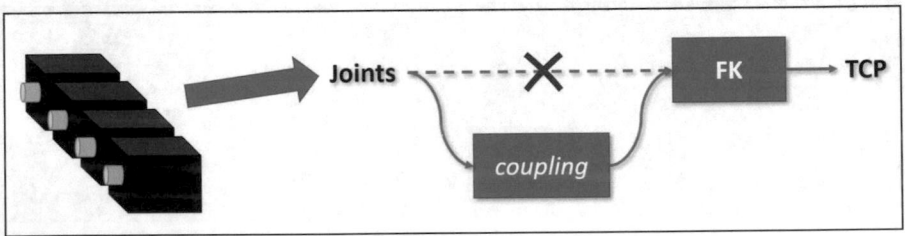

Figure 3-21. *Introducing mechanical coupling compensation before the forward kinematics*

Typically, six-axes robots have linear mechanical couplings between the wrist axes. The actual position of the fifth and sixth joints is affected by the position of the fourth axis according to the ratios c_{45} and c_{46}. In that case, the correct joint angles to consider are the sum of the encoder feedback values, plus the coupling to compensate for the movement of J_4. If J_4 is in its zero (home) position, then the actual values of J_5 and J_6 correspond to their own motor positions.

$$J_5 \leftarrow J_5 + J_4\, c_{45} \qquad\qquad (3\text{-}16)$$

$$J_6 \leftarrow J_6 + J_4\, c_{46} \qquad\qquad (3\text{-}17)$$

If we now feed these adjusted angles to the forward kinematics and calculate the TCP position, we find the correct value.

Summary

In this chapter, we described a step-by-step approach to solve the forward kinematics of a six-axes manipulator. Essentially, we built a homogeneous matrix for the whole robot, from its base to the TCP, by combining all the individual matrices of each consecutive joint. We also saw how to add a zero frame, a tool frame, and compensation for mechanical couplings, when required.

You can easily apply the same method to solve any generic serial kinematic chain. The solutions to some common industrial robots are shown in the Appendix.

CHAPTER 4

Inverse Kinematics

In the previous section, we derived the model for the forward kinematics: we calculated the TCP pose given the values of the joint axes. In this section, we solve the inverse problem: find the values of the joint axes given the position and orientation of TCP relative to the base frame (see Figure 4-1).

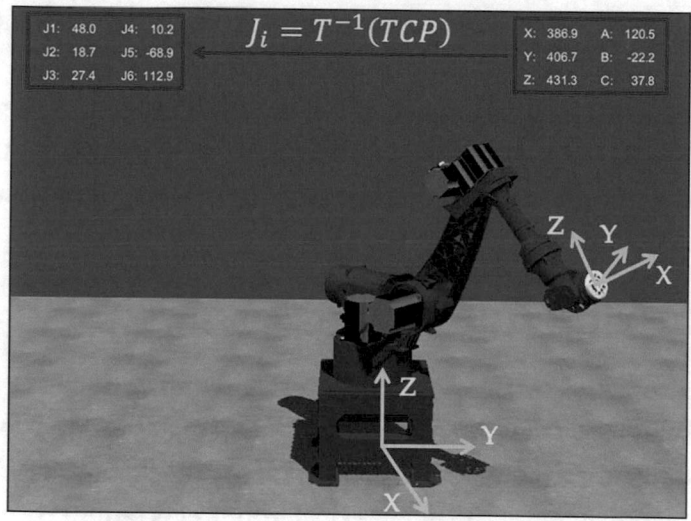

Figure 4-1. *IK: finding the joint angles from the TCP pose*

You might ask yourself: why do we need an inverse kinematic function if we already know the value of the joint axes from the motor's encoders?

© Fabrizio Frigeni 2023
F. Frigeni, *Industrial Robotics Control*, Maker Innovations Series,
https://doi.org/10.1007/978-1-4842-8989-1_4

While it is true that we can always read the current position of the joint angles, we also want to have the ability to predict where the joints need to be driven to, given a desired target pose for the robot's TCP. That information is essential for planning future movements and executing them, as we will see in the next chapters.

Robots are typically programmed by the operator in **path space** coordinates [X, Y, Z, A, B, C], which are intuitive to visualize and understand. Imagine, for instance, moving your own arm to grab something with your hand: you see the position of the target object in the path space (e.g., in front, left, above), but you have no idea what the required angles of your joints (shoulder, elbow, and wrist) are.

However, motion control software acts directly on the motors driving the joints of robots, so many internal calculations work in the **joint space** $[J_1...J_6]$. That is why we need a way to transform back from path values to joint angles.

For robots with a serial kinematic chain, just like the six-axes robot we are analyzing here, the problem of solving inverse transformations is usually more complicated than deriving the direct transformations.

Closed-Form Derivation

The approach we presented to solve the forward transformations results in a general solution, which can be directly applied to any kind of chain. For the inverse problem, there is no such general approach. Each chain must be analyzed individually, and geometrical intuition is required.

In fact, there are even mechanical structures for which it is impossible to solve the problem in a **closed form**. In other words, there are no equations that can be written to calculate the values of the joint axes. In those cases, a **numerical approach** is required.

For a six-axes manipulator, the condition that allows for a closed-form solution is the mechanical structure of the wrist.

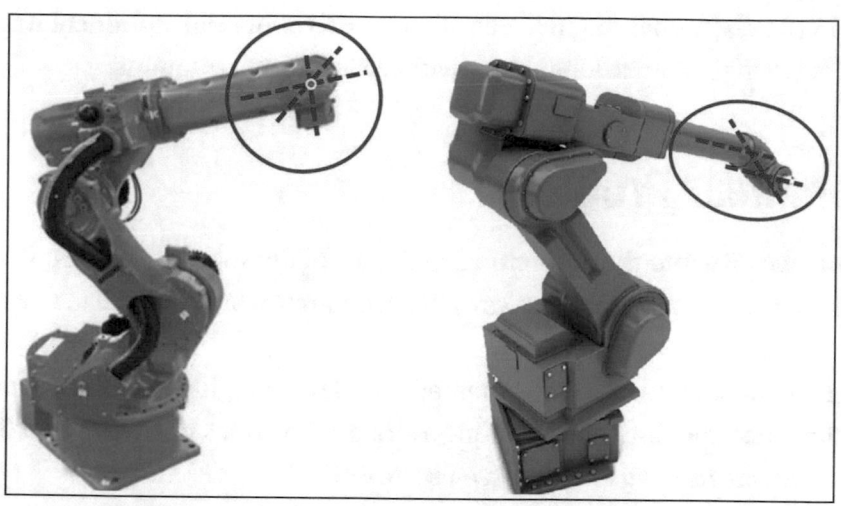

Figure 4-2. *Only a spherical wrist (left) can be solved in closed form*

There are two different kinds of wrists typically used in manipulators. The one on the left side of Figure 4-2 is called **spherical wrist**, because the three rotating axes $[J_4, J_5, J_6]$ all intersect in one point, the so-called wrist center point. This configuration is the most common in the industry and allows for a convenient closed-form solution.

On the right side of Figure 4-2, we see a robot with a different kind of wrist, whose rotation axes do not intersect in one point. Some people call this mechanical structure "elephant nose" for the vague resemblance with the animal's trunk. Critically, such configuration cannot be solved in closed form with fixed equations. Instead, it requires a numerical algorithm in the form of successive iterations in order to find approximate values for the joints. The process is mathematically more complex and computationally more expensive. It does offer other practical advantages though, namely, the fact that it does not suffer from singularity issues and can therefore always perform movements at constant speed. For that reason, this kind of wrist configuration is often used in painting applications, where a constant speed is strictly required to spray a uniform layer of paint over the workpiece.

In the rest of this chapter, we will focus on robots with spherical wrists and solve their inverse kinematics with closed-form equations.

Nonlinear Problem

Before we dive into the mathematical details of the solution, we need to understand a few characteristics of the problem we are about to tackle and why it can be so complicated.

First, we observe that the kinematic model is a highly nonlinear model. For instance, moving the TCP along a straight line does not mean that the joint axes are moving linearly (see Figure 4-3).

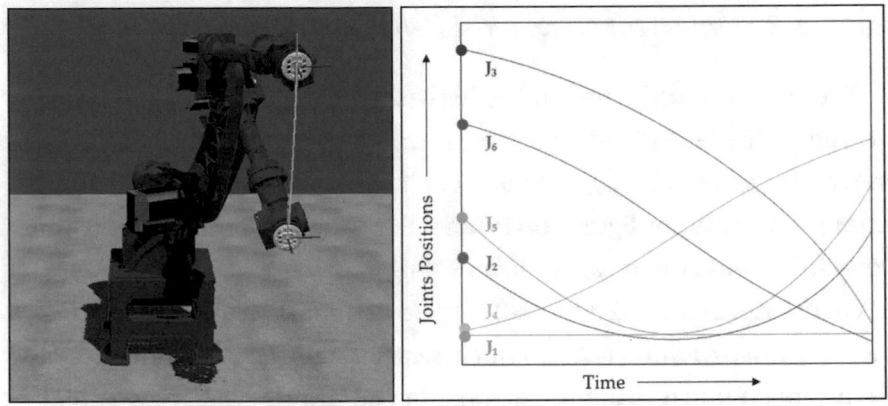

Figure 4-3. *A linear movement in the path space (left) is not linear in the joint space (right)*

The corresponding trajectories of the joints show that they are all performing nonlinear movements. The only axis that looks linear is the first joint, because it is in fact not moving at all, since the programmed TCP path only runs along the vertical Z axis and does not require a movement of the first joint. All other joints follow nonlinear position and speed profiles.

Nonlinearities can reach extreme effects around singularities, where minimal movements of the TCP require very large jumps of the joint axes. We will address this problem in detail when dealing with trajectory generation in Chapter 7.

Nonunique Solution

Another issue that makes solving inverse kinematics complicated is the fact that a generic target pose of the TCP might be reached by more than one configuration of the joint axes. In other words, the solution of the inverse transformation function is not always unique!

Let us consider a few examples. In Figure 4-4, we see a robotic arm reaching the exact same target position and orientation of its TCP using two different configurations of its joints:

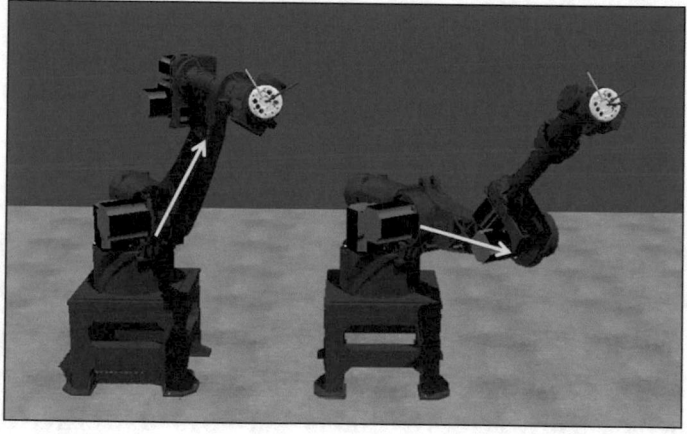

Figure 4-4. *UP and DOWN options to reach the same target*

We can call the two options UP and DOWN, because of the way the second and third links of the arm are aligned. These two solutions are both perfectly valid: the question is which one to select. The answer depends on a few conditions:

- *Safety*: One of the two solutions might now be acceptable because the resulting arm configuration would cause a collision with other objects in the environment.

- *Distance*: One of the two solutions is usually geometrically closer to the configuration that the joints have before starting the movement. In that case, we might as well select the closest option to minimize the movement time and reduce energy consumption.

- *Forced*: The operator might want to force either one of the two possible configurations for any reason related to the particular task the robot is performing. We need to provide that choice in our software interface.

Figure 4-5 shows another case of ambiguity between two possible valid solutions. We call these two configurations FRONT and BACK, according to the direction that the first joint points to. In the first case, the first joint is oriented toward the TCP, in the second case it points in the opposite direction, and the second joint needs to assume a very large negative value. It looks as a very uncomfortable position for the arm to work in. But it might be useful to access locations that are hard to reach from the top side.

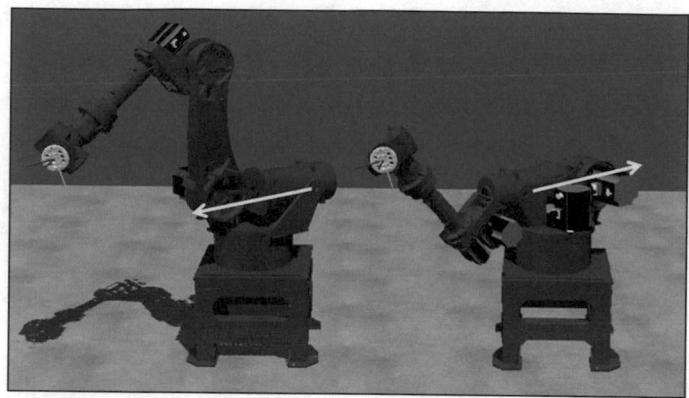

Figure 4-5. *FRONT and BACK options to reach the same target*

Finally, let's look at yet one other example in Figure 4-6. This one is more subtle.

Figure 4-6. *POSITIVE and NEGATIVE options to reach the same target*

At a first glance, the two joints' configurations look identical, but they are actually not. While the first three joints are in the same position and give the same pose to the main arm, the last three joints are oriented differently and generate two alternative wrist configurations.

The main difference is the value of the fifth axis, which has opposite values in the two cases: we call one solution POSITIVE, and the other NEGATIVE.

Singularities

From the examples shown so far, we can infer that there are always a few possible distinct solutions to choose from when solving the inverse kinematics problem for any generic TCP pose inside the workspace of the robot. However, it can be shown that there are extreme cases where the number of possible solutions is actually infinite.

These critical points are called **singularities** and occur for some specific target poses of the TCP. Let's look at some possible examples.

Figure 4-7. *Infinite solutions exist for a SHOULDER singularity*

In Figure 4-7, the target [X, Y] position of the TCP coincides with the origin of the robot [X=0, Y=0]. In this case, a fixed solution exists for most joints, except for J_1 and J_6, as those two axes are free to rotate synchronously without affecting the target pose of the TCP! This configuration is a special case of a shoulder singularity, which happens when the wrist point is aligned with the first axis. The value for J_1 can be chosen arbitrarily, and an infinite number of valid solutions exist.

In the particular case of Figure 4-7, we observe that the joints J_1 and J_6 are linearly dependent: we can calculate the correlation between them (i.e., their sum or their difference), but not their absolute values, which means we can choose among an infinite number of valid joint configurations for the robot.

The name singularity comes from linear algebra: a matrix is singular when its determinant is zero. The associate linear system has an infinite number of solutions because two or more variables are dependent on each other. In the case of the shoulder singularity, the values of J_1 and J_6 are the two dependent variables.

Another case of singularity for a six-axes robot is shown in Figure 4-8. Here the fifth joint axis J_5 is 0, which causes the joints J_4 and J_6 to be directly aligned with each other along the X axis of the TCP. The sum of J_4 and J_6 is therefore required to be equal to the orientation of the TCP around the X axis, but their individual values can be arbitrary. An infinite number of valid solutions exist.

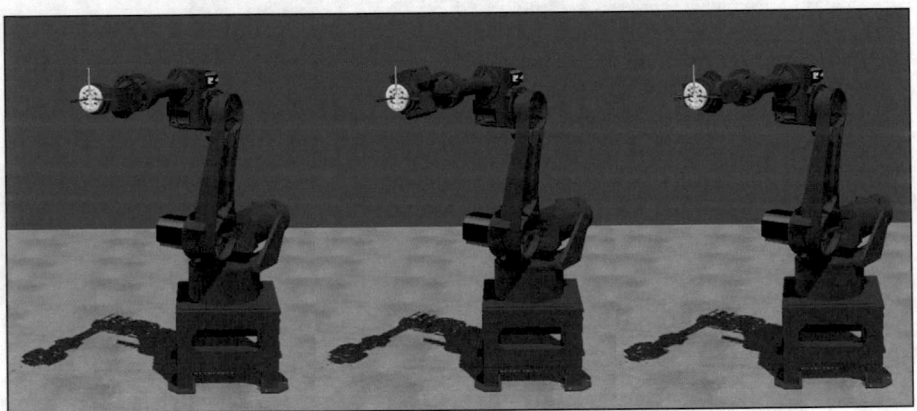

Figure 4-8. *The WRIST singularity*

Since this case involves the axes in the wrist of the robot, the configuration is called wrist singularity and is actually a very common trap for robots to fall into during normal operation, because as soon as the joint J_5 approaches its zero value, we hit the critical area. In fact, it even occurs at the home position of the robot, when all the joint axes have zero value.

In practice, solving the inverse kinematics in singular points is not problematic. It is enough to assign a specific value to one of the two dependent axes and immediately derive the value for the other joint.

The selected values are normally chosen to be close to the actual positions of the axes when approaching the singularity.

However, a far more complicated topic is handling singularities dynamically while planning a path and generating the trajectory for a movement. That is because the behavior of speeds and forces involved around singularities is highly nonlinear and their control becomes tricky. We will study the details in the Sections on Differential Kinematics and Dynamic Model.

It should be clear by now that the inverse kinematic model is more complex than forward one: not only it is more difficult to derive a solution, but sometimes there are many or even infinite possible solutions to pick from.

IK Step 1: Decoupling

Let's begin with the first step of the inverse kinematics: we start from the TCP, which is given, and we move back to the center of the spherical wrist to find the coordinates of the *wrist point* (WP).

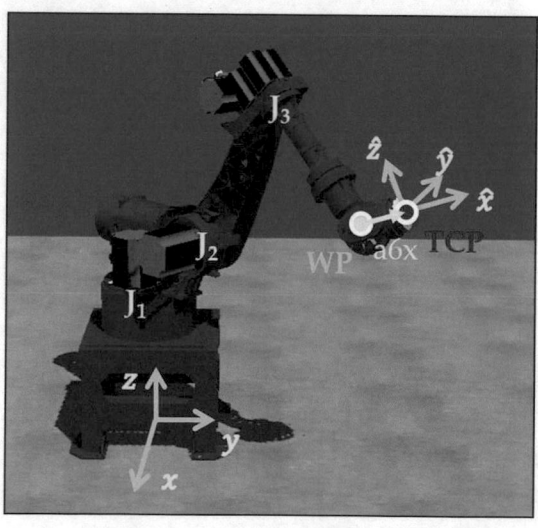

Figure 4-9. *Find the WP from the TCP*

By analyzing the kinematic chain in detail, we can observe that the position of the wrist point only depends on the first three joint axes of the robot $[J_1, J_2, J_3]$. Movements of the wrist axes $[J_4, J_5, J_6]$ *have no effect on the position of the wrist center point.*

This property allows for a convenient simplification of the inverse kinematics problem, by splitting the robot's body into two parts: the lower arm and the upper wrist. Given the position of the wrist point, we can first solve the lower arm and calculate the values of the first three joints. Then, using the previous result and the coordinates of the TCP, we solve the wrist and calculate the last three joint angles.

This operation is called **decoupling**: we decouple the wrist from the rest of the arm. Keep in mind that that this operation is only possible for robots with spherical wrists, because the movements of J_4, J_5, and J_6 do not modify the position of the wrist point.

Finding WP from TCP is a simple vector difference. Their distance is the mechanical offset $a6x$ along the X axis of the TCP frame, which we call $\hat{x}$ in the base frame (see Figure 4-9):

$$WP = TCP - a6x\,\hat{x} \tag{4-1}$$

Recall from the Column Vectors Section in Chapter 2 that $\hat{x}$ is simply the first column vector of the rotation matrix between the base frame and the TCP frame:

$$\hat{x} = RX = R\begin{bmatrix} 1 \\ 0 \\ 0 \end{bmatrix} = \begin{bmatrix} R_{11} & R_{12} & R_{13} \\ R_{21} & R_{22} & R_{23} \\ R_{31} & R_{32} & R_{33} \end{bmatrix}\begin{bmatrix} 1 \\ 0 \\ 0 \end{bmatrix} = \begin{bmatrix} R_{11} \\ R_{21} \\ R_{31} \end{bmatrix} \tag{4-2}$$

These coordinates express how the vector $X = [1\ 0\ 0]$ is seen from the base. The rotation matrix from base to TCP can be found by composing the orientation Euler angles of the TCP.

Putting all together, one step at the time,

1. Find R from [A, B, C]: $R = R_Z(C)\, R_Y(B)R_X(A)$

2. Extract $\hat{x}$ from R: $\hat{x} = \begin{bmatrix} R_{11} \\ R_{21} \\ R_{31} \end{bmatrix}$

3. Find WP from TCP: $\boldsymbol{WP = TCP - a6x\,\hat{x}}$

At this point, the robot is split in half, and the problem looks much easier to tackle. Let's solve the bottom half first.

IK Step 2: Solve the Arm

The subproblem we need to solve here is the following: given the position of a point WP, find the angles $[J_1, J_2, J_3]$ for which the three-axes arm reaches WP.

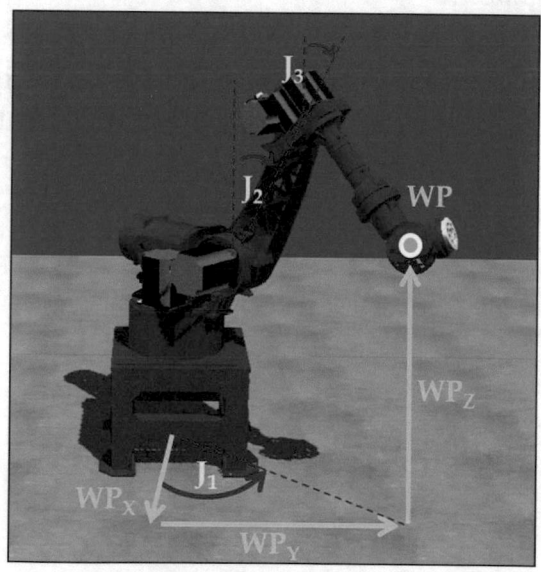

Figure 4-10. Finding $[J_1, J_2, J_3]$ from WP

Let's start from J_1. Since the first joint of the robot rotates around the base Z axis, it can only affect the WP_X and WP_Y coordinates of the wrist point, not its vertical position WP_Z.

The search for J_1 consequently reduces to a two-dimensional planar problem (see Figure 4-10) and is solved by observing that J_1 is the angle between the X and Y coordinates of WP:

$$J_1 = atan2\left(WP_Y, WP_X\right) \tag{4-3}$$

However, recall that two valid solutions exist, depending on the configuration that we select: either FRONT or BACK. In the first case, J_1 is the result of the arctangent function in Equation (4-3); in the second case, we need to add 180 degrees to rotate in the opposite direction.

Another issue is when both WP_X and WP_Y are 0. The wrist point is straight above the origin of the base frame: a shoulder singularity. Any position for J_1 would be a correct solution here. We can either use a value forced by the operator or hold the joint equal to its position just before reaching the singularity.

The next step consists in finding the angles J_2 and J_3. We need a few more geometrical intuitions. Figure 4-11 shows it all: it might seem daunting, but bear with me and we will solve it one step at the time.

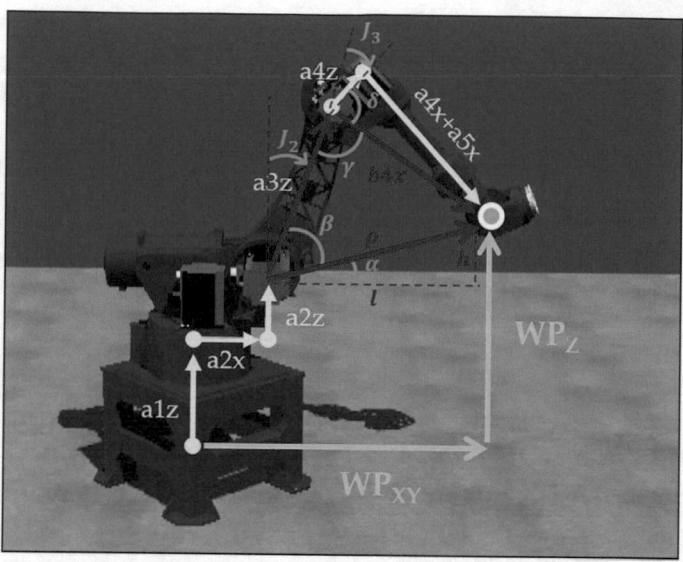

Figure 4-11. *Finding J_2 and J_3 requires some geometrical analysis*

Firstly, we observe that once J_1 is fixed, the movement of the arm caused by J_2 and J_3 is restricted to a vertical plane. We can simplify the problem by reducing it to a 2D geometry and focus our attention on the red triangle.

Let's start solving the triangle from its bottom side. We call its length ρ and its horizontal and vertical projections l and h.

Using the robot's base frame as reference, the start coordinates of ρ are $[a2x, a1z + a2z]$, while the end coordinates are $[WP_{XY}, WP_Z]$, where $WP_{XY} = \sqrt{WP_X^2 + WP_Y^2}$.

We can now quickly derive the values for l and h:

$$l = WP_{XY} \mp a2x \qquad (4\text{-}4)$$

$$h = WP_Z - a1z - a2z \qquad (4\text{-}5)$$

Whether to use a subtraction or a sum when finding l depends on the choice we made for J_1 in the previous step because the $a2x$ offset could be facing away from WP.

The length of the bottom side of the red triangle is then known:

$$\rho = \sqrt{h^2 + l^2} \tag{4-6}$$

The length of the left side of the triangle is also already known: it is the mechanical offset $a3z$.

Finally, the length of the top side of the triangle, which we call $b4x$, can also be quickly derived from its projections:

$$b4x = \sqrt{a4z^2 + (a4x + a5x)^2} \tag{4-7}$$

We have solved all the three sides of the triangle, but we need to make sure that the derived values are physically meaningful. The user could enter a TCP target position that is too far away for the robotic arm to reach, in which case our calculations would result in a large value for ρ. The arm cannot stretch far enough to reach that position, and no physical solution to the problem exists. The condition we need to impose for a realistic solution is as follows:

$$\rho \le a3z + b4x \tag{4-8}$$

On the other hand, the given TCP position should not be too close to the origin of the second joint; otherwise, the third link would not be able to reach it. The condition to impose in this case is as follows:

$$\rho \ge |a3z - b4x| \tag{4-9}$$

Mathematically speaking, Equations (4-8) and (4-9) represent the triangle inequality as a base condition to guarantee the triangle's existence.

Now that we know the length of the sides of the red triangle, we can calculate its angles by taking advantage of the law of cosines:

$$\cos\beta = \frac{\rho^2 + a3z^2 - b4x^2}{2\rho\, a3z} \tag{4-10}$$

You could directly use the arcsine to find beta, but a better approach to increase numerical stability is to use the arctangent: $\beta = \text{atan2}\left(\pm\sqrt{1-\cos^2\beta}, \cos\beta\right)$.

The two possible signs for the sine of β depend on the configuration we choose for the arm, either UP or DOWN, as previously discussed in the Section on Non-Unique Solution. In case we choose the DOWN pose, the second link will point downward and the third link upward. The resulting WP will be unchanged.

The value for α is simply $\alpha = \text{atan2}(h, l)$.

We now observe the 90 degrees angle centered in the J_2 joint stretching between the horizontal and vertical axes. The J_2 angle is as follows:

$$J_2 = \frac{\pi}{2} - \alpha - \beta \tag{4-11}$$

A couple of more steps are required to find J_3. We first apply the law of cosines one more time to find γ:

$$\cos\gamma = \frac{a3z^2 + b4x^2 - \rho^2}{2\,a3z\,b4x} \tag{4-12}$$

Then, we calculate δ from the size of the mechanical links:

$$\delta = \text{atan2}\left(a4x + a5x, a4z\right) \tag{4-13}$$

Finally, we derive J_3 by observing that the sum of γ, δ, and J_3 is always 180 degrees:

$$J_3 = \pi - \gamma - \delta \tag{4-14}$$

That was quite a bit of work, but we now have a solution for the first three joints of the robot. Let's move on to the wrist.

IK Step 3: Solve the Wrist

We need to find a value for the last three joints of the robot $[J_4, J_5, J_6]$ given the position and orientation of the lower arm $[J_1, J_2, J_3]$ and the TCP (see Figure 4-12). Let's focus on the orientation of those two parts.

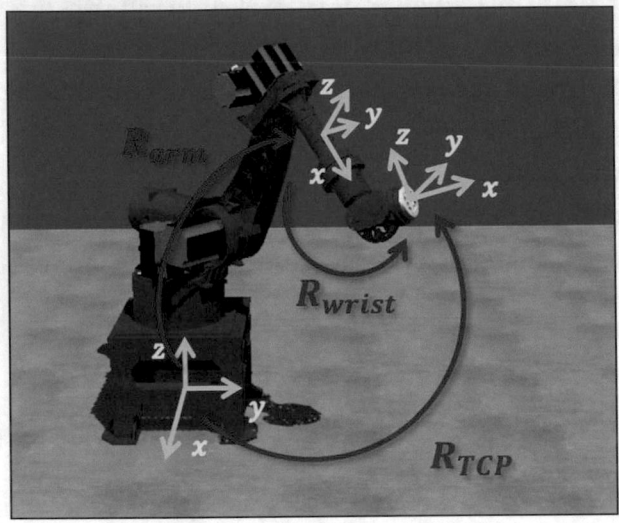

Figure 4-12. *Solving the wrist given the arm*

The orientation of the TCP with respect to the base frame is given by the Euler angles [A, B, C]. We can compose a rotation matrix R_{TCP} from them, as described in Chapter 2:

$$R_{TCP} = R_Z(C) R_Y(B) R_X(A)$$

(4-15)

The orientation of the lower arm with respect to the base frame is determined by the first three joint axes, which we just calculated in the previous step. We can compose a rotation matrix R_{arm} from them, keeping in mind their axes of rotation: J_1 rotates around Z, while both J_2 and J_3 rotate around Y.

$$R_{arm} = R_Z\left(J_1\right)R_Y\left(J_2 + J_3\right) = \begin{bmatrix} c_1 c_{23} & -s_1 & c_1 s_{23} \\ s_1 c_{23} & c_1 & s_1 s_{23} \\ -s_{23} & 0 & c_{23} \end{bmatrix} \qquad (4\text{-}16)$$

Note the compact notation $c_1 = \cos J_1$ and $c_{23} = \cos\left(J_2 + J_3\right)$.

The missing rotation between the lower arm and TCP is the rotation introduced by the wrist, which we call R_{wrist}. Recall that the product of rotation matrices is still a rotation matrix, and it is associative. We already observed in Equation (3-10) how the TCP orientation can be expressed by merging the two contributions from the arm and wrist:

$$R_{TCP} = R_{arm}R_{wrist} \qquad (4\text{-}17)$$

We already know R_{TCP} and R_{arm}. Finding R_{wrist} requires a pre-multiplication of both sides by R_{arm}^{-1}:

$$R_{arm}^{-1}R_{TCP} = R_{arm}^{-1}R_{arm}R_{wrist} \qquad (4\text{-}18)$$

$$R_{wrist} = R_{arm}^{-1}R_{TCP} = R_{arm}^{T}\, R_{TCP} \qquad (4\text{-}19)$$

Since R_{arm} is a rotation matrix, we take advantage of the property that its inverse is simply its transpose.

Once we know the total wrist rotation R_{wrist}, finding the individual axes $[J_4, J_5, J_6]$ is relatively straightforward. The rotation introduced by the wrist, starting from the lower arm and ending at the TCP, is solely due to the angles of those three joints (see Figure 4-13). All we need to do is decompose the wrist rotation matrix in the correct angles' notation.

Figure 4-13. *Finding J$_4$, J$_5$, and J$_6$ from the wrist rotation*

We recall that both J_4 and J_6 rotate around their local X axis, while J_5 rotates around its local Y axis. The matrix R_{wrist} is the combination of those three rotations:

$$R_{wrist} = R_X\left(J_4\right)R_Y\left(J_5\right)R_X\left(J_6\right) = \begin{bmatrix} c_5 & s_5 s_6 & s_5 c_6 \\ s_4 s_5 & c_4 c_6 - s_4 c_5 s_6 & -c_4 s_6 - s_4 c_5 c_6 \\ -c_4 s_5 & s_4 c_6 + c_4 c_5 s_6 & -s_4 s_6 + c_4 c_5 c_6 \end{bmatrix} \quad (4\text{-}20)$$

Extracting the individual angles from the matrix is a process we already learned in Chapter 2, when we decomposed a generic rotation matrix into Euler angles. The only difference is that the matrix was built from an X-Y-Z combination of rotations, while here we start from an X-Y-X form. The concept is absolutely the same though.

We indicate each element of the matrix with the symbol R_{ij}, where i is the row and j the column. From the very first element of the matrix, we can already derive the value of J_5:

$$J_5 = atan2\left(\pm\sqrt{1-\left(R_{11}\right)^2}, R_{11}\right) \quad (4\text{-}21)$$

As usual, take the arctangent for best numeric results. The ± sign depends on the desired configuration, either POSITIVE or NEGATIVE. Both are acceptable solutions and the final TCP pose will be identical, but the robot's joint configuration will look a bit different.

Once J_5 is fixed, we use the remaining elements on the first column to quickly derive J_4:

$$J_4 = atan2\left(\pm R_{21}, \mp R_{31}\right) \tag{4-22}$$

Similarly, from the elements on the first row, we derive J_6:

$$J_6 = atan2\left(\pm R_{12}, \pm R_{13}\right) \tag{4-23}$$

We have now calculated all the six joint values of the robot, and it looks like we are all set. But there is one more little catch: what if the first element of the wrist rotation matrix is 1 and therefore $J_5 = 0$? In that case, we are facing a wrist singularity, a condition we already described in the Section on Singularities.

When $J_5 = 0$, there is no rotation around the Y axis in the wrist, only two consecutive rotations around X, one from J_4 and one from J_6. We know what the final value of the total rotation needs to be, but there is no way to separate the individual contributions from J_4 and J_6. The matrix simplifies to the following:

$$R_{wrist} = \begin{bmatrix} 1 & 0 & 0 \\ 0 & c_4c_6 - s_4s_6 & -c_4s_6 - s_4c_6 \\ 0 & s_4c_6 + c_4s_6 & -s_4s_6 + c_4c_6 \end{bmatrix} = \begin{bmatrix} 1 & 0 & 0 \\ 0 & c_{46} & -s_{46} \\ 0 & s_{46} & c_{46} \end{bmatrix} \tag{4-24}$$

The symbols s_{46} and c_{46} represent the sine and cosine of $(J_4 + J_6)$, derived via standard trigonometric identities: $s_{46} = s_4c_6 + c_4s_6$.

Only the sum of the J_4 and J_6 angles is known, because the wrist in its singular configuration essentially reduces to a single rotation about the X axis:

$$J_4 + J_6 = atan2\left(R_{32}, R_{33}\right) \tag{4-25}$$

The solution is to manually fix one of the two angles, for example, holding J_4 equal to its current value right before hitting the singularity and then finding the other one accordingly.

One final tip: since angles are periodic, a 2π offset makes no difference to them. However, joint values are commanded to motors, which rotate the entire robot axes around. A 2π offset can cause a big rotation difference in practice, with possible consequences in the speed, energy, and safety of the resulting movement. We always try to minimize the robot's movements distances to reach a target pose, so make sure to add or subtract 2π when needed to achieve the most convenient solution.

Numerical Test

You now have a solution for the inverse kinematic model. We provide here a test set for you to compare your calculations, in case you are coding along.

The mechanical parameters are the same we used in Chapter 3 when testing the direct transformations (see Table 4-1).

Table 4-1. *Values of mechanical parameters for the example robot*

a1z	a2x	a2z	a3z	a4x	a4z	a5x	a6x
650	400	680	1100	766	230	345	244

We start from the homing position (see Figure 4-14). This time the input values are the TCP axes, and the outputs are the joint angles.

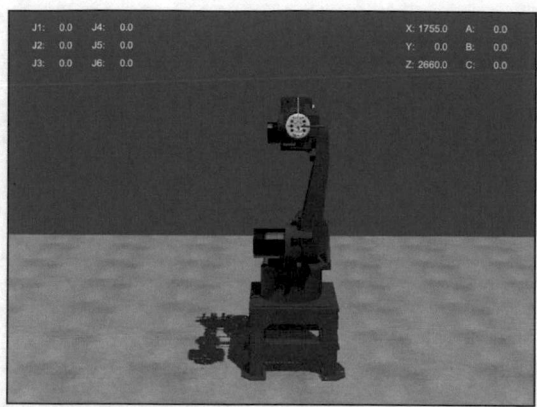

Figure 4-14. *Test Position #1*

You should find all the joints to be 0 for this particular position and orientation of the TCP (see Table 4-2).

Table 4-2. *Test Position #1 (home position)*

TCP		Joints	
X	1755	J_1	0
Y	0	J_2	0
Z	2660	J_3	0
A	0	J_4	0
B	0	J_5	0
C	0	J_6	0

Then we jog the robot to a few random positions in space as shown in Figure 4-15.

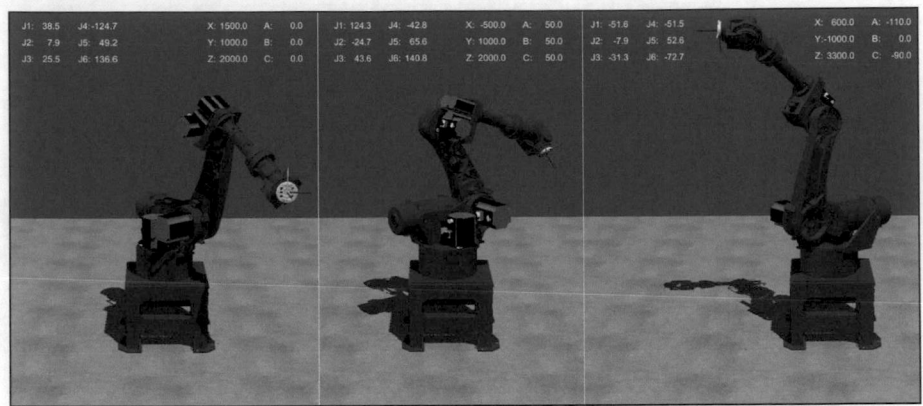

Figure 4-15. *Test Positions #2, #3, and #4*

The programmed path space poses and the resulting values in the joint space are reported in Table 4-3, Table 4-4, and Table 4-5.

Table 4-3. *Test Position #2*

	TCP		Joints
X	1500	J_1	38.5
Y	1000	J_2	7.9
Z	2000	J_3	25.5
A	0	J_4	-124.7
B	0	J_5	49.2
C	0	J_6	136.6

Table 4-4. *Test Position #3*

TCP		Joints	
X	-500	J_1	124.3
Y	1000	J_2	-24.7
Z	2000	J_3	43.6
A	50	J_4	-42.8
B	50	J_5	65.6
C	50	J_6	140.8

Table 4-5. *Test Position #4*

TCP		Joints	
X	600	J_1	-51.6
Y	-1000	J_2	-7.9
Z	3300	J_3	-31.3
A	250	J_4	-51.5
B	0	J_5	52.6
C	-90	J_6	-72.7

Keep in mind that the results for the joint axes are not unique. They depend on the various possible pose configurations you select (UP/ DOWN, FRONT/BACK, POSITIVE/NEGATIVE) and also on multiples of $\pm 2\pi$ offsets you might add along.

Zero Frame

The solution we derived for the inverse kinematics assumes that the input TCP position and orientation are given with respect to the robot's own frame base, also known as machine coordinate system (MCS).

In practical applications, especially when working with multiple robots in the same cell, it is more natural to assume the target pose specified with respect to a global coordinate system (GCS), as shown in Figure 4-16.

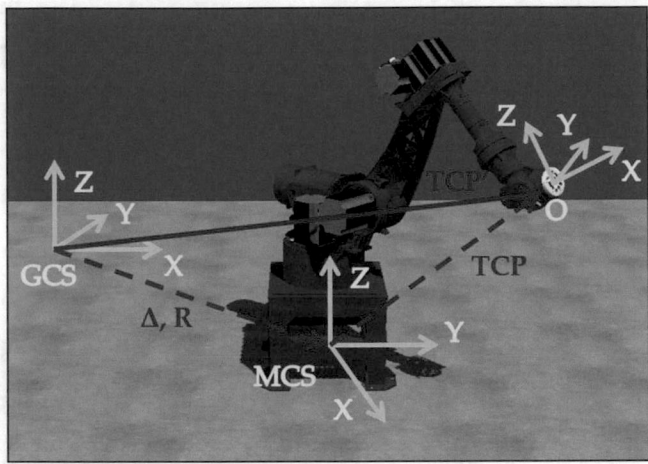

Figure 4-16. *Removing the zero frame*

In that case, we first need to move back to our robot's base frame and then solve locally. In other words, we need to find the actual TCP coordinates in the local frame, given the input TCP coordinates in the global frame. The difference between the two frames is given by a translational offset Δ and a rotation matrix R, as usual:

$$T_{Base} = \begin{bmatrix} R & \Delta \\ 0 & 1 \end{bmatrix} \tag{4-26}$$

Removing a frame requires an inverted transformation:

$$TCP = T_{Base}^{-1} \, TCP'$$

(4-27)

The procedure to invert a homogeneous transformation was described in the Section on Inverted Transformation in Chapter 2:

$$T_{Base}^{-1} = \begin{bmatrix} R^T & -R^T \Delta \\ 0 & 1 \end{bmatrix}$$

(4-28)

Finding the position of the TCP as seen from the local frame is then given by the following:

$$TCP_{XYZ} = T_{Base}^{-1} \, TCP'_{XYZ}$$

(4-29)

To find the orientation, we only need the rotation part of the matrix, whose inverse is equal to its transpose:

$$TCP_{ABC} = R^T TCP'_{ABC}$$

(4-30)

Tool Frame

A similar concept applies to the tool. Given the TCP, we first need to find the mounting point by removing the additional tool frame and then from there solve using the standard inverse transformation function (see Figure 4-17).

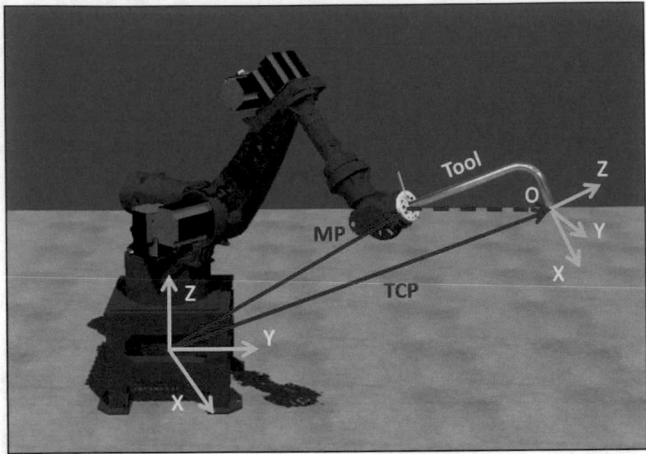

Figure 4-17. *Removing the tool frame*

Removing the tool frame also requires inverting the homogeneous transformation for the tool position and orientation. However, since the tool comes after the MP, we need to post-multiply the matrices:

$$MP = TCP \, T_{Tool}^{-1} \qquad (4\text{-}31)$$

Finding the position of the MP requires the entire homogeneous matrix:

$$MP_{XYZ} = T_{TCP} \, T_{Tool}^{-1} \begin{bmatrix} 0 \\ 0 \\ 0 \\ 1 \end{bmatrix} \qquad (4\text{-}32)$$

To find the orientation, we only need the rotation part of the matrices:

$$R_{MP} = R_{TCP} \, R_{Tool}^{-1} = R_{TCP} \, R_{Tool}^{T} \qquad (4\text{-}33)$$

Mechanical Coupling

We introduced the concept of mechanical coupling when we solved the forward kinematics: the movement of one axis can affect the position of a different axis regardless of whether its joint motor is moving or not. We used coupling coefficients to compensate for that effect in the calculations. Here we use the same process in reverse.

The output of the inverse kinematics is the position of the real joints of the robot. Normally, those positions are passed directly to the motors to control the robot.

However, if a mechanical coupling is present, we need to correct the calculated values, because the real joint positions do not correspond to the actual motors' angles anymore (see diagram in Figure 4-18).

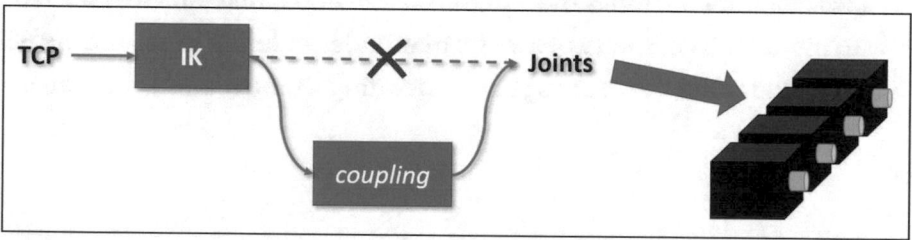

Figure 4-18. *Introducing mechanical coupling compensation after the inverse kinematics*

Mathematically, we adjust the output of the inverse transformations by subtracting the effects of the coupling coefficients. For example, in a standard six-axes robot, the set position of the fifth and sixth axes that we need to send to the motors is the difference between the target position of the joints calculated by the IK and the coupling offsets caused by the movement of the fourth axis:

$$J_5 \leftarrow J_5 - J_4 \, c_{45} \tag{4-34}$$

$$J_6 \leftarrow J_6 - J_4 \, c_{46} \tag{4-35}$$

Summary

Congratulations if you made it this far: you have now mastered the technique of solving the inverse kinematic model of a standard six-axes robot. The journey was not simple: we saw that the problem is highly nonlinear and that it often degenerates into singularities.

The critical step in allowing for a closed-form solution was the decoupling between the arm and wrist. If the geometry of the robot does not allow for that, then you are stuck with numerical approximations.

Just as in the case of forward kinematics, we also saw how to add a zero frame, a tool frame, and compensation for mechanical couplings. These practical details are often needed in real applications, and you should integrate them into your software control library.

This chapter concludes the first part of the book, in which we learned how to model the geometry of a static manipulator. In the next part, we will add life to the robot by planning and generating its movements in space.

PART II

Robot Movements

The kinematic model analyzed in the previous chapter allows us to calculate individual positions of the robot: for example, two fixed points A and B. What we are missing now is a way to smoothly move the robot from A to B without a sudden jump of its axes, as shown in Figure II-1.

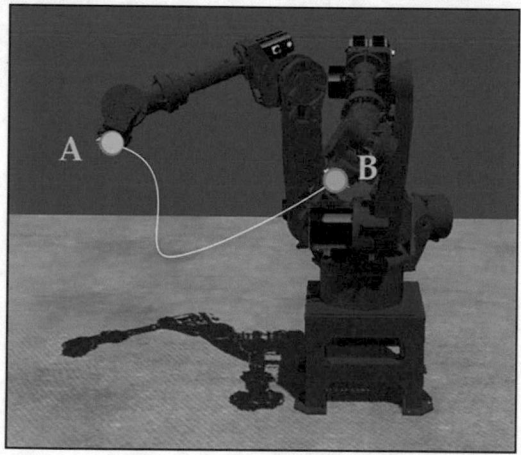

Figure II-1. *Smooth movement between two points in space*

This second part of the book is all about planning the geometrical path of a movement, making sure it is safe to execute, and then dynamically generating the correct speed for the joint motors in order for the TCP to move along the planned path.

CHAPTER 5

Path-Planning

A **path** is the geometrical description of the robot's movement in space. Given two points A and B, there are many different (actually infinite) possible paths to connect them. A straight line is the simplest path, but any kind of other curve could be programmed (see Figure 5-1).

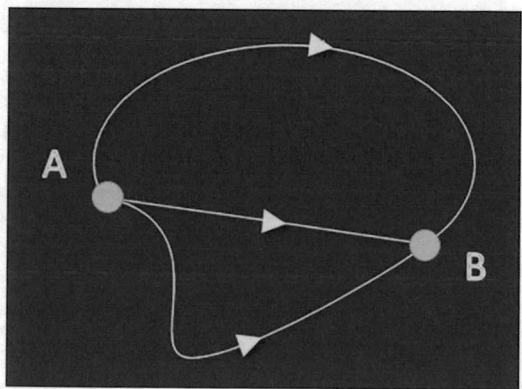

Figure 5-1. *The robot can cover different paths between start and end point*

By path-planning, we mean deriving the equation of the curve that connects the starting and ending points of a movement. When the operator chooses a curve for the robot to follow, we translate that curve into a mathematical equation and then execute it with a specific time-dependent **trajectory**. Path and trajectory are two separate topics, although they are closely related to each other. We talk about paths in this chapter, while we will study trajectories in Chapter 7.

© Fabrizio Frigeni 2023
F. Frigeni, *Industrial Robotics Control*, Maker Innovations Series,
https://doi.org/10.1007/978-1-4842-8989-1_5

The coordinates of the target point for each movement can be expressed either in the joint space $[J_1...J_6]$ or in the path space [X, Y, Z, A, B, C]. Similarly, the movement to reach that target point can be planned either in the joint or in the path space.

As we saw in the Nonlinear Problem Section in Chapter 4, the two spaces are not linearly dependent with each other. A linear interpolation of the joint axes to reach the target point will not result in a linear movement of the TCP in the path space. Vice versa, a line in the path space will appear as a line to an external observer looking at the TCP but will generate seemingly random movements when monitoring the motor's velocity profiles.

The former kind of movement is called **PTP** (point-to-point) and is planned in the joint space. The latter kind is a **path-interpolated movement** in the path space. Figure 5-2 shows the resulting paths in the two cases.

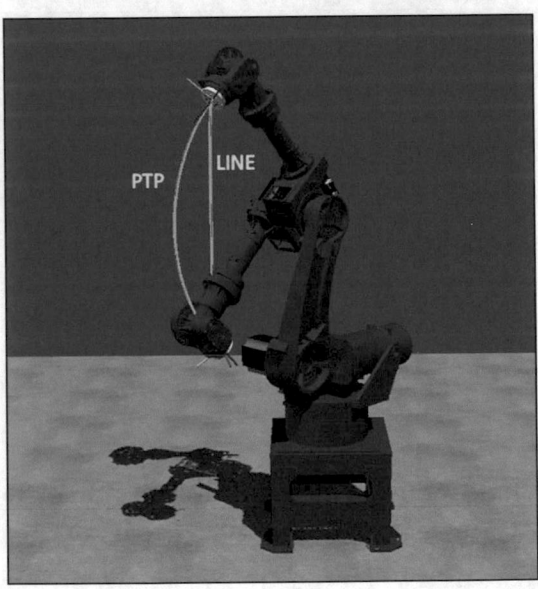

Figure 5-2. *Movements can be planned in the joint or in the path space*

Since the operator is free to decide what movement to command the robot to execute, we need to understand and be able to plan both kinds.

PTP Movements

A point-to-point movement is a **linear interpolation** of the joint axes of the robot.

Imagine we start from point P_0, where all the joints have certain defined values $[J_1...J_6]$, and we want to reach point P_1, where the joints have different values $[J_1'...J_6']$. Interpolating between P_0 and P_1 means that we linearly increase or decrease the joint angles, from their starting values to their target ones, as shown in Figure 5-3.

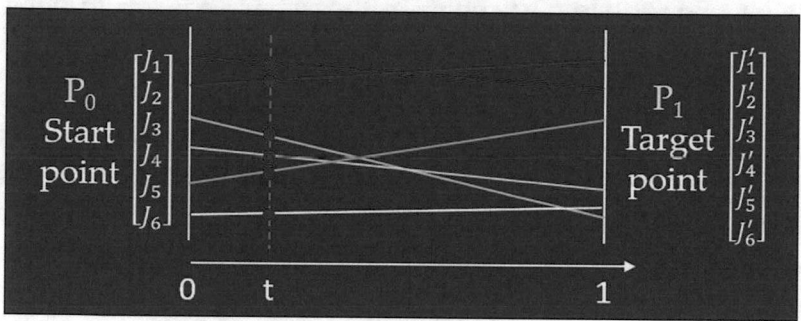

Figure 5-3. *Linear interpolation of the joint axes*

Mathematically, a linear interpolation is described using the following parametric equation:

$$P = P_0(1-t) + P_1 t \quad t \in [0,1] \tag{5-1}$$

The parameter t sweeps the entire movement, starting from P_0 for $t = 0$ all the way to P_1 for $t = 1$.

A key point to understand here is that the parameter t does not represent time. The variation of t from 0 to 1 describes the geometrical path in space, but not how the path is covered in time. The time variable only belongs to trajectories and is the topic of Chapter 7. At this stage, we only plan the geometry, not the speed of the movement.

A second important concept to grasp is that the described linear interpolation happens entirely in the joint space. The joint angles move linearly from their starting position to their target value. However, this linear movement of the joints does not translate at all into a linear movement of the TCP. Imagine the first joint of the robot moving linearly from 0 to 360 degrees: the TCP actually moves around a whole circle in space!

The major consequence of this observation is that PTP movements offer no control over the actual path of the end effector of the robot. These movements are very simple to program and very fast to execute, but they are inherently unsafe because it is hard to intuitively predict their behavior in the path space.

At planning time, we do not know what the position and orientation of the TCP will look like. During the movement, we can cyclically call the direct transformations to find out, in case we need to.

Let's look at a practical example: a linear interpolation of the first joint from 0 to 360 degrees. Figure 5-4 shows the positions of the robot at $t = 0, \frac{1}{3}, \frac{2}{3}, 1$.

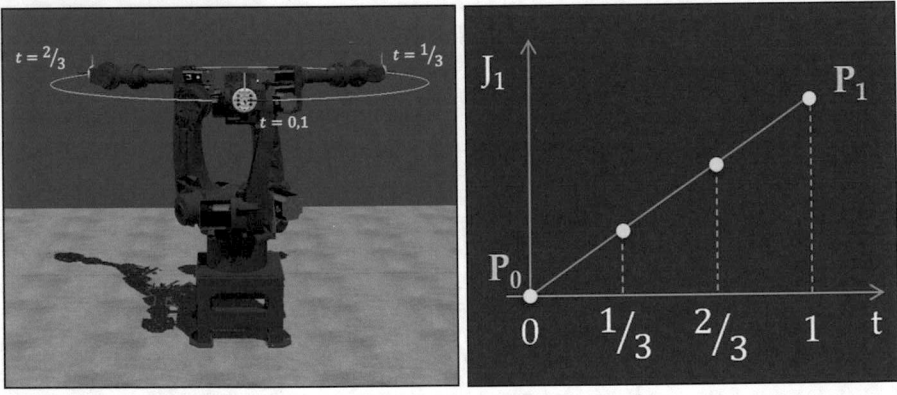

Figure 5-4. *A PTP movement of the robot*

Clearly, the resulting TCP movement is not a line. Actually, if we had planned a linear interpolation in the path space for the TCP using the same starting and ending points, *the robot would have not moved at all!* That is because the two points $P_0 = [0, 0, 0, 0, 0, 0]$ and $P_1 = [360, 0, 0, 0, 0, 0]$ (in joint coordinates), while being very distant from each other in the joint space, are actually the same coincident point in the path space.

Remember that t is not time. While t increases from 0 to 1, the influence of P_0 over P decreases, while the influence of P_1 increases, until eventually, at $t = 1$, P reaches the target point and completely forgets about the starting point.

Notice that the starting and ending points of a PTP movement must be expressed in joint coordinates in order for the interpolation to be executed. In case the operator only provides the point coordinates in path space, we first need to transform them using the inverse kinematic function and then proceed to plan the movement.

PTP movements are typically used for jogging individual joint axes, with the intention of moving one single link of the robot without caring about the actual path of the TCP.

A characteristic of PTP movements is that they are time-optimal, in the sense that they always take the shortest possible time to complete from start to end given the speed limits of the motors. Achieving the same feature in the path space is possible but not easy, as we will learn in the Section on Time-Optimal Movements in Chapter 7.

Finally, and most importantly, *a PTP movement is the only movement that allows a modification of the actual joint configuration.* Figure 5-5 shows a robot reconfiguring from the UP to the DOWN pose. Attempting to do that with a path-interpolated movement would not be possible, as it would lead the TCP to leave the programmed path.

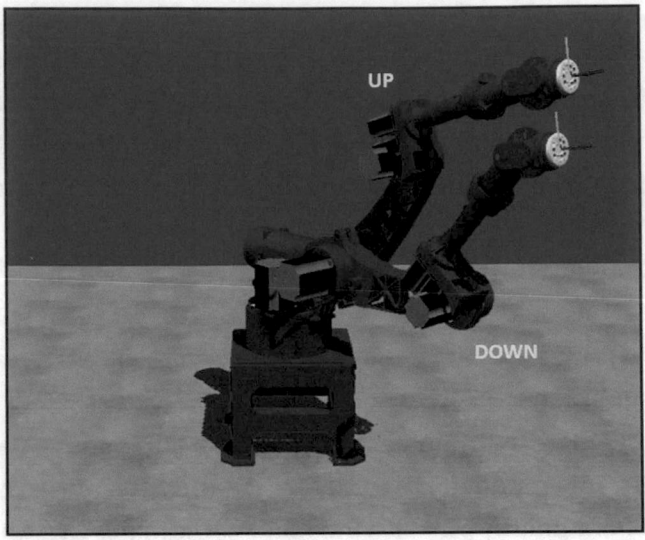

Figure 5-5. *Only PTP movements allow for a reconfiguration of the joints*

A PTP movement allows for total control over the joint values. Path-interpolated movements cannot do that, since the configuration stays constant along the programmed path. For instance, there is no way to replicate the movement of Figure 5-5 while forcing the TCP to follow a line in space.

On the other hand, while point-to-point movements are nice-to-have features in some cases, in the vast majority of practical situations, the operator wants the end effector of the robot to exactly follow a specific path. For that reason, we now turn our attention to path-interpolated movements.

Path Movements

Path-interpolated movements are entirely planned in the path space. The robot we are working with has six degrees of freedom, so when describing a target pose for the TCP in space, we need to use six coordinates: three for the position [X, Y, Z] and three for the orientation [A, B, C].

Both the starting and ending points of a path-interpolated movement must be expressed in the path space coordinates in order for the interpolation to be executed. In case the operator only provides the point coordinates in joint space, we first need to transform them using the forward kinematic function and then proceed to plan the movement.

The simplest way to interpolate between two points is a linear interpolation. Given that we are now working in the path space, the resulting movement of the TCP will be a line.

You might be tempted to think that we can simply apply the same formula (Equation (5-1)) that we used for PTP to the two points expressed in path space coordinates. It turns out that you can do that only for the position part of the coordinates, not for the orientation (Figure 5-6). As we mentioned in the Expressing Rotations Section in Chapter 2, Euler angles do not interpolate linearly.

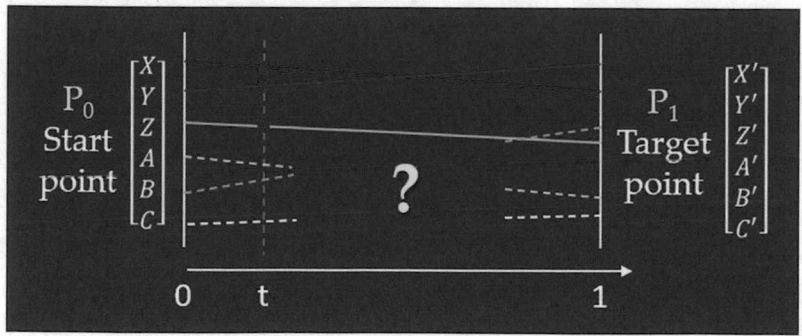

Figure 5-6. *Only position coordinates can be directly interpolated in the path space*

Positions can be directly interpolated because we are working in a Euclidean space, where vectors add up linearly. Think of it this way: if you start from your current position, then move a step left, then one forward, then one right, and finally one backward; you end up going back to the initial position.

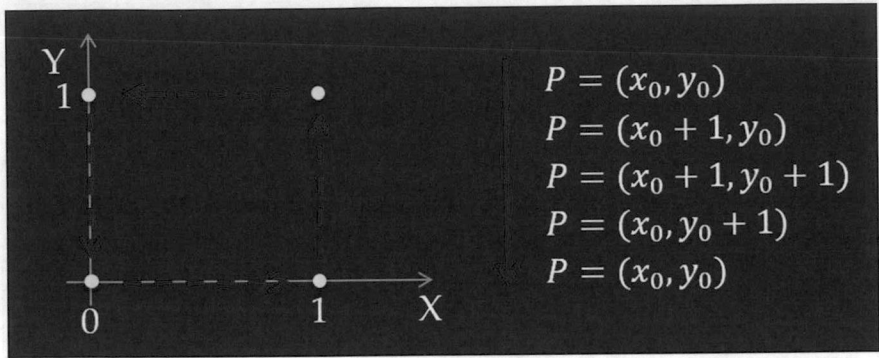

Figure 5-7. *Vectors add up linearly in a Euclidean space*

The same simple thought experiment does not work for Euler orientation angles: *vectors on a sphere do not add up linearly*. A sphere is not a Euclidean space, unlike the three-dimensional [X, Y, Z] space, and measuring angular distances requires a different kind of geometrical framework.

Let's look at a simple example, as shown in Figure 5-7. We take the yellow frame as fixed reference and rotate the colored axes with the sequence A+90, C+90, A-90, and C-90, where the Euler angle A is a rotation around the fixed X axis and the angle C is a rotation around the fixed Z axis. If angles would add up linearly as positions do, we would end up going back to the same initial orientation. However, the test shown in Figure 5-8 clearly shows a different result.

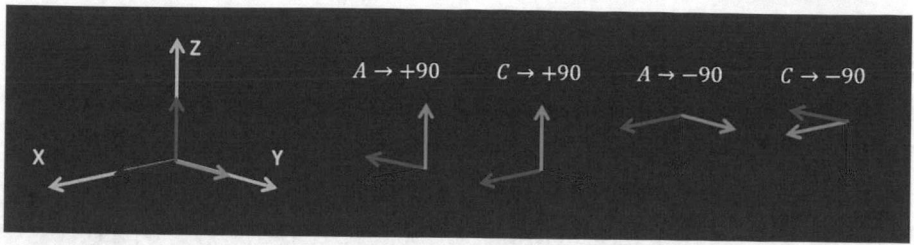

Figure 5-8. *Euler rotation angles do not add up linearly*

We actually already know that the correct way of finding the target rotated frame is to pre-multiply by an additional rotation matrix. However, matrices are not suitable for continuous interpolations: they cannot be parameterized with one single variable running from 0 to 1.

The solution is to introduce a new way of expressing rotations: **quaternions**. They are more convenient to handle than matrices, and they are able to interpolate linearly between two given orientations. We have to briefly introduce quaternions and study how to interpolate them, before being able to continue planning path movements.

Quaternions

Quaternions are simply another way to express orientations, just like Euler angles and rotation matrices. We are not going to study all quaternions algebra here, because it would take too long and is not strictly required for our purposes. We only present some properties relevant to robotics movements.

The general expression for a quaternion q is given by the following:

$$q = x\mathbf{i} + y\mathbf{j} + z\mathbf{k} + w \tag{5-2}$$

$i, j,$ and k are imaginary numbers, specifically square roots of -1:

$$i^2 = j^2 = k^2 = -1 \tag{5-3}$$

115

The first part of a quaternion $(x\boldsymbol{i} + y\boldsymbol{j} + z\boldsymbol{k})$ is a vector, while the second part (w) is a scalar. *You can intuitively imagine a quaternion as a vector pointing along a direction in space, plus a scalar rotation around that direction to uniquely define an orientation.*

The name quaternion comes from the number 4 (*quater* meaning "four times" in Latin), representing the four elements [x, y, z, w] it takes to describe an orientation, as opposed to three elements for Euler angles and nine for a rotation matrix.

When working with quaternions, it is customary to use them in a normalized form, called *unit quaternion*:

$$\boldsymbol{q} = \frac{q}{\|q\|} \tag{5-4}$$

The norm of the quaternion is simply the Euclidean size of the vector of elements [x, y, z, w]:

$$\|q\| = \sqrt{x^2 + y^2 + z^2 + w^2} \tag{5-5}$$

The first and most important property we learn is that any given orientation in space can be uniquely represented by one unit quaternion q (see Figure 5-9), along with its negative $-q$, which also represents the same orientation. This uniqueness is unlike Euler angles, with which many representations are possible for the same orientation.

In fact, given an axis of rotation v and an angle θ, we can uniquely build the corresponding unit quaternion with the following formula:

$$q = cos\frac{\theta}{2} + \left(v_x\boldsymbol{i} + v_y\boldsymbol{j} + v_z\boldsymbol{k}\right)sin\frac{\theta}{2} \tag{5-6}$$

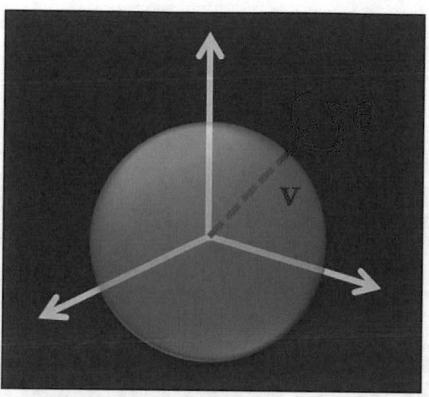

Figure 5-9. *Building a quaternion from a vector and an angle*

Similar to rotation matrices, quaternions can be multiplied by each other to add consecutive rotations. Consider the example in Figure 5-10. Given a starting frame with orientation q_1, and applying a rotation q_2 (around the Y axis), the resulting orientation can be found as follows:

$$q = q_2 q_1 \tag{5-7}$$

We build q_2 according to Equation (5-6) given the direction and angle of the desired rotation.

The resulting product of two quaternions is still a quaternion, and its expression is a lengthy formula to write down. We can simplify the notation by separating the vector and scalar part of each quaternion as follows:

$$q_1 = \boldsymbol{q_1} + w_1 \tag{5-8}$$

Then, the vector and scalar parts of the product $q = q_2 q_1$ are given by the following expressions:

$$\boldsymbol{q} = w_1 \boldsymbol{q_2} + w_2 \boldsymbol{q_1} + \boldsymbol{q_2} \times \boldsymbol{q_1} \tag{5-9}$$

$$w = w_2 w_1 - \boldsymbol{q_2} \cdot \boldsymbol{q_1} \tag{5-10}$$

117

Note that the product is *not commutative*: the order of rotation is always important. Rotating q_1 by q_2 is a totally different operation than rotating q_2 by q_1.

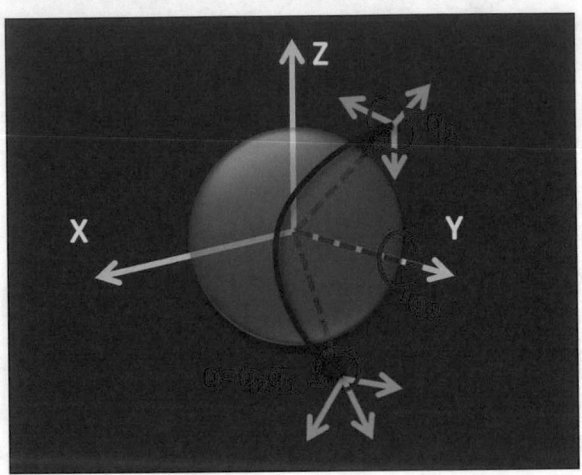

Figure 5-10. *Rotating a frame using a multiplication of quaternions*

Quaternions have several advantages over the other orientation representations we studied before. Compared to Euler angles, *quaternions do not suffer from singularities*: they offer a unique undisputable way to identify rotations. Also, as we will learn very soon, they offer an easy way to *linearly interpolate between orientations*. These are two massive advantages for robotic applications and come at the cost of only one additional parameter.

Compared to rotation matrices, quaternions are more compact: they only use four elements instead of nine. Matrices are also impossible to interpolate, and *they are not numerically stable*. It is possible for a rotation matrix that undergoes several computations to turn out non-orthogonal anymore because of numerical approximations. In that case, the matrix is very difficult to recover. On the other hand, a quaternion can be renormalized at any time using Equation (5-4) and is therefore *very robust against approximations*.

SLERP

At last, we come to study how quaternions interpolate, which is actually the main reason we introduced them in the first place.

Given two orientations, represented by the two quaternions q_1 and q_2 as shown in Figure 5-11, we need to find a quick and simple way to smoothly transition between them. In other words, we are looking for a parametric formula to calculate the generic quaternion q in between q_1 and q_2, as a function of a single parameter t varying from 0 to 1.

You can imagine q_1 being the current orientation of a robot and q_2 the target orientation we want to reach at the end of the movement.

There are actually infinite ways to connect two points on a sphere, but we are interested in the shortest solution shown in yellow. It is sometimes called *torque-optimal* solution because it requires the least amount of energy to move from the starting to the ending orientation.

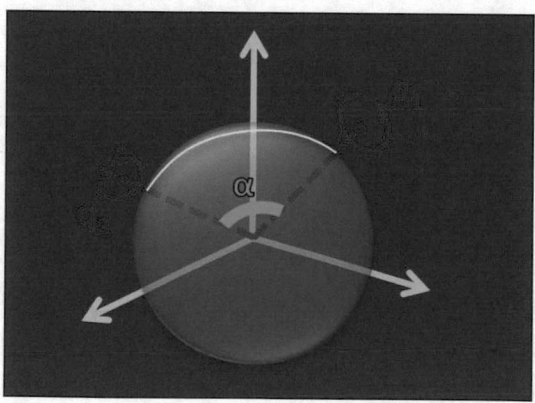

Figure 5-11. *Interpolating between q_1 and q_2*

The optimal solution is given by the following equation:

$$q = \frac{sin(\alpha(1-t))}{sin\,\alpha}q_1 + \frac{sin(\alpha t)}{sin\,\alpha}q_2 \qquad (5\text{-}11)$$

This kind of interpolation is called **SLERP**, which stands for *Spherical Linear intERPolation*.

By sliding the parameter t from 0 to 1, we cover all the quaternions between the initial orientation q_1 and the final value q_2. The formula is essentially equivalent to a linear interpolation in the Euclidean space, except that this time the movement takes place on a sphere instead of a plane.

Similar to what we observed when interpolating positions, the parameter t does not represent time. The same path can be traversed at different speeds depending on how t varies in time. For example, rotating the TCP at a constant angular speed requires a linear increase of t in time. The details will be discussed when introducing trajectories in Chapter 7.

The other unknown variable in the formula is α, the angle between the two quaternions. Calculating α is analogous to deriving the angle between two vectors on plane:

$$\alpha = \cos^{-1} \frac{q_1 \cdot q_2}{\|q_1\| \|q_2\|} \tag{5-12}$$

The numerator is a scalar value given by the dot product between the two quaternions:

$$q_1 \cdot q_2 = x_1 x_2 + y_1 y_2 + z_1 z_2 + w_1 w_2 \tag{5-13}$$

The arccosine function returns two different angles: either the shortest, highlighted in green in Figure 5-11, or the opposite angle around the back side of the sphere. Using either angle for the interpolation would lead to the same final orientation at the end of the movement. However, the rotation process during the movement would be different.

To guarantee that we always use the shortest angle when planning a movement, we need to make sure that the dot product between the two quaternions at the numerator is positive before taking the arccosine. If we happen to have a negative dot product, then we should invert it by inverting one of the two quaternions. For example, we can use the target orientation $-q_2 = -x_2\boldsymbol{i} - y_2\boldsymbol{j} - z_2\boldsymbol{k} - w_2$. The negative of a quaternion still represents the same exact orientation as the original, but the rotation path taken to reach it lies on the opposite side of the quaternion sphere shown in Figure 5-11.

Another potential issue we need to address is the case when the starting and ending orientations of the movement are very similar, which causes the two quaternions to come next to each other. A small angle between them forces the denominators in Equation (5-11) to approach 0, introducing numerical instabilities in our code. To avoid any complication, we simply observe that a small arc on the sphere essentially degrades into a linear segment, so that we can safely fall back to a standard linear interpolation between q_1 and q_2 as a good approximation and guarantee a more robust calculation:

$$q = q_1(1-t) + q_2 t \tag{5-14}$$

The result of the interpolation should be normalized to make sure the final quaternion is valid. This simplified interpolation is often referred to as NLERP (normalized linear interpolation).

Figure 5-12 shows two examples of the SLERP interpolation in practice. The position of the TCP during these movements is held fixed, while the orientation axes are rotated. The projection of the movement at a fixed distance from the TCP describes a circle in space.

Figure 5-12. *Some SLERP rotations in practice*

A step-by-step procedure to program a robot's rotation works as follows (see Figure 5-13):

- The robot is in the current orientation $[A_1 \quad B_1 \quad C_1]$, and the operator provides a target orientation $[A_2 \quad B_2 \quad C_2]$. Euler angles are convenient when interfacing with the operator, but they cannot be interpolated directly. We need to transform them into quaternions first.

- The easiest way to go from Euler angles to quaternions is to take an extra step in between and use rotation matrices. We first compose the rotation matrix from the given Euler angles as we learned in Chapter 2. Then, we decompose the rotation matrix into the corresponding quaternion. There are a few different ways to do that; one possible formula is the following (where R_{ij} is the element of the rotation matrix on row i and column j):

$$\begin{cases} w = \dfrac{1}{2}\sqrt{1 + R_{11} + R_{22} + R_{33}} \\[2mm] x = \dfrac{1}{4w}\left(R_{32} - R_{23}\right) \\[2mm] y = \dfrac{1}{4w}\left(R_{13} - R_{31}\right) \\[2mm] z = \dfrac{1}{4w}\left(R_{21} - R_{12}\right) \end{cases} \qquad (5\text{-}15)$$

- Interpolate the starting and ending quaternions q_1 and q_2 with the SLERP formula given in Equation (5-11) to find the desired orientation q according to the current value of t.

- The calculated orientation needs to be fed to the motors in the joint space. However, the inverse kinematic function normally takes Euler angles as inputs, so we first need to transform the quaternion back into a rotation matrix and from there extract the final Euler angles. To compose a rotation matrix from a quaternion $q = x\boldsymbol{i} + y\boldsymbol{j} + z\boldsymbol{k} + w$, we use the following expression:

$$R = \begin{bmatrix} 1 - 2y^2 - 2z^2 & 2xy - 2zw & 2xz + 2yw \\ 2xy + 2zw & 1 - 2x^2 - 2z^2 & 2yz - 2xw \\ 2xz - 2yw & 2yz + 2xw & 1 - 2x^2 - 2y^2 \end{bmatrix} \qquad (5\text{-}16)$$

This is a standard formula, although some minor optimizations for numerical stability and computational speed could be added. Also, depending on how you implemented the IK function, you might be able to pass the rotation matrix directly as input value, without the need to decompose it into Euler angles.

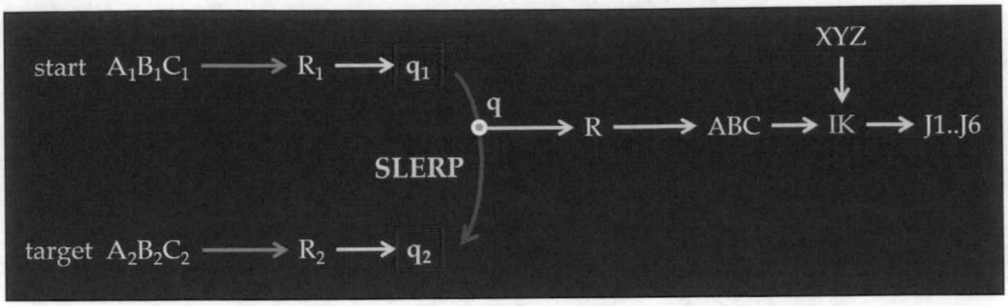

Figure 5-13. *Interpolating between start and target orientation requires several in-between steps*

There are more involved ways to directly transform orientations between Euler angles and quaternions without having to step through rotation matrices. However, since you already know how to handle rotation matrices and have probably implemented that function already, you can easily reuse your code and only add the small extra step for quaternions.

With this new powerful function to interpolate orientations in our toolbox, we can now proceed and start implementing all sorts of path-interpolated movements.

Line

The simplest interpolation in the path space is a line. Figure 5-14 shows the robot's TCP starting from point P_0 with coordinates $\begin{bmatrix} x_0 & y_0 & z_0 & A_0 & B_0 & C_0 \end{bmatrix}$ and reaching point P_1 with coordinates $\begin{bmatrix} x_1 & y_1 & z_1 & A_1 & B_1 & C_1 \end{bmatrix}$ along a line in space. *Notice that both position and orientation of the TCP change during the movement.*

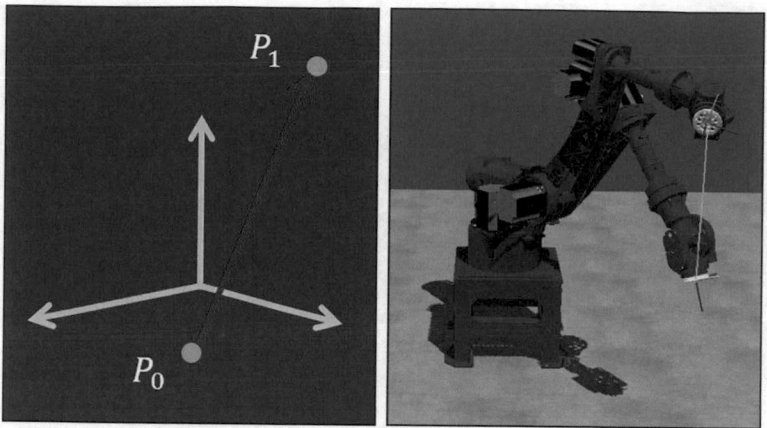

Figure 5-14. *The robot moving along a line*

Position coordinates are easily interpolated linearly with the following three equations:

$$\begin{cases} x = x_0(1-t) + x_1 t \\ y = y_0(1-t) + y_1 t \\ z = z_0(1-t) + z_1 t \end{cases} \qquad (5\text{-}17)$$

Orientation coordinates are interpolated using SLERP. The two interpolations run in parallel at the same time using the same value for the parameter t sweeping from 0 to 1.

Circle

The next movement we want to plan is a circle. Unlike a line, two points are not enough to define a circle in space. We need three of them: the starting point P_0 and two other points P_1 and P_2, where P_2 could also be the final target point (see Figure 5-15).

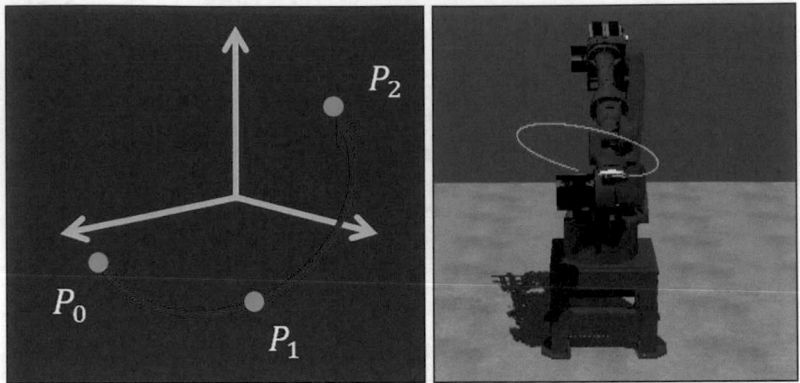

Figure 5-15. *The robot moving along a circle*

We focus on interpolating the position coordinates first, so for each point we only consider the [X, Y, Z] coordinates. It is possible to describe the position of a generic point P on a circle in space using a parametric equation with a single parameter t:

$$P = C + r \cos t\, \boldsymbol{U} + r \sin t\, \boldsymbol{V} \tag{5-18}$$

We call C the center point of the circle and r its radius. U is the normalized vector from the center of the circle to the starting point of the path. V is a vector perpendicular to U, lying on the circle plane (see Figure 5-16).

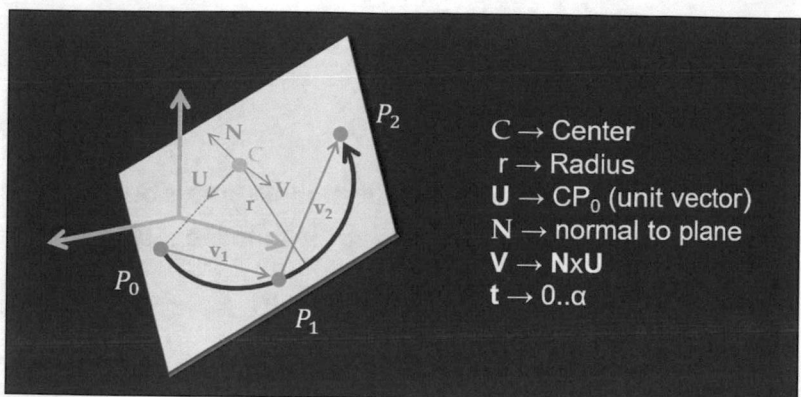

Figure 5-16. *Showing all the parameters needed to parameterize a circle*

By sliding the parameter t between 0 and the ending angle, we let the point P run around the circle.

Let's see how to find all those values, one step at a time.

First of all, we need to make sure that the three points P_0, P_1, and P_2 are not collinear; otherwise, no circle can be defined. Collinear means that the points lie on the same line in space.

We take any two vectors defined by the three points, for example, $v_1 = P_0P_1$ and $v_2 = P_1P_2$, and we calculate the vector N normal to the plane as the cross product between v_1 and v_2:

$$N = v_1 \times v_2 \qquad (5\text{-}19)$$

The cross product of any two vectors lying on a plane is always perpendicular to the plane itself, no matter what the vectors are. Using different vectors changes the magnitude of N but not its direction. However, the magnitude goes to 0 in case we multiply collinear vectors. Therefore, it is enough to check that $|v_1 \times v_2|$ is nonzero to make sure that the three points define a valid circle.

Remember to normalize N after the product, so that we can use it later for the next steps.

Once we have N, we can calculate V as the cross product between N and U (both normalized):

$$V = N \times U \tag{5-20}$$

Next, we need to find the radius r. Given three points on a circle, the length of its radius can be calculated using the following formula:

$$r = \frac{|P_0 - P_1||P_1 - P_2||P_2 - P_0|}{2|(P_0 - P_1) \times (P_1 - P_2)|} \tag{5-21}$$

When coding the calculations, remember that all those points are actually three-dimensional vectors with [X, Y, Z] position coordinates.

Finding the center point C is a bit more complicated. It can be proven that C is a linear combination of any three points on the circle:

$$C = c_0 P_0 + c_1 P_1 + c_2 P_2 \tag{5-22}$$

The coefficient c_0 is given by following formula:

$$c_0 = \frac{|P_1 - P_2|^2 (P_0 - P_1) \cdot (P_0 - P_2)}{2|(P_0 - P_1) \times (P_1 - P_2)|^2} \tag{5-23}$$

You can quickly modify the indexes to find c_1 and c_2 as well.

The last parameter we need to find is the angle α, so that we know where to stop the interpolation. We have two distinct cases here. The first is when the angle length is specifically forced by operator: e.g., by programming a rotation of 720 degrees to execute two passes around the circle. The second option is when no specific angle is given and it is assumed that the circular movement will end exactly at P_2. In that case, we can calculate α from the start and end vectors $V_0 = CP_0$ and $V_2 = CP_2$:

$$\alpha = cos^{-1} \frac{V_0 \cdot V_2}{\|V_0\| \|V_2\|} \tag{5-24}$$

However, since the arccosine operator cannot distinguish between angles larger and smaller than π, we need to manually adjust the result of this equation.

The way to do it is to check whether the vector N is normal to the plane (which defines the direction of travel of the circle), and the cross product between start and end vectors $V_0 \times V_1$ has the same sign. If they do, then you can safely take the result of the arccosine. If they have opposite signs, then the correct angle is 2π minus the result of the arccosine.

At this point, we have numerical values for all the required parameters of the interpolation in Equation (5-18), and we can calculate the position of the point on the circle. Keep in mind that that is in fact a group of three individual equations, one in x, one in y, and one in z.

Note that P_2 is not necessarily the final target point of the movement: we can use the angle α to define the circle length. We can stop the TCP at P_2, but we can also stop before P_2 or even before P_1 for small values of α. Alternatively, for very large value of α, we can run several revolutions around the circle. Additionally, using negative values for t, we could even run along the same circle but in reverse direction.

Along with the interpolation of the position, we will synchronously run the interpolation of the orientation axes using the SLERP function. According to the programmed angle length α, we might have to run up to three separate SLERP sections: from P_0 to P_1, from P_1 to P_2, and from P_2 to P_0.

Spline

We now introduce a new type of movement, more general and powerful than just basic lines and circles.

Factory workers who operate robots do not usually master complex math. They often require the robot to move from a starting point P_0 to a target point P_1 in a smooth way, but they do not have the ability to describe that curve mathematically. It is our task to offer the most convenient interface possible to simplify their life: we only ask them to provide two points, and we automatically generate the path in between.

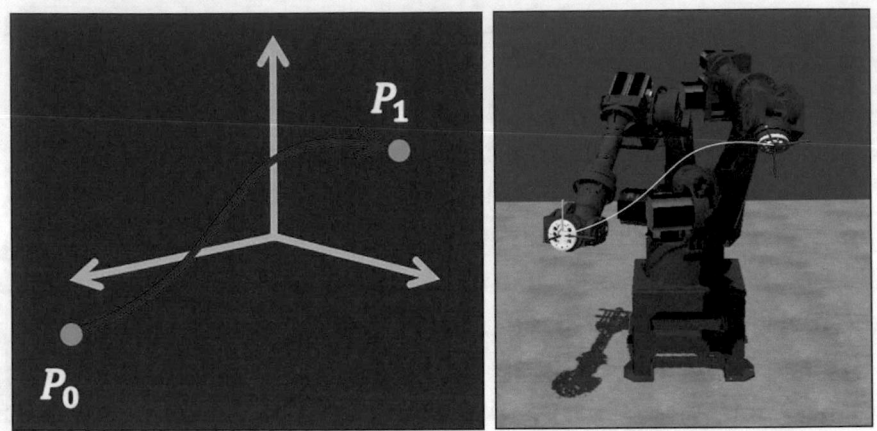

Figure 5-17. *Splines can smoothly connect two or more points in space*

One simple way to mathematically describe the red curve in Figure 5-17 is to use Bezier curves, which are a particular kind of **splines**.

Splines are functions built by composing piece-by-piece polynomials. Several different kinds of splines exist, each with its own set of properties. Here we focus on the family of Bezier splines, which are commonly used in computer graphics and path-planning. More specifically, we use the following cubic spline expression:

$$\boldsymbol{P} = \left(1-t\right)^3 P_0 + 3\left(1-t\right)^2 tP_1 + 3\left(1-t\right)t^2 P_2 + t^3 P_3 \qquad (5\text{-}25)$$

This curve describes the position of a generic point P moving along a cubic spline, starting from P_0 and ending at the target point P_3, according to the value of the parameter t, which ranges from 0 to 1. As in all the other interpolation we learned in the previous sections, t is just a dimensionless parameter and does not represent time.

You noticed that the equation includes four points: P_0 is the starting point and P_3 the target point, while P_1 and P_2 are two intermediate *control points* usually not touched by the curve but are critical in defining in what direction and with what magnitude the curve moves away from the two edge points.

There are a total of four points in the equation because we decided to use a *cubic spline*: t^3 is the highest power of t in the formula. Higher order splines are also possible but require more control points.

Modifying the position of the two middle control points leads to completely different paths between the start and end points: Figure 5-18 shows some examples.

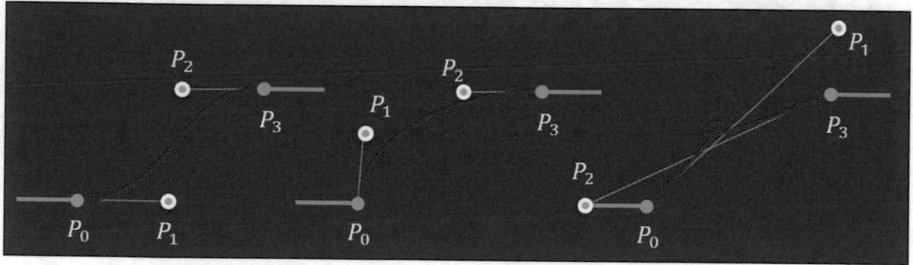

Figure 5-18. *The effect of the control points over the spline path*

While you can virtually place the control points anywhere in space, in practice you need to be careful when picking their location, because the resulting curve could show unexpected behaviors. For instance, the last example in Figure 5-18 (on the right) shows a loop in the path, which is probably not how the operator would expect the robot to move. We need to formulate a clear and safe way to place control points.

Another key feature that we must provide is a smooth transition between consecutive curves. In the second example in Figure 5-18 (middle), the path segments before and after P_0 do not connect smoothly with each other. The operator would not like that behavior either.

The ideal path between P_0 and P_3 is the one depicted in the first example of Figure 5-18 (left). There are no loops in the curve, and all transitions at the beginning and end of the movements are smooth. That is most likely what the operator would expect. Let's see how to place the middle control points in practice to achieve that result.

The first rule to follow is that the middle points need to be positioned along the directions tangent to the segments of path preceding and following the spline. See Figure 5-19 for some examples.

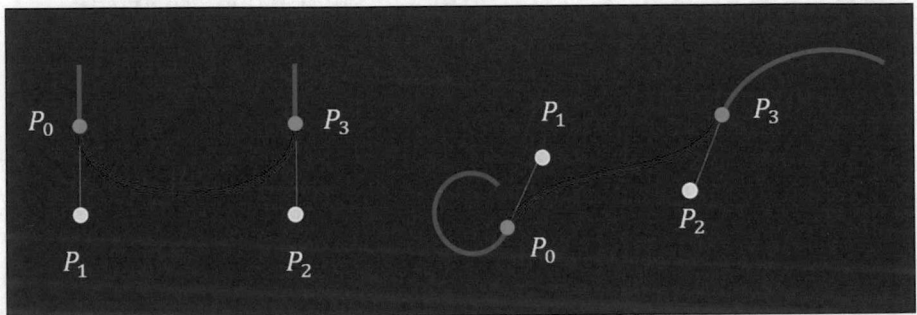

Figure 5-19. *Place the middle control points along the tangents of the path*

Placing the control points along the tangential directions ensures that the connection has a continuous first derivative. Geometrically, both the spline and the connected path segment are oriented along the same line. Later we will also introduce higher degrees of smoothness.

The path segments to be connected with a spline are not restricted to lines. You can connect virtually any geometrical shape: circles, other splines, or even PTPs. Once you find the tangents and place the control points, then you can calculate the spline immediately.

One might wonder how to find the tangent direction out of an individual point, in case we start a movement from a position in space without any knowledge of the path followed by the TCP to arrive there. In other words, if we have no previous geometry to connect to, there is no tangential direction to follow. In such situations, it does not really matter where the beginning of the spline points to, but we can always use the target point as a way to fix the initial tangent (see Figure 5-20).

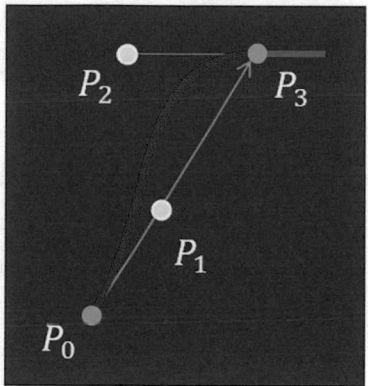

Figure 5-20. *Starting a spline from a lonely point*

One additional question we need to address is how far away from the edge points should we place the two middle control points. There is no uniquely correct answer to that question. All the generated splines with different distances between the edge and middle points are mathematically correct. Some are more useful than others though (see Figure 5-21).

A very short distance between the starting point P_0 and the first control point P_1 could result in a sharp corner, depending on the position of the second control point P_2. That is often an undesirable behavior because it forces the robot to slow down considerably during the movement.

On the other hand, a long distance between the starting point P_0 and the first control point P_1 results in a large initial jump of the spline. If that jump overshoots the second control point P_2, then the spline is very likely to temporarily move away from the target; in extreme cases, it will build a loop in its path. That is something we definitely want to avoid.

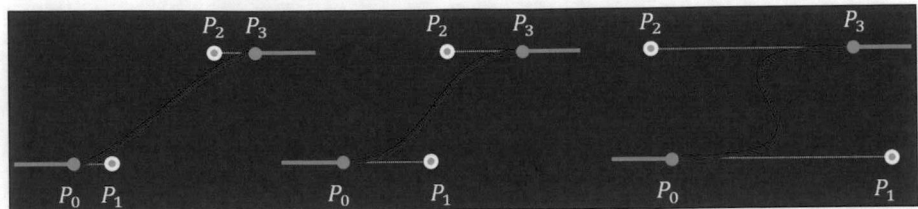

Figure 5-21. *Sensible placing of the middle control points is critical to achieve a smooth movement of the robot*

A common rule of thumb is to place the middle control points so that their distance from the corresponding edge point is about 1/3 of the linear distance between start and end point. That is the rule we followed in all the examples shown in the figures of this section and generally results in safe and smooth curves without loops.

Following all the rules we described so far, we can generalize the procedure and build cubic Bezier curves to smoothly connect any number of points in space. Figure 5-22 shows an example with three set points: A, B, and C. While it is usually clear how to build the tangents at start and end points (A and C), we have not yet defined a way to define tangents for all the other points in between (B in this case). A common method is to force the tangents going through a set point in the middle of a movement to be parallel to the imaginary line connecting the previous and next points.

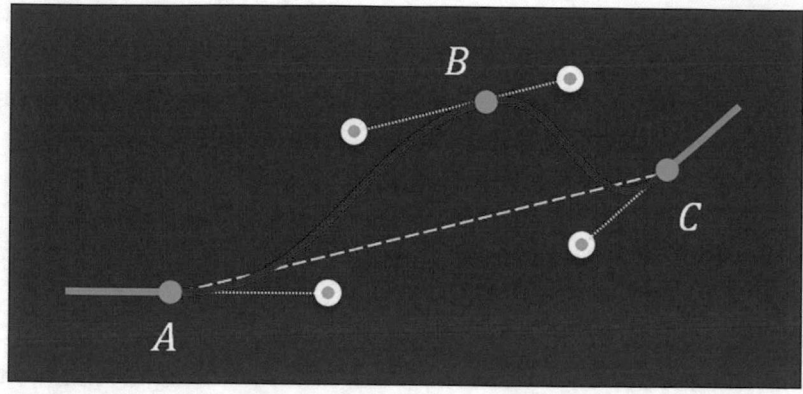

Figure 5-22. *Merging multiple splines to generate a path between several points*

This is a valuable function for robots' operators: they can simply teach the position of certain points where the robot is required to go through, and then our software will automatically generate the whole path in between, by connecting multiple splines with each other.

A typical example is shown in Figure 5-23. The operator needs to move something from point A to point B without colliding with the obstacle in between. A simple PTP movement is ruled out because it would certainly cause the robot to hit to obstacle. A spline is the easiest way around: set either one or two points around the obstacle, and the robot will smoothly transition between them.

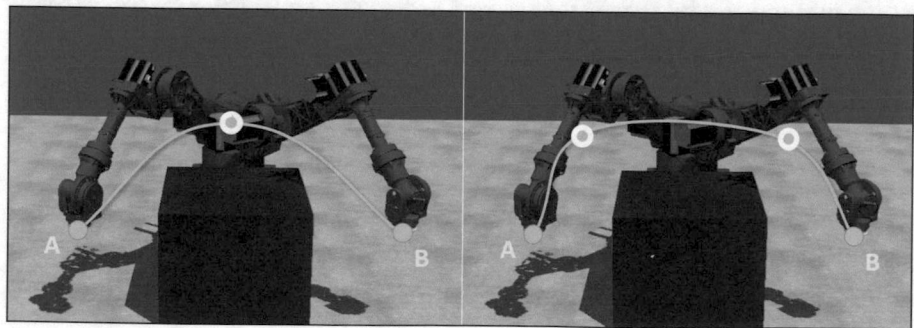

Figure 5-23. *Automatically generated splines between set points*

Finally, remember that we use splines only to interpolate the position coordinates of the TCP. In parallel to that, we also need to use the SLERP algorithm to interpolate the orientation coordinates between the different orientations of all the set points.

De Casteljau's Algorithm

Coding the cubic Bezier spline interpolation could be done by directly applying Equation (5-25). Given two points to be joined by a spline, we first derive the position of the two middle control points using the rules explained in the previous section and then swipe the parameter t from 0 to 1 to calculate all the intermediate points along the curve.

This method, however, suffers from numerical instabilities because it involves taking quadratic and cubic powers of very small numbers, which can result in low accuracy when the planned path is long. The problem is especially critical when working with higher order splines, for example, quartics, which we will use later on for other applications.

We present here a much better alternative method that only involves linear operations: the De Casteljau's algorithm. It might be slightly less efficient, because it requires more operations, but it is definitely more accurate.

The underlying idea is simple: instead of interpolating between all four points at once, we take only two points at the time and linearly interpolate between each pair. Let's follow the step-by-step procedure shown in Figure 5-24.

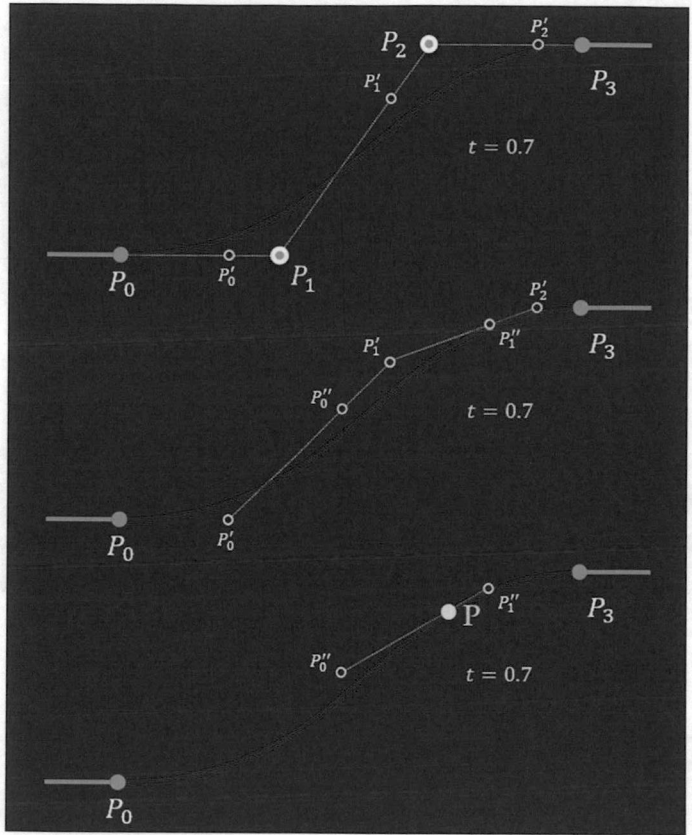

Figure 5-24. *Step-by-step De Casteljau's algorithm*

We know the position of the four points P_0, P_1, P_2, P_3. Imagine we want to calculate the position of point P for $t = 0.7$.

- First, using $t = 0.7$, we linearly interpolate between P_0 and P_1 to find P_0', between P_1 and P_2 to find P_1', and between P_2 and P_3 to find P_2':

$$\begin{cases} P_0' = (1-t)P_0 + tP_1 \\ P_1' = (1-t)P_1 + tP_2 \\ P_2' = (1-t)P_2 + tP_3 \end{cases} \tag{5-26}$$

137

- Then, starting from these three new points, we repeat the same algorithm (again with $t = 0.7$) to find P_0'' and P_1'':

$$\begin{cases} P_0'' = (1-t)P_0' + tP_1' \\ P_1'' = (1-t)P_1' + tP_2' \end{cases} \tag{5-27}$$

- Finally, using these two new points, we run a last linear interpolation and find the target point P:

$$P = (1-t)P_0'' + tP_1'' \tag{5-28}$$

The algorithm is simple and can be programmed recursively. It takes a few more steps than the direct equation, but it offers much higher computational stability because it only uses linear calculations instead of power operations of very small numbers.

Round Edges

So far, we have studied different geometries to connect two points in space: lines, circles, and splines. Now we need to study how to connect those geometries with each other. The point of contact between two consecutive segments of a path is called a **transition**.

The first issue we encounter when analyzing transitions is that there might be a sharp angle between two adjacent segments. Sharp edges are usually undesirable when planning paths for robots because they cause a considerable drop in speed: the robot has to decelerate, almost come to a stop at the transition, and then start accelerating again along the next section. Such a behavior is forced by the maximum dynamic limits allowed by the mechanical structure of the arm. The result is a suboptimal movement performance and also much stress induced on the mechanics, which reduces their life. A smoother movement would be much more desirable and beneficial, both for the application process and for the robot itself.

One typical solution is to introduce a round edge as a path smoothing transition effect. The result is a more fluid curve as shown in Figure 5-25. The round edge allows for higher travelling speed, since no stop is required at the transition point. It also reduces stress and wear on the mechanics. The size of the round edge shown in this example is purposely exaggerated to make it clearly visible. Typically, a radius of a few millimeters suffices to achieve smooth dynamics.

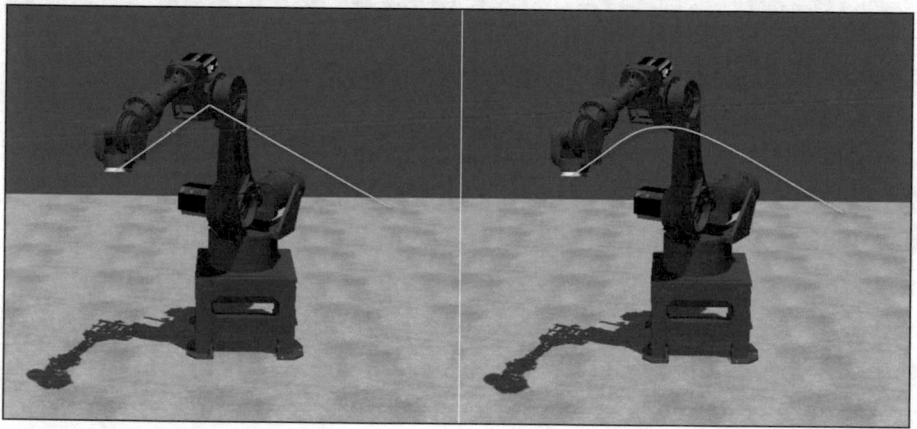

Figure 5-25. *Introducing a round edge at a path transition*

There are specific applications where a sharp 90 degrees corner between two edges is strictly required. In those cases, we obviously cannot add a round edge to the path, and we must accept that the robot will come to a complete stop at the transition point. As always, a sensible compromise between path accuracy and traveling speed must be found according to the specific application requirements.

Let's now see how to describe a round edge mathematically. One possible way is by means of a Bezier curve, the same kind we used to generate splines between two points in space. Splines require control points: we need to devise a way to place control points given two adjacent edges and a desired radius R for the rounding, as shown in Figure 5-26.

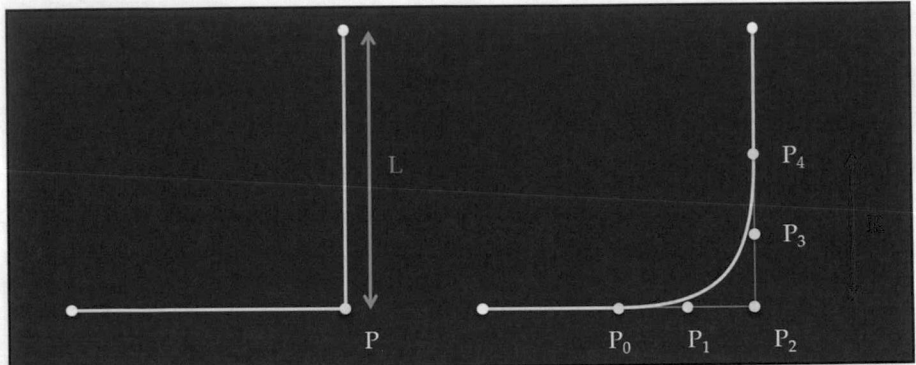

Figure 5-26. *Constructing a round edge with a Bezier curve*

Note that we need to automatically introduce limitations for the size of R. First of all, R cannot be longer than the length of the segment itself: that should be clear because the round edge cannot be longer than the entire segment anyway (L in Figure 5-26). More often, we also need to limit R to half of the segment length, because another round edge might be required at the next transition, and we need to leave space for that.

Once the size of R is fixed, we can already define the first two control points for the spline: the starting and ending points P_0 and P_4 equidistant from the transition point P.

We could now add just two middle control points and generate a cubic spline, as we learned earlier in this Chapter, but given the symmetry of the problem, it is more convenient to use three additional control points and generate a quartic spline.

We end up with a total of five control points: the aforementioned P_0 and P_4, one at the transition point $P_2 \equiv P$, and the final two (P_1 and P_3) in the middle of the two edge segments. The resulting equation for the quartic spline is as follows:

$$P = (1-t)^4 P_0 + 4(1-t)^3 tP_1 + 6(1-t)^2 t^2 P_2 + 4(1-t)t^3 P_3 + t^4 P_4 \quad (5\text{-}29)$$

It can be shown that such a configuration ensures geometrical continuity of the spline with the original path up to the second derivative: the two sides of the transition have same direction (first derivative) and also same curvature (second derivative). The transition between the programmed path and the inserted round edge is "smooth." In the next section, we will define that smoothness mathematically.

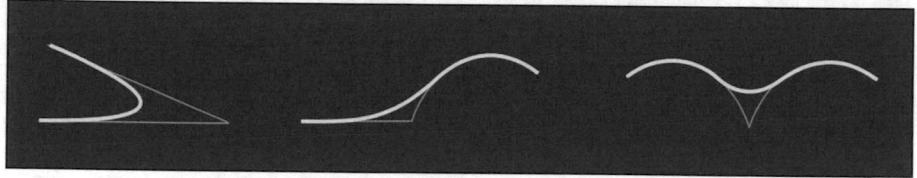

Figure 5-27. *Examples of round edges between different geometries*

As shown in Figure 5-27, smooth transitions can be added between lines, circles, and technically even PTPs. All we need to do is find the direction of the path tangents at the transition points. Since PTPs are not planned in the path space, a bit more work is needed, and some numerical approximation must be accounted for.

Transitions

It is possible to mathematically define the **degree of continuity C^k** between two adjacent path sections.

If two curves simply connect at one point, these are said to be C^0 continuous. The ending position of the first curve is equal to the starting position of the second curve (see Figure 5-28 left). However, in that case, the directions of the two tangents (blue in the drawing) are not the same. The edge is sharp, not smooth, and a robot travelling along that path would come to a halt at the transition point. This is called a *nontangential transition*.

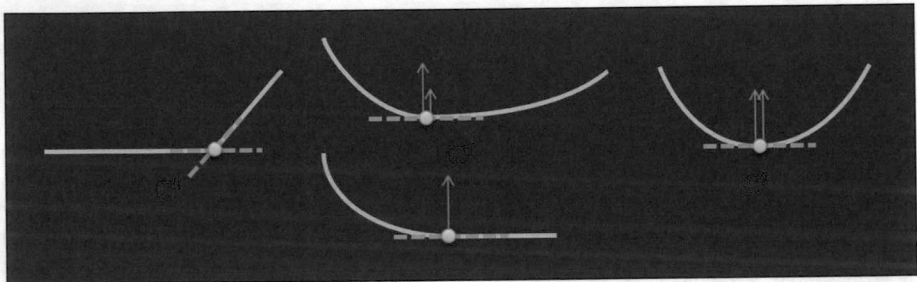

Figure 5-28. *Examples of transitions with different degrees of continuity*

In the second example (Figure 5-28 middle top), two curves touch at the transition point, and also their tangents are coincident along the same blue line. The two path segments have C^0 and also C^1 continuity. A movement is not required to stop at this point and the transition is called *tangential*.

However, taking the second derivative of the curves returns different values (green in the drawing) for each side of the transition. Geometrically, the second derivative of a curve represents its curvature.

In fact, if you were to feel with your hand such a transition between two curves, you would notice an edge, not a really smooth feeling. That is because the curvature of the two sides is not the same, with the left side having a higher curvature than the right one. Therefore, this transition only reaches C^1 degree of continuity. It is better than C^0, but not yet entirely satisfying for a high-quality movement.

A similar example, although not always so immediate to recognize, is the transition between a circle and a line (Figure 5-28 middle bottom). You can think of the line as circle with infinite radius: a line has zero curvature and therefore a second derivative equal to zero. Since the two sides have different curvature, this transition is not C^2 continuous.

That is precisely the reason why we did not use circles for round edges in the previous section. It would have been much simpler conceptually and mathematically, but it would have not guaranteed the same high degree of continuity of the quartic spline we adopted instead. *The radius of curvature of the spline is not constant*: it starts at zero, increases in the middle, and goes back to zero at the end, so that the transitions with the previous and next segments are smooth.

What does that mean in practice exactly? Why is the degree of continuity so important? A C^1 continuity between consecutive path sections translates into the same movement speed on both sides of the transition. However, if the transition is not C^2 continuous, the two sides require different axial accelerations to hold the same path speed.

For instance, we mentioned that the transition between a line and a circle is not C^2 continuous. If we let the robot travel at constant speed across the two curves, the acceleration along the axis perpendicular to the path changes instantaneously: from zero along the line to a sudden finite positive value when entering the circle. If you run that movement with a real machine, you can easily hear a hard "click" on the axes caused by the abrupt change in acceleration. On the long term, that is harmful to the mechanics.

Consequently, the continuity between two curves must be satisfied at higher order of differentiation in order to have a smooth dynamic profile and avoid slowing or halting the movement. It certainly helps to increase production speed and decrease wear on the mechanics.

That should now clarify why we required C^2 continuity in round edges between path sections. The two curves at the right side of Figure 5-28 are also C^2 continuous as they share the same tangent directions and the same curvature.

Let's now see what it means mathematically to have a C^2 transition.

We start from a simple linear segment P_0P_1 as the simplest example. Its parametric equation is as follows:

$$P = (1-t)P_0 + tP_1 \qquad (5\text{-}30)$$

Taking the first derivative with respect to the parameter t gives the following:

$$P' = P_1 - P_0 \qquad (5\text{-}31)$$

That is the vector that points from P_0 to P_1 and represents exactly the direction of this segment. The result is quite obvious, as the tangent to a line is the line itself.

If we further differentiate that vector with respect to t, we simply get 0. That is also what we expected, because a line has zero curvature.

$$P'' = 0 \qquad (5\text{-}32)$$

Let's now move on to the quartic spline, which has five control points:

$$P = (1-t)^4 P_0 + 4(1-t)^3 tP_1 + 6(1-t)^2 t^2 P_2 + 4(1-t)t^3 P_3 + t^4 P_4 \qquad (5\text{-}33)$$

We take the first and second derivatives with respect to t:

$$P' = 4(1-t)^3 (P_1 - P_0) + 12(1-t)^2 t(P_2 - P_1) + \\ 12(1-t)t^2 (P_3 - P_2) + 4t^3 (P_4 - P_3) \qquad (5\text{-}34)$$

$$P'' = 12(1-t)^2 (P_2 - 2P_1 + P_0) + 24(1-t)t(P_3 - 2P_2 + P_1) + \\ 12t^2 (P_4 - 2P_3 + P_2) \qquad (5\text{-}35)$$

The important result to observe out of these formulas is that both first and second derivative are not constant: they vary over the interval from $t = 0$ to $t = 1$. Physically, it means that the direction and curvature of the round edge change along the path from start to end.

While that is obvious for the direction, it might be unintuitive for the curvature. However, that is exactly the feature we require in a smooth round edge transition: a varying curvature means that we have full control over the amount of acceleration required from the physical axes during the movement. That is unlike the case of a simple circle, where the constant curvature over the path forces a constant centripetal acceleration, leading to discontinuities at the transitions with linear path segments.

Let's now focus at the transition point P_0, which is the starting point of the spline ($t = 0$). Since we require the spline to have zero curvature at the edges, the condition to satisfy is as follows:

$$P''|_{t=0} = 12\left(P_2 - 2P_1 + P_0\right) = 0 \tag{5-36}$$

The solution is as follows:

$$P_1 = \frac{P_2 + P_0}{2} \tag{5-37}$$

That is, the point P_1 should be placed in the middle point between P_0 and P_2.

Similarly, by forcing the curvature to be zero at P_4 ($t = 1$), we find that P_3 must be placed in the middle between P_2 and P_4.

That explains why, when constructing the round edge in the previous section, we decided to place the control points of the quartic spline exactly in the middle of the two segments. With that configuration, we achieve a geometrically and physically smooth transition between the two path sections.

Path Length

We have defined different kinds of paths and transitions between them. Now we are interested in measuring their Cartesian length. The requirement comes from the trajectory generation task: when the operator asks the robot to move at a specific speed along a path, we need to know how long that path is, in order to find out how much time it takes the parameter t to slide from 0 to 1. We will discuss trajectory generation in Chapter 7, where we will define a relation between the parameter t and the actual time.

Finding the path length for lines and circles (as in Figure 5-29) can be easily done analytically.

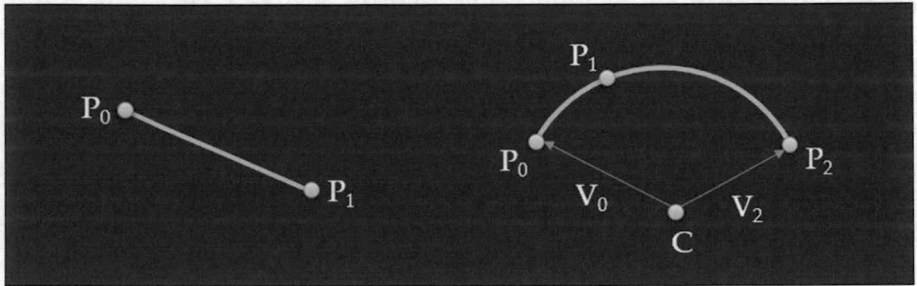

Figure 5-29. *The length of lines and circles is easy to calculate*

The length of a line P_0P_1 is given by the following:

$$L = \sqrt{\left(x_1 - x_0\right)^2 + \left(y_1 - y_0\right)^2 + \left(z_1 - z_0\right)^2}$$

(5-38)

The length of a circle is given by the product of its radius and the angle of rotation:

$$L = R\alpha$$

(5-39)

Recall from the Circle Section that the angle between the starting and ending vectors of a circle is given by the arccosine of their normalized dot product.

$$\alpha = cos^{-1} \frac{V_0 \cdot V_2}{\|V_0\| \|V_2\|} \tag{5-40}$$

Just as we noticed earlier, the arccosine operator cannot distinguish between angles larger and smaller than π, so we need to manually adjust the result of this equation based on the direction of the cross product between V_0 and V_2.

Calculating the path length of splines and point-to-point movements is not as straightforward. A simple solution is to use a numerical approximation by segmenting the given movement into a large number of subsections, as shown in Figure 5-30. The individual sections can then be linearized and their lengths added up to find an approximate value for the total path length.

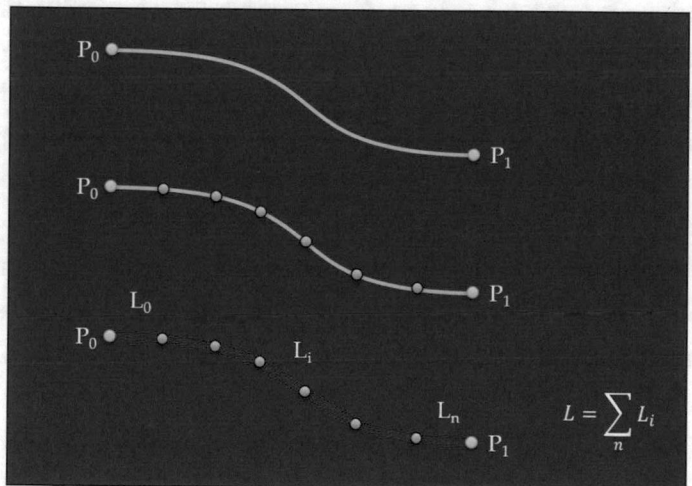

Figure 5-30. *Segmentation and linearization of a curve to calculate its length*

Calculating the actual positions of all the intermediate points in space is relatively simple. For PTP movements, we use the forward transformations, because the original movement is planned in the joint space. For splines, we can evaluate the curve directly using the De Casteljau's algorithm.

The density of segmentation depends on the required accuracy and on the available computing power.

External Path Corrections

Typically, the path-planning procedure takes place well before the actual movement is executed. Planning ahead of time allows the robot to "see" the path farther ahead and safely reach higher speeds. However, there are situations where the planned path must be modified at runtime according to external signals that are not available at planning time.

A very common example in many production lines is conveyor tracking, where the robot has to perform an action on a moving target: e.g., picking incoming cookies and placing them in a box or filling moving water bottles. These tasks all happen while the cookies, boxes, or bottles are moving on a conveyor. The result is a much higher productivity of the line, but from a path-planning point of view, things get a bit more complicated.

Another example is when the robot needs to adjust its own path to avoid a sudden obstacle along the way. Taking a detour before reaching the planned destination increases the path length and therefore affects the execution time.

All these situations happen at runtime: they cannot be planned ahead of time because the target positions or correction offsets are unknown until they are provided by a camera or a sensor, and the exact behavior of the path cannot be predicted in advance. Critically, the actual path length is not known until it is physically executed.

When implementing a practical solution for online path corrections, you should keep in mind a couple of potential issues. As far as the path itself is concerned, if the target position suddenly changes, then the planned path for the next movement will also be affected because its starting point is not the one it used to be. As a consequence, the planning procedure needs to be restarted entirely.

Safety is also critical: unplanned paths need to be monitored closely at runtime to prevent position limits, violations, and collisions. We will describe workspace monitoring in detail in Chapter 6.

Finally, if the path length changes significantly from the originally planned one, then the actual speed of the movement needs to be recalculated in real time to avoid violating speed and acceleration limits. We will describe dynamic limits when talking about trajectory generation in Chapter 7.

Summary

The first step of moving a robot from A to B is to plan a geometrical path between those two points. Paths for kinematic chains can be planned either in the joint space (PTP movements) or in the path space (path-interpolated movements).

In this chapter, we described several alternative path geometries, and we derived the required equations to interpolate both the position and the orientation of the robot's TCP during its movement. In particular, we introduced quaternions to allow for smooth orientation changes.

The most advanced path geometry we introduced is the spline, which gives you a powerful tool to easily plan complex paths and also to round the edges of otherwise nontangential transitions.

In the next chapter, we discuss how to monitor the safety of the path sections we just generated before physically executing them.

CHAPTER 6

Workspace Monitoring

In the path-planning chapter, we have learned how to describe several different curves in space using parametric equations. However, while all those formulas can generate mathematically valid paths, it is not always physically possible for the robot's TCP to move along those curves.

There are essentially two distinct reasons why a given position in space would not be acceptable. The first is that the robot's own mechanics cannot stretch enough to reach that point. Verifying such a condition is straightforward: we solve the inverse kinematic function for the desired TCP pose and check if it returns any error. For example, when solving the lower arm for the six-axes robot, we imposed the conditions in Equations (4-8) and (4-9) in order for the solution to be physically acceptable. Every kind of robot has a finite reachable workspace that depends on its mechanical size, which is always described in the manufacturer's datasheet (see, e.g., Figure 6-1).

© Fabrizio Frigeni 2023
F. Frigeni, *Industrial Robotics Control*, Maker Innovations Series,
https://doi.org/10.1007/978-1-4842-8989-1_6

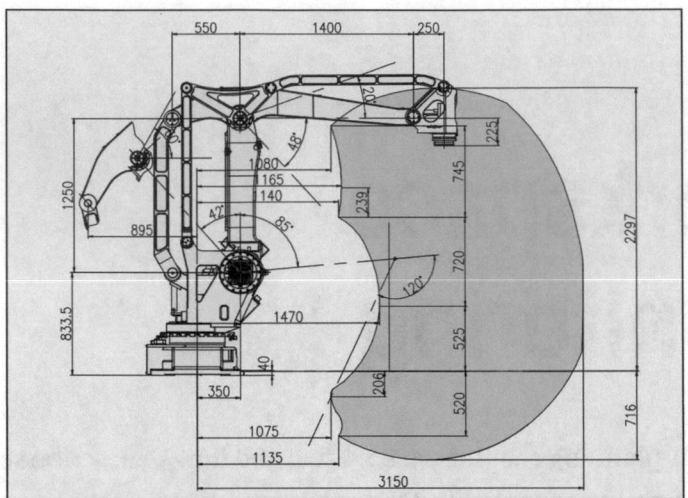

Figure 6-1. *Example of a robot's mechanical workspace (with permission from EFORT, adapted from ER180-4-3200 datasheet)*

The second reason why a robot might not be able to reach a given position, even if inside its mechanical workspace, is safety. Each practical application of a robotic arm usually requires a further restriction of the allowed workspace to avoid collisions between the robot and the surrounding environment.

Collisions can happen with fixed objects (e.g., cabinets, conveyors), with other moving robots, or even with the robot's own body itself! To handle the first case, we need to introduce a fixed workspace monitoring function using forbidden and safe zones of space. To avoid collisions with other moving machines, we deploy a dynamic workspace monitoring function using exclusive zones. Finally, a self-collision detection function prevents the robotic arm to hit itself.

A **safe zone** is defined to keep the entire robot inside a virtual cage, where it is safe for it to move without colliding with other elements of the environment (see Figure 6-2 left). The rule is that no part of the robot's body can be outside a safe zone at any time.

On the other hand, a **forbidden zone** precludes the robot from entering certain areas of space (see Figure 6-2 right), because they might already be occupied by other objects.

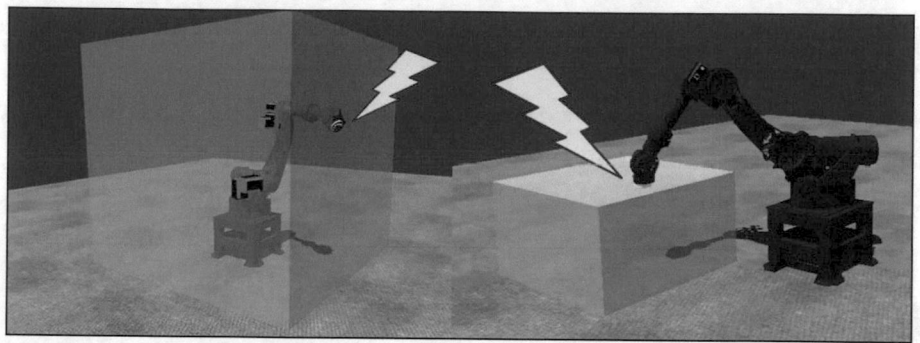

Figure 6-2. *SAFE zones (left) and FORBIDDEN zones (right) are used to keep the robot inside or outside specific regions of space*

Safe and forbidden areas can be combined as logical elements in order to describe complex space geometries and offer the operator a flexible and convenient interface to define the allowed workspace. For instance, a forbidden area could be placed inside a safe area: the resulting allowed workspace would be the difference between the safe and the forbidden area.

When implementing workspace monitoring functions in practice, the earlier we are able to predict a collision, the better. In most cases, we do not need to wait for the robot to come in contact with a safety area to trigger a movement stop. It is much wiser and safer to *monitor the path while planning it, not while executing it*. In that case, a movement can be aborted even before it starts. This practice should overrule all programming errors by the operator, making sure that the robot will not try to execute dangerous movements.

However, there are also cases when monitoring in advance is not possible, for example, when the position of other robots in the same workspace is commanded by other controllers and the information about their planned path is not available to our control software. Another example is when we deal with external additive corrections commanded by sensors or cameras. Those corrections are added at real time, after the path has been planned, and therefore require dynamic monitoring of all actual positions.

In the next sections, we will describe a number of functions that can be implemented to monitor and guarantee a safe behavior of the robot both while planning and while executing a movement.

Linearization

When monitoring the safety of a given path, it is not enough to check only the start and end points of the corresponding movement: both points might be inside a safe area, but the actual path could intersect a forbidden region of space (see Figure 6-3). The entire length of the path needs to be closely monitored.

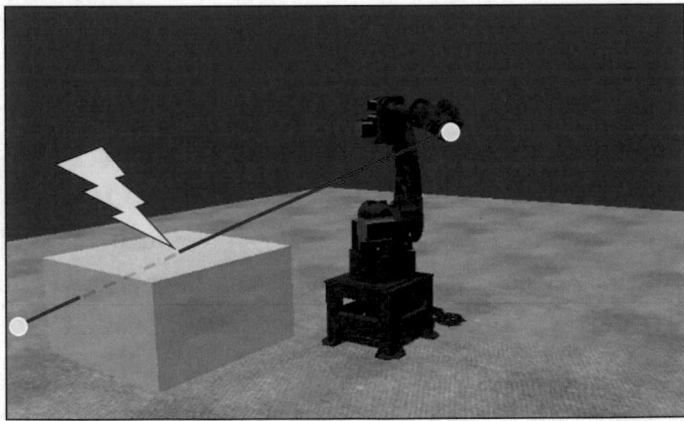

Figure 6-3. *A path might intersect a forbidden area even though both start and end points are in safe space*

However, detecting the intersection between a generic path and a generic area of space can be quite complex and computationally demanding. We need to introduce two practical simplifications to make the problem easier to define and solve: we linearize both the path and the space.

We start by linearizing the space using a combination of 3D shapes with six rectangular faces, technically called **cuboids**. More complicated geometrical shapes would require much more computationally expensive evaluations. Since the control software runs in real time, we need to optimize resources when possible. Intricated regions of space can always be approximated by the sum of individual cuboids of different sizes.

Another, more important reason to rely on cuboids is that complex geometries are not easy to define and configure in the robot's visual interface. Conversely, *a cuboid requires just two points in space* and can be quickly defined by the operator. In particular, it is enough to provide any pair of opposite vertexes to uniquely identify a cuboid (see Figure 6-4).

Figure 6-4. *Any pair of opposite vertexes can be used to define a cuboid*

The second object we linearize is the movement profile of the robot. While it is relatively easy to predict collisions if the path is straight line, things get much more complicated and computationally expensive if the path is a spline.

A typical approach is to decompose the path in linear segments, as shown in Figure 6-5, similarly to what we did in the Path Length Section in Chapter 5 in order to calculate its length. Note that the higher the segmentation density, the higher the accuracy of our predictions, but also the higher the computational load. As usual, the optimal compromise must be found according to the application requirements and available computational power.

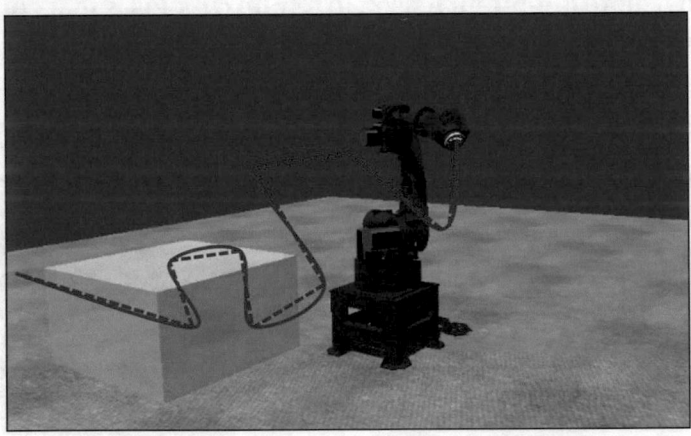

Figure 6-5. *Linearizing a complex path to simplify collision computations*

By linearizing the path and the space zones, the problem of detecting collisions reduces to the much easier geometrical problem of finding the intersection between a line and a cuboid. Let's see some examples in detail.

Safe Zones

Safe zones require the entire movement to stay inside the defined space. The solution is fairly easy: if either one or both of the edge points of the line are outside the zone, then part of the movement is violating the safety area for sure.

The only acceptable solution is that both start and end points are inside the safety area. Given the simple geometry of the problem, that condition is enough to guarantee that the entire movement is safe (see Figure 6-6).

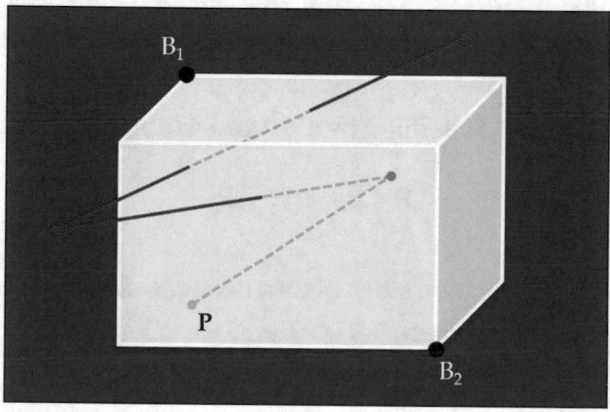

Figure 6-6. *The entire line must stay inside a safe zone*

Mathematically, saying that a point *P* is inside a cuboid is equivalent to say that its coordinates are within those of any two opposite vertexes (B_1 and B_2):

$$\begin{cases} P < max\left(B_1, B_2\right) \\ P > min\left(B_1, B_2\right) \end{cases} \qquad (6\text{-}1)$$

Since P, B_1, and B_2 are all three-dimensional vectors, the condition must be verified along all three axes [X, Y, Z].

Forbidden Zones

Forbidden zones require the entire movement to stay outside the defined space. Geometrically, the condition to verify is that the line should never intersect the cuboid.

There are different methods to solve this problem mathematically, some of which are highly optimized for graphic applications, but not very easy to understand. We present a simpler method here, which you can then further optimize if needed.

To make the explanation more intuitive, we start from the 2D equivalent problem: we check if a line intersects a rectangle. The solution to the 3D problem is essentially the same, only with the addition of one coordinate.

A line from P_0 to P_1 is defined by a parametric equation with $t = 0...1$:

$$P = P_0 + t\left(P_1 - P_0\right) \tag{6-2}$$

We first consider the two sides of the rectangle closer to the point P_0 (highlighted in green in Figure 6-7 left), and we calculate the values of t for which the line intersects them. There will be two intersection points, one for each side of the rectangle ($x = x_B$ and $y = y_B$), which will result in two values for t:

$$\begin{cases} t_x = \dfrac{x_B - x_0}{x_1 - x_0} \\[2mm] t_y = \dfrac{y_B - y_0}{y_1 - y_0} \end{cases} \tag{6-3}$$

We take the largest one of the two, which represents the latest intersection between the line and the closest edges of the rectangle, and we call it t_0.

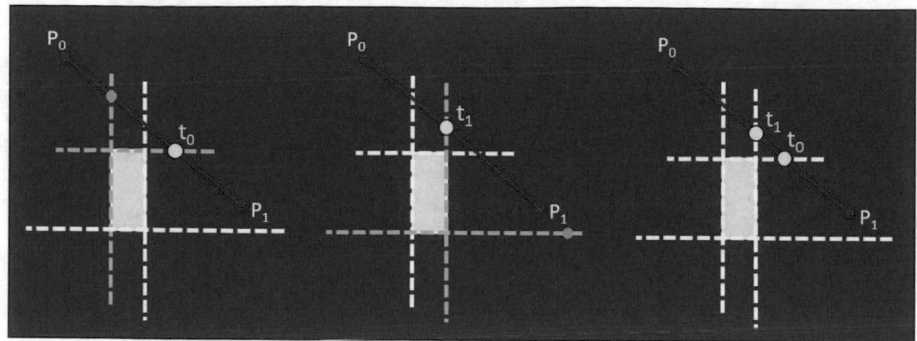

Figure 6-7. *Simple procedure to detect if a line intersects a rectangle*

Now we repeat the same technique for the two sides farthest away from P_0 (see Figure 6-7 middle). We calculate the two intersecting values and take the smallest one, which represents the earliest intersection between the line and the furthest edges of the rectangle. We call it t_1.

It can be shown, and should be easy to see without proof, that the line intersects the rectangle if $t_0 \le t_1$; otherwise, it does not (see Figure 6-7 right).

However, since we want to consider only the segment P_0P_1 and not the whole unbound line, we also need to check that the intersection happens within the interval $t = 0...1$. In other words, either t_0 or t_1 must be inside the 0...1 section.

The same exact procedure also works for a line and a cuboid in 3D (Figure 6-8), with the only addition of one coordinate. When calculating the values of t_0 and t_1, there are three distinct intersection points to consider (t_x, t_y, t_z). The condition for intersection is the same: $t_0 \le t_1$.

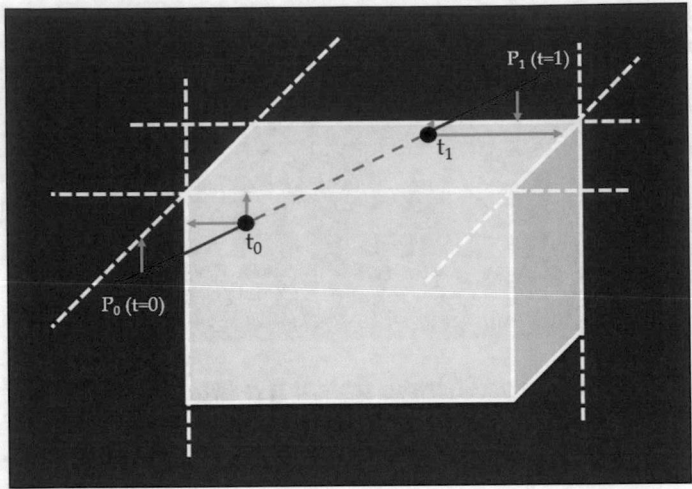

Figure 6-8. *Extension to the 3D case*

In practical implementations, one must also check for the special cases in which the line is parallel to the cuboid along one axis. Calculating the values for t_0 and t_1 for that specific axis is then not possible anymore because the denominator in Equation (6-3) becomes 0.

We need to distinguish two possible cases: if the coordinate of the line along that parallel axis is inside the cuboid, then we only use the other two axes to test for collision; if, on the other hand, the parallel coordinate is outside the cuboid, then we already know that the line cannot intersect the cuboid, and no further test is necessary.

Wire-frame Model

Monitoring the position of the robot's TCP and checking its position with respect to safe and forbidden zones are not always enough to guarantee safety during movements. The reason is that the TCP might lie well inside a safe area, while other parts of the robot's body are outside of it (see Figure 6-9 for an example). Similarly, the robot could find itself colliding with a forbidden area with parts of the body located away from the TCP.

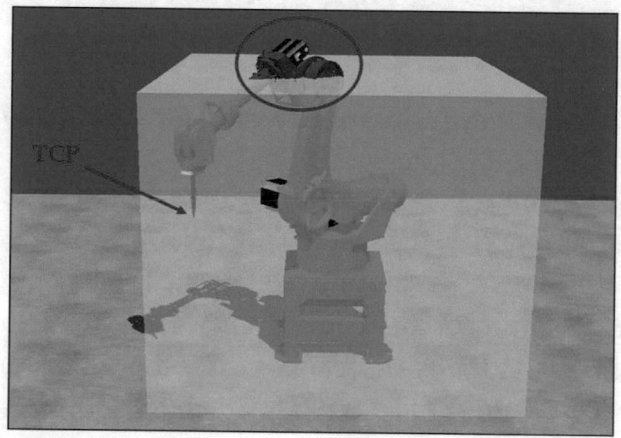

Figure 6-9.   *The whole body of the robot needs to be monitored!*

However, monitoring the movement of the entire body along with the TCP can be quite complicated, depending on the shape of the mechanics. A solution commonly adopted in the industry is to simplify the robot's body to a **wire-frame model**, by connecting a few key points in its structure with linear segments (see Figure 6-10). Monitoring only the position of those points and the lines between them instead of the entire body volume simplifies the problem considerably with no significant drop in accuracy.

Figure 6-10. *Example of wire-frame model for a six-axes robot*

The number of the points needed to generate the wire-frame depends on the desired accuracy, on the shape of the robot, and on the actual application requirements. A standard model usually uses one point per joint. A minimalistic model could simply remove the points and lines that are not expected to come into contact with other objects in the environment. On the other hand, if the environment is unknown and more robots are working together, a more sophisticated model with a larger number of key points might be required.

Since the positions of the points selected for the wire-frame model depend on the actual values of the joint axes, they must be cyclically calculated using the direct kinematics and constantly updated during movements. For example, the position of the point centered in J_3 depends on the value of the first two joints and can be calculated as follows:

$$F_3\left(X,Y,Z\right)=T_{01}T_{12}T_{23}O=T_{03}\begin{bmatrix}0\\0\\0\\1\end{bmatrix} \tag{6-4}$$

Once all the positions are known, we can use the line to cuboid intersection function to check whether the wire-frame model (both points and connecting lines) collides with forbidden zones or moves outside safe zones.

Keep in mind that linear movements of the TCP do not correspond to linear movements of the points in the wire-frame model. Therefore, the planned path always needs to be segmented, so that we can check the position of all the individual points along the entire movement.

This feature is obviously very demanding in terms of computational power and should only be activated when really needed and on powerful enough hardware.

Safe Orientation

For most applications and for most robotic structures, monitoring the position of the TCP along with the entire wire-frame is enough to guarantee safe movements. However, there are some situations in which the orientation of the TCP also needs to be monitored.

A typical case is when a six-axes robot is deployed in laser cutting applications: the fact that the laser pointer mounted on the TCP stays within a safe zone does not prevent the laser from pointing upward in the air and maybe toward the operator with possibly dangerous consequences. The orientation of the laser pointer must also be monitored to make sure it remains safe during the entire path.

A safe zone for orientation is called a **safe cone** and is shown in Figure 6-11.

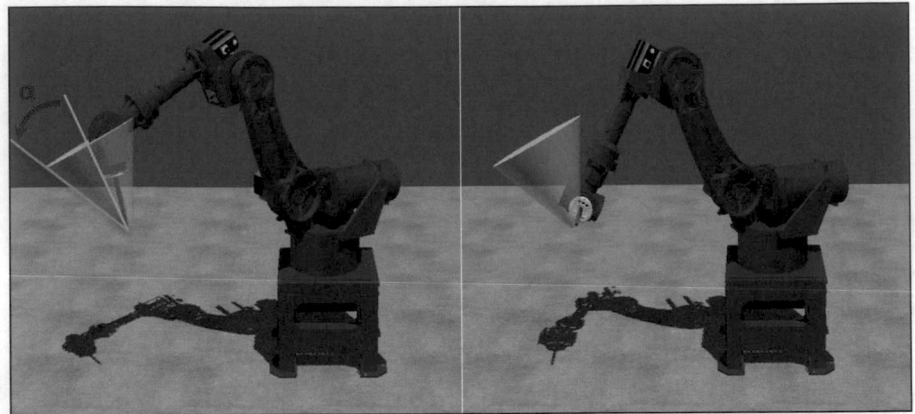

Figure 6-11. *Definition (left) and violation (right) of a safe cone*

The operator needs to define a central working orientation (as shown with the yellow line in Figure 6-11 left) and then a maximum allowed deviation angle α. The resulting geometry is a cone, outside of which the orientation of the TCP is considered unsafe (see Figure 6-11 right).

Recall that when expressing orientations with quaternions, the angle δ between two orientations q_1 and q_2 is given by the following formula:

$$\delta = cos^{-1} \frac{q_1 \cdot q_2}{\| q_1 \| \| q_2 \|} \tag{6-5}$$

Using the central working orientation as reference (q_1), we can monitor whether the current TCP orientation (q_2) lies inside the safe cone ($\delta < \alpha$), both at planning time and during the execution of a movement at real time.

Self-Collision

All the forbidden zones defined by the operator are normally located in the space next to the robot. However, there is one additional area that the TCP must not enter under any circumstance, and that is the body of the robot itself. In other words, we want to monitor and prevent self-collisions.

The risk for self-collision increases if the attached tool is particularly large, as shown in Figure 6-12.

Figure 6-12. *Self-collision of the robot with its own body*

The solution is quite simple and consists in automatically generating forbidden zones around each of the body links (usually corresponding to the wire-frame model), starting from the base up to the TCP. These bounding regions will not correspond exactly to the body of the robot, but they will cover it safely and prevent self-collisions.

For the shape of the bounding regions, we choose **capsules** because they are computationally inexpensive. The operator only needs to specify a numerical radius for each arm, and we then use the direct kinematics to automatically generate the positions of the zones given the actual position of the joints (see Figure 6-13).

The only difference with standard forbidden zones is that the position of the body capsules changes over time and must be dynamically updated.

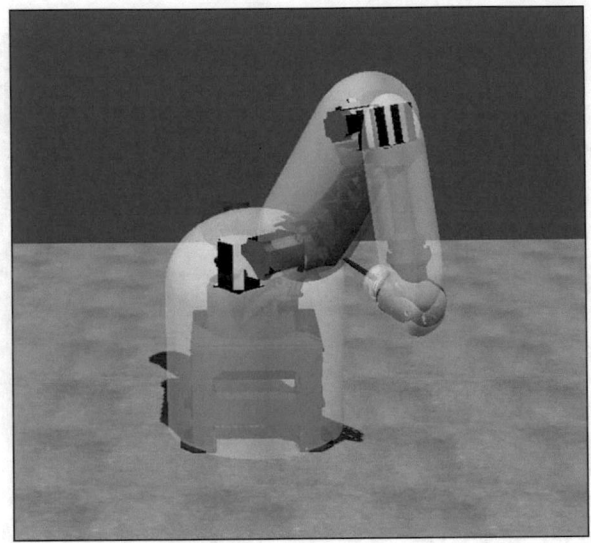

Figure 6-13. *Virtual protection capsules built around each link of the robot*

In order to verify if a point in space is inside or outside the capsule, it is enough to calculate its distance from the capsule originating segment. We will describe how to do that in the next section.

Capsules

A capsule is geometrically defined as the set of points located at the same distance from a given segment. It is composed of a cylinder with two semispheres at the extremes. You can imagine sliding a sphere along the segment to find all the points at the same distance r (see Figure 6-14).

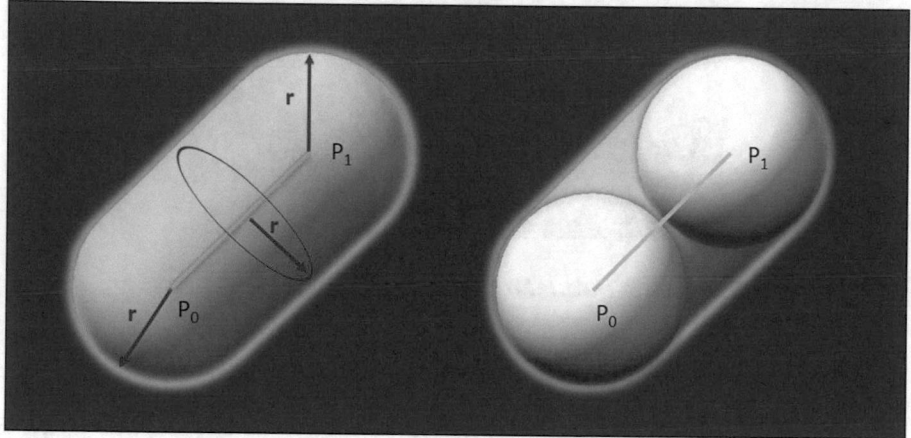

Figure 6-14. *Geometrical description and construction of a capsule*

The radius r is normally defined by the operator, based on the size of the mechanical arm of the robot.

The problem we want to solve is finding the distance between any given point Q in space and the segment P_0P_1. If that distance is smaller than r, then a collision has occurred.

We refer to Figure 6-15 and solve the problem one step at a time. The segment P_0P_1 can be described parametrically as a function of $t = 0...1$. We project the point Q onto P_0P_1 and call the intersection P.

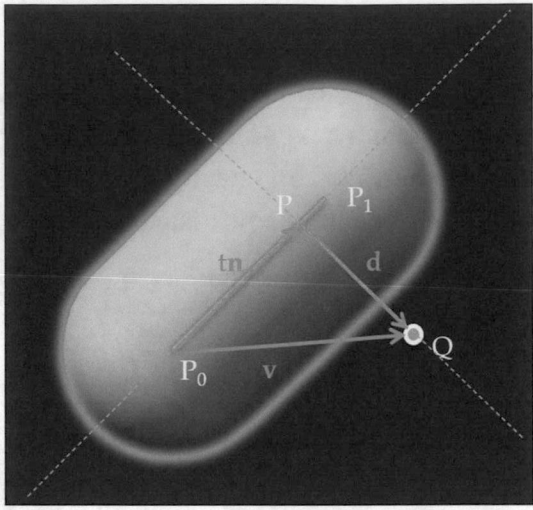

Figure 6-15. *Distance between point and capsule*

We then define the following basic vectors:

$$\overline{P_0P_1} = \boldsymbol{n} \tag{6-6}$$

$$\overline{P_0P} = t\boldsymbol{n} \tag{6-7}$$

$$\overline{PQ} = \boldsymbol{d} \tag{6-8}$$

$$\overline{P_0Q} = \boldsymbol{v} \tag{6-9}$$

Observing the blue triangle in Figure 6-15, we can express d as the difference between the vectors v and tn:

$$\boldsymbol{d} = \boldsymbol{v} - t\boldsymbol{n} \tag{6-10}$$

We already know v, because the positions of both P_0 and Q are known. We are missing P. The solution is to impose the condition that d and n are perpendicular to each other. Mathematically, that is equivalent to forcing their dot product equal to 0:

$$d \cdot n = 0 \tag{6-11}$$

Merging Equation (6-10) into Equation (6-11) and solving for t leads to the following:

$$(v - tn) \cdot n = 0 \tag{6-12}$$

$$v \cdot n - t(n \cdot n) = 0 \tag{6-13}$$

$$t = \frac{v \cdot n}{\|n\|^2} \tag{6-14}$$

We can safely assume that n is not 0; otherwise, no capsule would exist at all.

Once we find t, we can first calculate the position of P using Equation (6-7) and then immediately the distance between P and Q, which is the quantity we need to evaluate to detect a possible collision between the point and the capsule.

Note that this procedure works as long as t lies in the interval between 0 and 1, which means that the point P lies on the segment P_0P_1. In case we find a value for t smaller than 0 or larger than 1, it means that the perpendicular from Q onto the line n falls outside the range P_0P_1, either before P_0 (for $t < 0$) or after P_1 (for $t > 1$). In those cases, the distance between the point Q and the segment P_0P_1 is simply given by the distance between Q and either P_0 or P_1, whichever one is closer (see Figure 6-16).

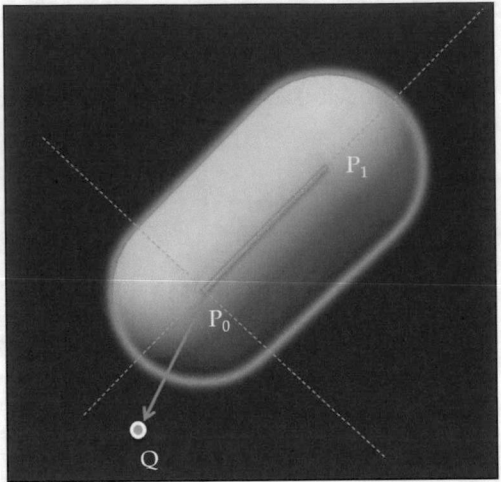

Figure 6-16. *Distance between point and capsule when t is outside the 0...1 interval*

Exclusive Zones

So far, we have introduced functionalities to monitor the position and orientation of a single robot. However, it can happen that several robots work together in the same environment and share their workspaces.

That is actually a very common situation in practical applications. One example is when multiple robots work on an automotive assembly line and they perform different operations on the same workpiece. One robot might be laser cutting, a second robot welding, another one driving the screws, and one more doing the painting.

Pick-and-place applications are also a typical example: several delta robots are lined up to pick products off a common conveyor. It is critical to monitor their position relative to each other in order to avoid collisions (see Figure 6-17).

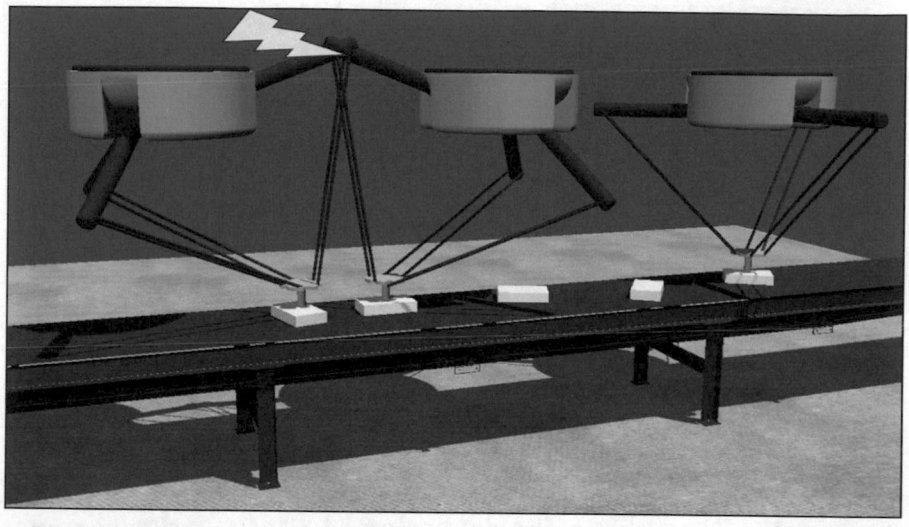

Figure 6-17. *Multi-robot applications require sophisticated collision detection functionalities*

One way to perform workspace monitoring for multiple robots is to introduce **exclusive zones**: regions of space that only one robot (or at least its TCP) can occupy at any given time.

From the control software point of view, each exclusive zone is associated to a flag bit, which is visible to all robots and warns of the presence of a robot inside that space. If the TCP of a second robot is programmed to enter the same exclusive zone, it will have to check the corresponding status flag first and will only be allowed to enter if the zone is free (see Figure 6-18). Otherwise, the movement is temporarily paused until the first robot leaves the exclusive zone. At that point, the second robot can resume its movement and can enter the region. As a consequence, the region immediately becomes locked again, and no other robot is allowed inside.

Figure 6-18. *The robot on the right cannot enter the exclusive zone (yellow) until the robot on the left leaves it*

In case that more than two robots are sharing the same exclusive zone, there might be a few of them waiting in line to enter the zone. Typically, the first robot in line gets the green light and has access to the zone, locking it. Alternatively, each robot might have a specific priority according to the importance of the task they perform. In that case, the robot with the highest priority gets to enter the unlocked zone first.

The main advantage of using exclusive zones is that they are easy to configure. However, they do not guarantee the safety of all robots in the entire workspace. For instance, the operator typically configures an exclusive zone around the workpiece but rarely thinks of possible collisions away from it. In the next section, we introduce a more sophisticated function to provide additional safety to the robots.

One important observation when working with a multi-robot system is that the collision monitoring functions only work correctly if all robots are referenced to the same global frame. To achieve that condition, the zero frame described in Chapters 3 and 4 must be activated.

Collision Detection

Collision detection consists in monitoring the possible intersection between the bodies of two robots sharing a workspace. The function is conceptually simple to understand, but a bit tricky to implement mathematically.

The idea is to build bounding capsules around the joints of each robot, just like we did for the self-collision monitoring function, and evaluate all possible collisions between each capsule of a robot against all the capsule of the other robots.

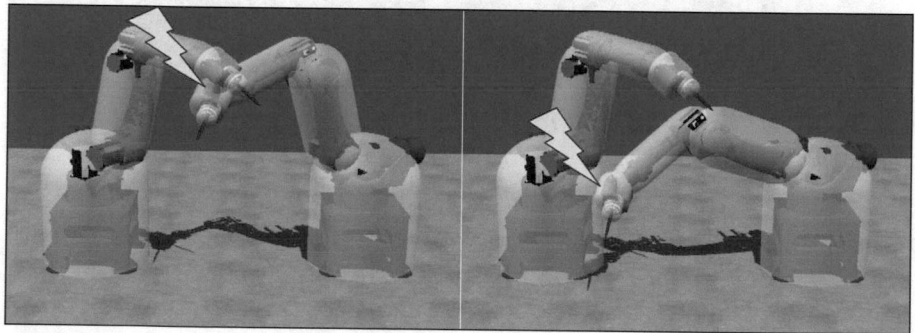

Figure 6-19. *Using capsules around the body links to detect collisions between robots*

Using capsules is much more convenient than cuboids or cylinders because the collision test is extremely quick and computationally inexpensive: *two capsules intersect if the distance between the capsules' generating segments is smaller than the sum of their radii.*

Essentially, we need to implement a function to compute the distance between two segments in space. Notice that, since the robots move independently running separate programs, we cannot perform collision detection monitoring during planning time: the planned paths might intersect in space but not in time. The monitoring needs to run at real time, while the movements are being executed.

We take Figure 6-20 as reference and start to compute the distance between two segments P_0P_1 and Q_0Q_1, lying, respectively, on the lines L_1 and L_2.

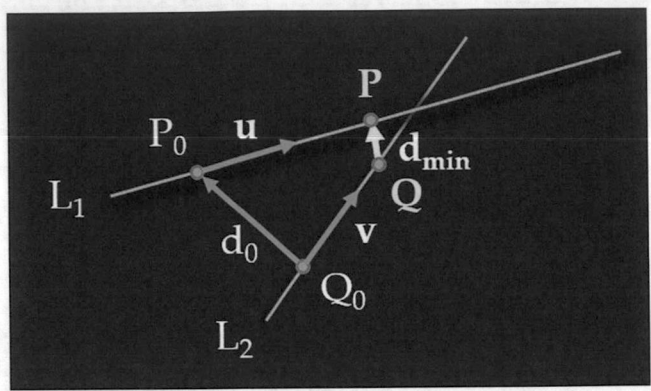

Figure 6-20. *Calculating the distance between two segments*

We use the direction vectors u and v for the two lines and the parameters s and t to describe the segments in the interval $s, t = 0...1$. The two parametric equations for the lines are as follows:

$$L_1 : P = P_0 + su \qquad\qquad (6\text{-}15)$$

$$L_2 : Q = Q_0 + tv \qquad\qquad (6\text{-}16)$$

We call d_0 the distance between P_0 and Q_0, where the two segments begin. Then we observe that the generic distance between the lines L_1 and L_2 is given by the following:

$$d = P - Q = P_0 + su - Q_0 - tv = d_0 + su - tv \qquad\qquad (6\text{-}17)$$

In other words, d is a function of the two parameters s and t, and our goal is to find its minimum value, which is the minimum distance between the two lines.

One way to proceed is with simple calculus. We derive the squared norm of d and observe that the result is a paraboloid in s and t:

$$\| d \|^2 = d \cdot d = (d_0 + su - tv) \cdot (d_0 + su - tv) = (d_0 \cdot d_0) + s^2 (u \cdot u) +$$
$$t^2 (v \cdot v) + 2s (d_0 \cdot u) - 2t (d_0 \cdot v) - 2st (u \cdot v) \tag{6-18}$$

This function has exactly one minimum value (d_{min}) in correspondence of (s_{min}, t_{min}) as shown in Figure 6-21, unless, of course, the two lines are parallel, in which case their distance is always constant.

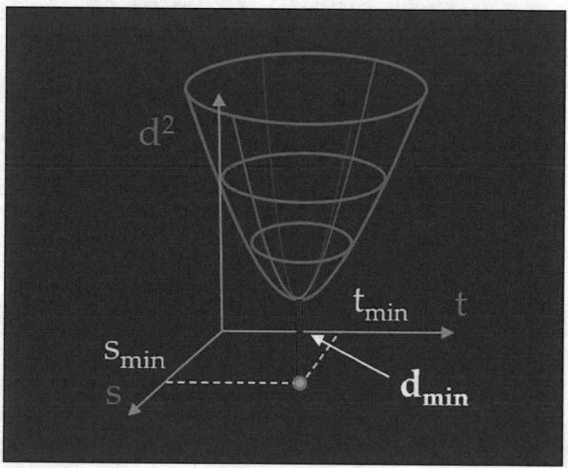

Figure 6-21. *The distance between two lines is a quadratic function in two variables*

Finding the minimum of the quadratic function is solved by forcing its gradient to 0: we write out the partial derivatives with respect to s and t and set both of them equal to 0.

$$\begin{cases} \dfrac{\partial \left(\| d \|^2 \right)}{\partial s} = 2s (u \cdot u) + 2 (d_0 \cdot u) - 2t (u \cdot v) = 0 \\[2mm] \dfrac{\partial \left(\| d \|^2 \right)}{\partial t} = 2t (v \cdot v) - 2 (d_0 \cdot v) - 2s (u \cdot v) = 0 \end{cases} \tag{6-19}$$

Interestingly, we could have also derived the same equations by geometrical intuition: since the minimum of d is a vector perpendicular to both L_1 and L_2, the two dot products $d \cdot v$ and $d \cdot u$ must both be 0, which leads again to Equation (6-19).

Solving for s and t we find the two values that minimize the function d:

$$s_{min} = \frac{(d_0 \cdot v)(u \cdot v) - (d_0 \cdot u)(v \cdot v)}{(u \cdot u)(v \cdot v) - (u \cdot v)^2} \qquad (6\text{-}20)$$

$$t_{min} = \frac{(d_0 \cdot v)(u \cdot u) - (d_0 \cdot u)(u \cdot v)}{(u \cdot u)(v \cdot v) - (u \cdot v)^2} \qquad (6\text{-}21)$$

This solution is always valid except when the denominator is 0, which happens when the two lines are parallel to each other and their distance is constant.

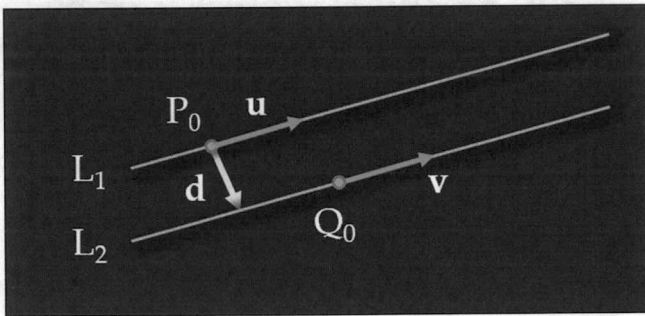

Figure 6-22. *Choose any point on parallel lines to calculate their distance*

To handle this particular case, we can pick any point on one line and then evaluate the distance from there to the other line (see Figure 6-22). In other words, we fix an arbitrary value for s (e.g., $s_{min} = 0$) and solve the second equation of Equation (6-19) to find the corresponding value of t:

$$t_{min} = \frac{d_0 \cdot v}{v \cdot v} \tag{6-22}$$

Once we have the values for s_{min} and t_{min}, we can find the corresponding points P and Q, where the two lines are closest to each other. The distance between them is simply $|P - Q|$.

However, recall that we are specifically looking for the minimum distance between the two finite segments inside the capsules, not just between the two infinite lines. Therefore, the solution we just derived is only valid if both s_{min} and t_{min} lie in the interval 0...1. If they do not, then either P or Q lie outside the original segments, which means that the lines L_1 and L_2 get closer to each other outside the capsule than they do inside it (see Figure 6-23). To find the actual distance between the capsules in this case, we need to work a bit further.

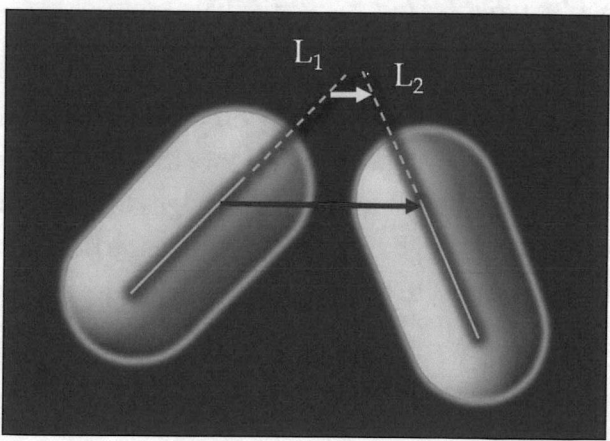

Figure 6-23. *The actual distance between the capsules might be larger than the distance between the lines*

Let's look again at the paraboloid representing the squared distance $\|d\|^2$ between the two lines as a quadrating function of s and t (see Figure 6-24). The global minimum might fall outside the 0…1 square region. In that case we need to find the tuple (s, t) that minimizes the paraboloid locally, in the region generated by the projected square on its surface.

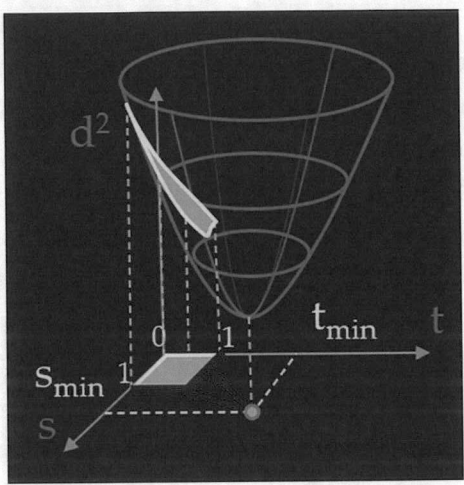

Figure 6-24. *Solving for a local minimum*

The solution is to notice that the local minimum will necessarily lie on the edges of the (0…1) square, specifically on the ones closer to the global minimum point (s_{min}, t_{min}). We need to evaluate the partial derivatives of the paraboloid, but this time only on the two edges of the square.

We start with the first edge $s = 1$ and minimize the partial derivative of the paraboloid with respect to t:

$$\frac{\partial\left(\|d\|^2\right)}{\partial t} = 2t(v\cdot v) - 2(d_0\cdot v) - 2s(u\cdot v) = 0 \qquad (6\text{-}23)$$

Solving for $s = 1$ we find the minimum value for t:

$$t_{min} = \frac{(d_0 \cdot v) + (u \cdot v)}{(v \cdot v)} \qquad (6\text{-}24)$$

If $0 \leq t_{min} \leq 1$, then the minimum distance between the segments is $s = 1$ and $t = t_{min}$.

Otherwise, we need to clamp t_{min} in the interval 0...1 and go check on the other edge of the square, where the local minimum will be located.

Once we derive the valid solutions for s_{min} and t_{min}, we can calculate P and Q and finally calculate their distance, which is the distance between the two capsules. We then compare that distance with the sum of the two radii and find out whether the two robots are colliding.

Summary

Technically, not all geometrical paths are safe to be executed by the robot. This chapter on workspace monitoring provides a detailed description of all the safety functions we should add to our control software in order to make sure that the planned path does not violate the imposed spatial constraints.

In particular, an individual robot needs to stay inside predefined safe zones and outside of forbidden zones while keeping its TCP orientation within limits and also avoiding collisions with its own body.

When more robots are brought to work together, then collision detection is extended to a multitude of bodies, and exclusive zones are introduced to prevent dangerous situations.

CHAPTER 7

Trajectory Generator

In the previous chapters, we have learned how to describe a path geometrically, by means of parametric equations. Now, we need to learn how to let the robot run along those curves.

Given a planned path s, we introduce the concept of **trajectory** $s(t)$, defined as *a function that samples points from the path s along the time t.*

A trajectory transforms a static path into a dynamic function of time. Importantly, any number of different trajectories can be generated from the same path. This is easy to understand, as you can move along a line from A to B using different speed profiles.

For example, we can sample points from a path at a constant rate in time (see Figure 7-1). We call the time interval Δt, say 1ms. Since every 1ms we cover a constant amount of space, we are traveling at constant speed v_1.

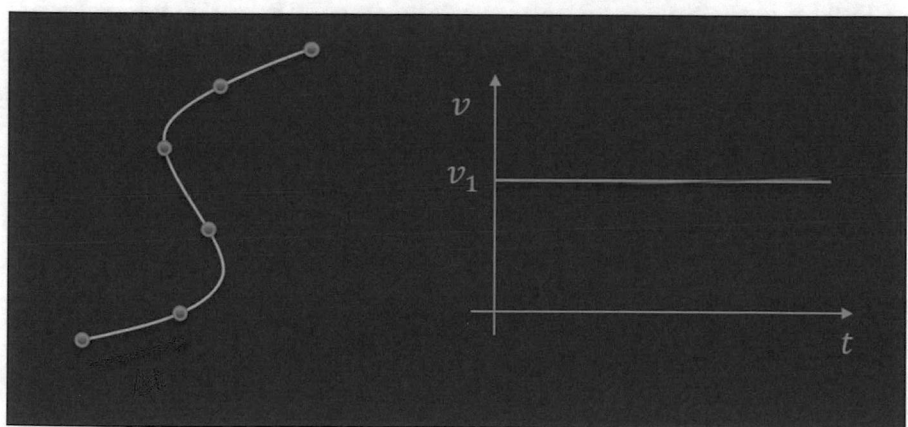

Figure 7-1. *Sampling points from a path generates a trajectory*

© Fabrizio Frigeni 2023
F. Frigeni, *Industrial Robotics Control*, Maker Innovations Series,
https://doi.org/10.1007/978-1-4842-8989-1_7

If we decide to increase the density of points, we cover a smaller amount of distance in the same time Δt, which means we are now travelling over the same path, but with a lower speed $v_2 < v_1$. The result is a different trajectory, even if the path is the same (see Figure 7-2). The geometrical curve in space is identical, but the relation between space and time has changed.

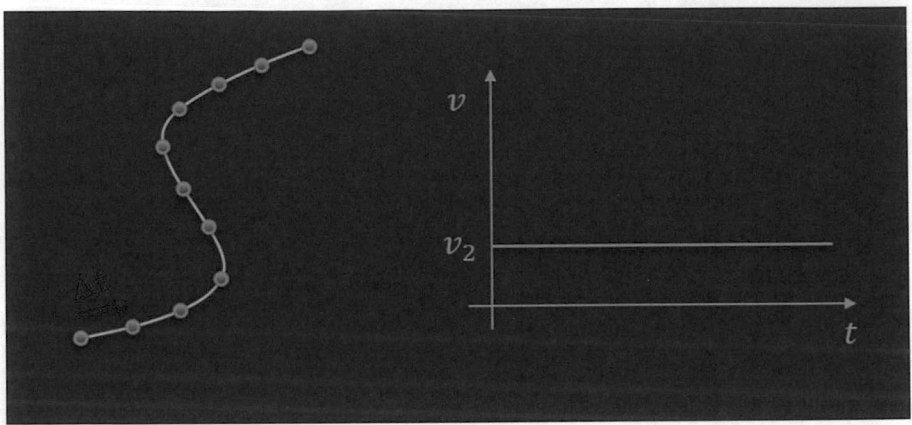

Figure 7-2. Increasing the sampling rate reduces the speed of the movement

Finally, we sample a different set of points at different distances from each other: first closer, then farther apart, and finally closer again (see Figure 7-3). This trajectory is yet another different way to traverse the same original path. This time we use a nonuniform speed: we start slowly, then accelerate to maximum speed, and finally slow back down to 0.

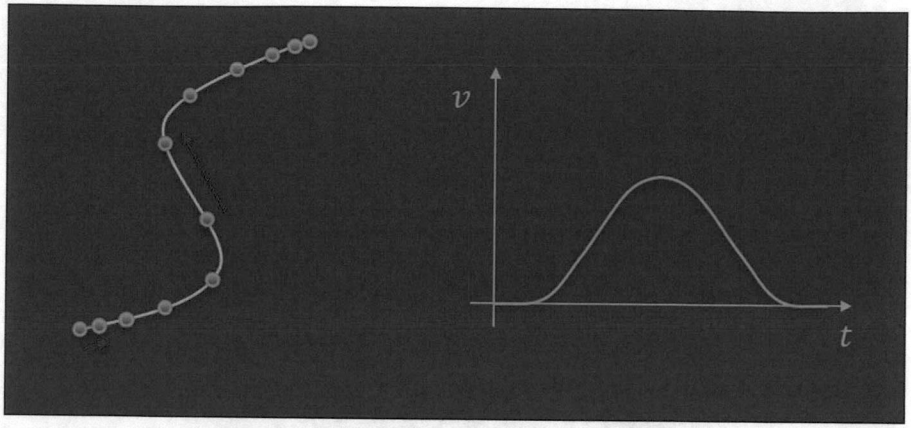

Figure 7-3. *Sampling at nonuniform speed*

The message to take out from this simple example is that we can visualize the trajectory generator as a sampler of points from the given path, where the sampling frequency decides the instantaneous movement's speed.

In the specific case of a six-axes robot, the trajectory generator picks a point from the planned path at every time tick of the controller (e.g., at 1kHz frequency). Then it calculates the set joint values via inverse kinematics and sends them to the motor drives.

If the points are all equidistant from each other, the path speed is constant. If we move those points closer together, we achieve lower speed. Vice versa, spreading them apart generates higher speed movements, because in the same amount of time we force the robot to cover more space.

S-Curve Profile

Generating a trajectory with constant speed from start to end is mathematically valid, but not physically realistic. A body starting from standstill state would need an infinite positive acceleration to instantaneously move at a constant finite speed and an infinite negative acceleration (or simply deceleration) to stop abruptly at the end of the movement (see Figure 7-4).

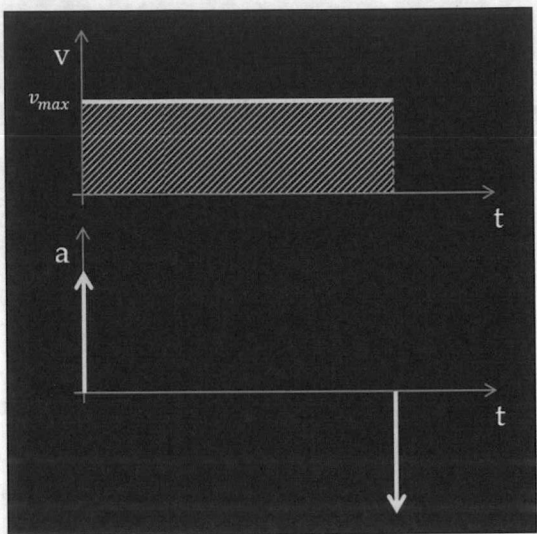

Figure 7-4. *Reaching cruise speed instantaneously requires infinite acceleration*

From a physical perspective, asking a robotic arm to accelerate instantaneously would require an infinite amount of torque from its motors, which is clearly impossible. Given a limited maximum torque, it will take some time to accelerate to cruise speed and some time to slow down to a halt. Taking into account the acceleration limit a_{max}, we can modify the speed profile accordingly (see Figure 7-5).

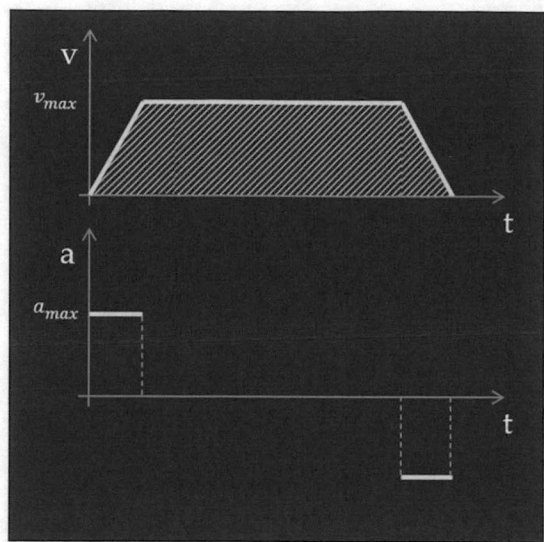

Figure 7-5. *Introducing an acceleration limit in the trajectory*

The path length to cover is the same, but the movement will take more time to execute because we have decreased its speed in the initial and final phases.

While an acceleration-limited profile is more realistic to execute, it is still far from being ideal. In practice, asking a robot to change its acceleration so abruptly increases wear and stress on the mechanics, reducing the life of the parts. A better solution is to further smooth the speed profile by also limiting the changes in acceleration. The derivative of acceleration over time is called **jerk**. Introducing a jerk limit j_{max} produces a smoother profile (see Figure 7-6), which is the typical solution adopted in the industry.

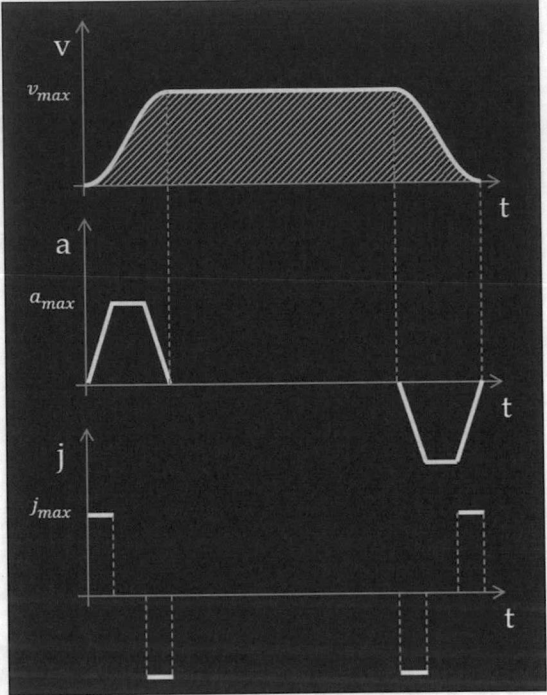

Figure 7-6. *Introducing a jerk limit generates a smoother trajectory*

The resulting speed profile with limited acceleration and jerk is normally called an **S-curve**, because its gradual speed changing behavior is shaped like an "S". The profile is composed of seven distinct zones:

- An initial increase of acceleration with limited jerk

- A phase of limited acceleration

- A decrease of acceleration with limited jerk to reach maximum speed

- A central area of constant limited speed

- Three more zone similar to the initial three, reducing speed to 0 with limited acceleration and jerk

We can now plot together the three different speed profiles that we have analyzed so far (see Figure 7-7). The length of the covered path is the same in all three cases and is given by the area under each speed profile:

$$L = \int v(t)\,dt \tag{7-1}$$

A constant-speed trajectory will complete in the shortest time, while the introduction of acceleration and jerk limitations will automatically increase the total time required to complete the movement. The first profile (red) is impossible to achieve in practice. The second (yellow) is essentially valid, but not recommended. The third (green) is the best choice.

Limiting the jerk is a very important requirement when generating trajectories in practice. By tuning the imposed limits (a_{max} and j_{max}) according to the motors' power and robot's mechanical strength, we can adjust the movement to complete with optimal timing.

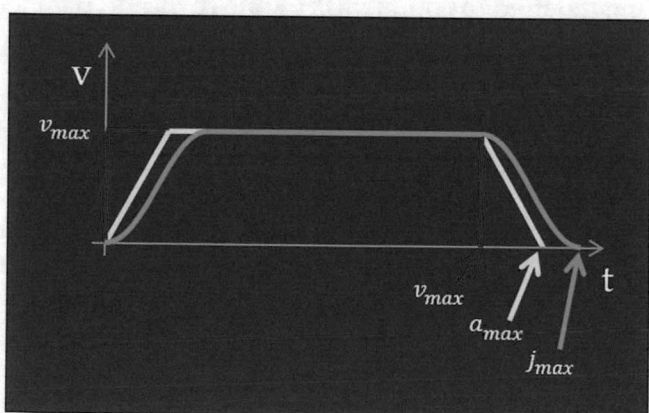

Figure 7-7. *Comparison between trajectories with different dynamic limitations*

The equations describing an S-curve depend on three main parameters: $[v_{max}, a_{max}, j_{max}]$. Sometimes these parameters double in number, in case the user decides to set different values for positive and negative speed, acceleration, and jerk. The concept does not change in either case, so we present here the equations in their simplest form.

Also, we consider here the simple case of starting and ending with zero speed and zero acceleration. This is typically the case when planning a trajectory between two non-tangential transition points (see Section on Transitions in Chapter 5), where the path speed is required to halt. If the transition between two consecutive blocks is tangential, then we do not need to stop, and we can merge the two blocks together into a single large movement. Upon reaching the next non-tangential transition, we stop the current movement and generate a new trajectory for the next block starting again from zero speed.

We now analyze all the seven zones of the S-curve shown in Figure 7-8: for each of them ($i = 1...7$), we derive the *time-dependent equations* for speed (v_i), time required to complete the zone (Δt_i), and path length covered (d_i). To simplify notation, we use $[v, a, j]$ in the formulas in place of $[v_{max}, a_{max}, j_{max}]$.

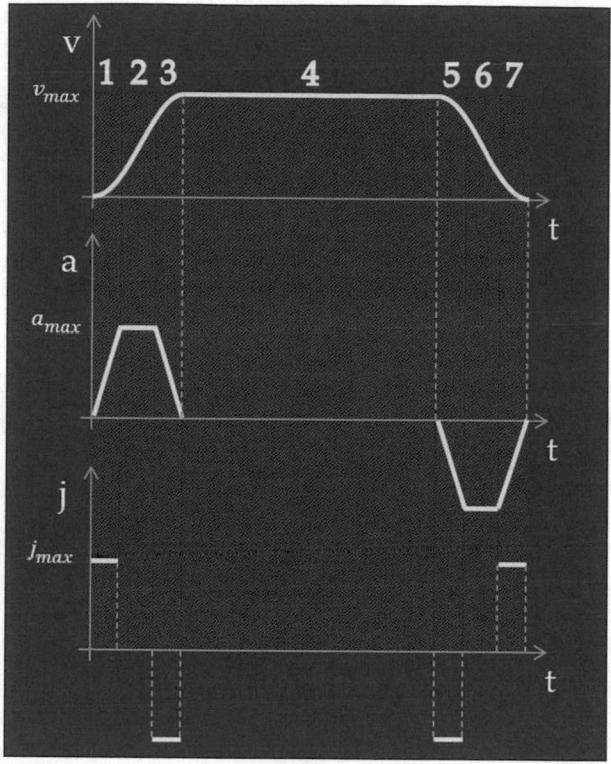

Figure 7-8. *The seven zones of the S-curve*

Zone 1: The acceleration increases linearly, and the speed increases very quickly as a square function of time, only limited by the maximum jerk *j*:

$$v_1 = \frac{1}{2} j t^2$$

(7-2)

The time duration of this section is as follows:

$$\Delta t_1 = \frac{a}{j}$$

(7-3)

In fact, a higher jerk limit allows to reach the target acceleration more quickly.

The covered path length can be obtained by integrating the movement speed v_1 over the time interval Δt_1:

$$d_1 = \frac{1}{6} j \Delta t_1^{\,3} \tag{7-4}$$

Zone 2: We have now reached maximum acceleration, and speed cannot grow quadratically anymore. The speed increase becomes linear, until we reach the point where we need to start reducing acceleration; otherwise, we overshoot the target maximum speed.

$$v_2 = v_1 + \boldsymbol{a}t \tag{7-5}$$

The time duration of this second section depends on the speed limit $\boldsymbol{v}$:

$$\Delta t_2 = \frac{v}{a} - \frac{a}{j} \tag{7-6}$$

$$d_2 = v_1 \Delta t_2 + \frac{1}{2} a \Delta t_2^{\,2} \tag{7-7}$$

Zone 3: The acceleration reduces to 0 so that the speed can plateau at its programmed limit value:

$$v_3 = \boldsymbol{v} - \frac{1}{2} j \left(\frac{a}{j} - t \right)^2 \tag{7-8}$$

The time duration of this segment is the same as the first zone. Note the symmetry of the curve, caused by the maximum and minimum jerk to be equal to each other, as is usually the case in practice.

$$\Delta t_3 = \frac{a}{j} \tag{7-9}$$

$$d_3 = v \Delta t_1 - \frac{1}{6} j \Delta t_1^{\,3} \tag{7-10}$$

Zone 4: The fourth central section is the easiest to solve, because its speed is constant:

$$v_4 = v \tag{7-11}$$

Typically, this is the longest part of a movement because it covers the whole path length, except for the small segments of the acceleration and deceleration phases:

$$d_4 = L - \left(d_1 + d_2 + d_3 + d_5 + d_6 + d_7 \right) \tag{7-12}$$

Its duration in time also depends on the total movement length: the longer the movement, the longer this zone takes to complete, while all the others remain unaffected.

$$\Delta t_4 = \frac{d_4}{v} \tag{7-13}$$

Zone 5–7: The last three sections are responsible for the deceleration phase, and they are symmetrical to the first three zones:

$$v_5 = v - \frac{1}{2} j t^2 \tag{7-14}$$

$$\Delta t_5 = \Delta t_3 \tag{7-15}$$

$$d_5 = d_3 \tag{7-16}$$

$$v_6 = v_5 - at \tag{7-17}$$

$$\Delta t_6 = \Delta t_2 \tag{7-18}$$

$$d_6 = d_2 \tag{7-19}$$

$$v_7 = \frac{1}{2} j \left(\frac{a}{j} - t \right)^2 \tag{7-20}$$

$$\Delta t_7 = \Delta t_1 \tag{7-21}$$

$$d_7 = d_1 \tag{7-22}$$

If you want to generate a trajectory using different values for positive and negative speed, acceleration, or jerk, then the preceding formulas need to be adjusted accordingly.

There are two special cases we need to analyze in detail: it can happen that the length of the movement is very short and the trajectory does not have enough time to reach the maximum speed or acceleration.

In particular, calculating the length of the central zone d_4 using Equation (7-12) could result in a negative value, losing physical meaning. In that case, the path is not long enough to allow the programmed speed v to be reached. The solution is forcing the maximum speed value to a lower value by imposing the condition $d_4 = 0$ and recalculating the acceleration and deceleration sections.

In other words, as soon as the new target speed is reached after the acceleration phase, the deceleration section starts immediately. The new generated profile does not have a central zone with constant speed (see Figure 7-9).

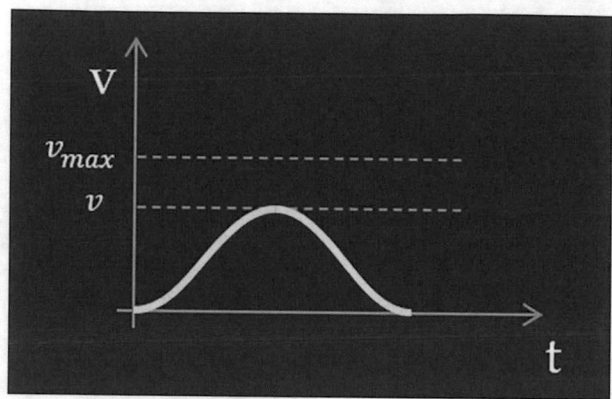

Figure 7-9. *A short movement is not always able to reach the limit speed*

In extreme cases, when the path is very short and/or the limit acceleration is set very high, even the maximum acceleration value a cannot be reached. The result is that the value calculated for Δt_2 in Equation (7-6) is negative. The solution is to impose the condition $\Delta t_2 = 0$ and recalculate the new lower acceleration limit. The resulting movement will only be jerk up and jerk down, with the S-curve reduced to a four-zone profile: zones $(1, 3, 5, 7)$ are still present, while $(2, 4, 6)$ disappear (see Figure 7-10).

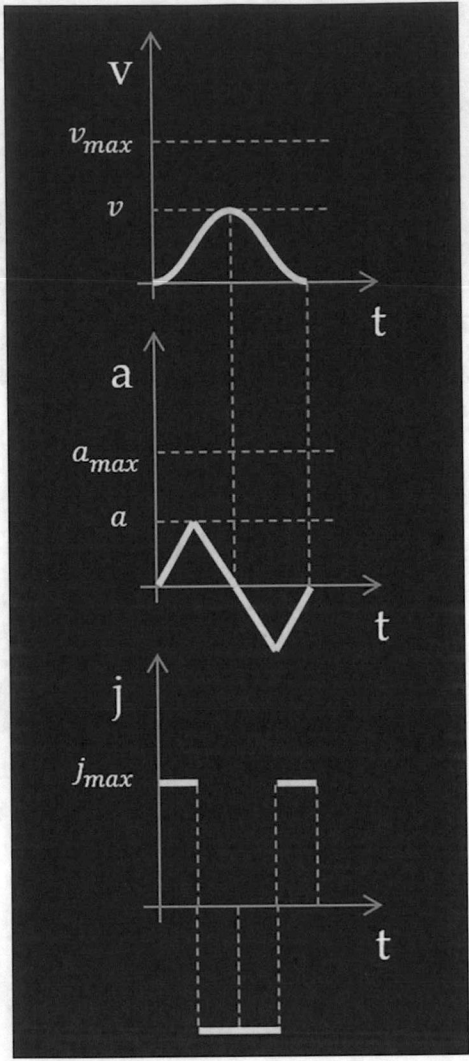

Figure 7-10. *A very short movement does not even reach the limit acceleration*

Sinusoidal Profile

The S-curve we just examined is the most common solution to generate a speed profile because it allows for complete control over all positive and negative speed, acceleration, and jerk values. However, it is not the only possible solution. There are other alternatives worth exploring, either because they are usually simpler to implement with models relying on fewer parameters or because they have interesting features that can outperform the S-curve in some cases.

For example, while the S-curve is jerk-limited, it is not limited in its higher derivatives. The acceleration profile is trapezoidal, and the jerk jumps to maximum limit values instantaneously. The derivative of jerk (the fourth derivative of path displacement over time) results in an infinite pulse.

We might face requirements to generate extra-smooth movements for certain applications. In that case, it could be useful to use a **sinusoidal profile**, so that its derivatives are limited to a higher degree.

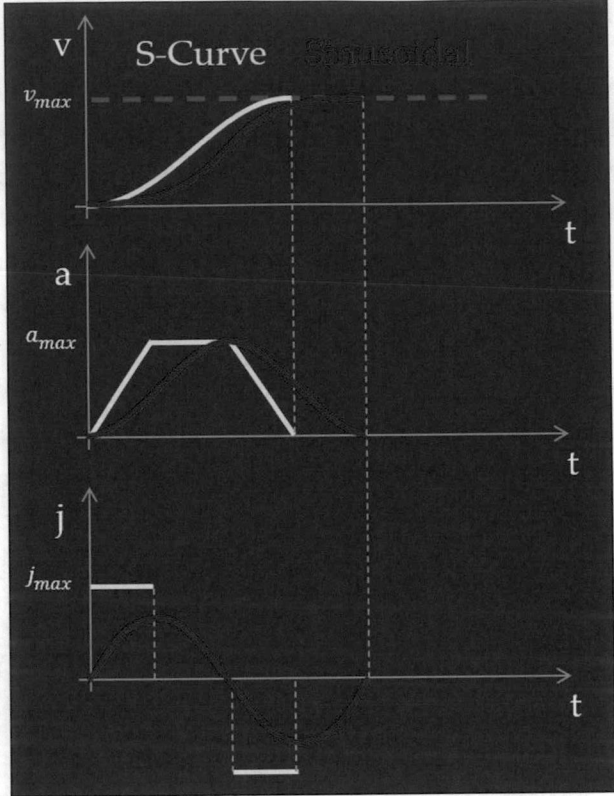

Figure 7-11. *Comparison between the S-curve and a sinusoidal profile*

The red curve in Figure 7-11 is constructed with one single sinusoidal segment to reach maximum speed. Both the acceleration and the jerk profiles exhibit a much smoother behavior when compared to the standard S-curve. That is a consequence of the limited harmonic content of sinusoidal curves. The maximum value reached by either the acceleration or the jerk is actually lower, depending on the set limits for each. It can also be proven that the power peak required by this curve is lower than the S-curve. On the other hand, the time to reach maximum speed is longer, under the constraint of equal dynamic limits.

Another major advantage of the sinusoidal profile is that one single segment is enough to smoothly change speed. Instead of a seven-segment curve for a complete profile, we would only need a three-segment curve: one zone to accelerate to the programmed speed, one linear zone to cruise at constant speed, and one final zone to decelerate to a halt.

The resulting code is much simpler to write. A minor drawback is that the computational load is slightly higher, because transcendental functions are used instead of simple multiplications. For modern CPUs, that is not really an issue.

The equations describing the first profile zone, where the movement accelerates from standstill to the target speed v, are as follows:

$$v_1 = k\left(\omega t - sin\ \omega t\right) \tag{7-23}$$

$$a_1 = k\omega\left(1 - cos\ \omega t\right) \tag{7-24}$$

$$j_1 = k\omega^2 sin\ \omega t \tag{7-25}$$

The parameter k determines the amplitude of the speed sinusoid and is fixed by the value of the target speed v:

$$k = \frac{v}{2\pi} \tag{7-26}$$

The parameter ω determines the period of the sinusoids and is constrained by the limit values for acceleration a and jerk j, whichever is smaller:

$$\omega = min\left(\frac{a}{2k}, \sqrt{\frac{j}{k}}\right) \tag{7-27}$$

The time needed to reach the target speed is as follows:

$$\Delta t_1 = 2\pi / \omega \qquad \qquad (7\text{-}28)$$

Integrating the speed profile over the elapsed time gives the path distance covered during the acceleration phase:

$$d_1 = \frac{v^2}{a} \qquad \qquad (7\text{-}29)$$

The second zone is a simple constant-speed profile, while the third zone is a perfect mirror of the first. In a similar way to what we already described for S-curves, there are certain conditions that must be verified to make sure that the generated sinusoidal profile is physically valid. Specifically, it is not guaranteed that the programmed limit values can be reached if the path length is too short. In that case, the k and ω parameters need to be adjusted to respect the newly calculated limits.

Let's compute a numerical example to highlight the difference between the two profiles. We consider a movement from standstill to target speed $v = 1$ m/s, imposing the maximum limit for acceleration $a = 10$ m/s^2 and jerk $j = 100$ m/s^3.

Computing the total acceleration time for the polynomial S-curve results in $\Delta t_1 = 0.2$ s. Conversely, the sinusoidal profile will take $\Delta t_1 = 0.25$ s. Notice, however, that the actual maximum acceleration reached by the sinusoidal curve is only about $a = 8$ m/s^2. The smother profile of the sinusoidal acceleration *is gentler on the mechanics at the cost of a slightly longer time to execute.*

Bezier Profile

A second alternative to a standard S-curve is a **Bezier profile**. We already introduced these curves in the path-planning chapter to generate smooth position profiles. We can use them here as well to generate smooth speed profiles.

The easiest Bezier curve we can build is a cubic spline with four control points to connect the starting standstill state with the target speed:

$$P = (1-t)^3 P_0 + 3(1-t)^2 tP_1 + 3(1-t)t^2 P_2 + t^3 P_3 \qquad (7\text{-}30)$$

We can let P move from P_0 to P_3 by sliding the parameter t in the interval $[0...1]$. The two middle control points P_1 and P_2 must be placed at the same speed of the end points, so that the initial and final accelerations are both 0. They also need to be placed at constant time distance, a third of the way each, so that the spline is traversed uniformly in time (see Figure 7-12). These two conditions also conveniently guarantee the C^2 continuity at the border points when joined in series to similar profiles.

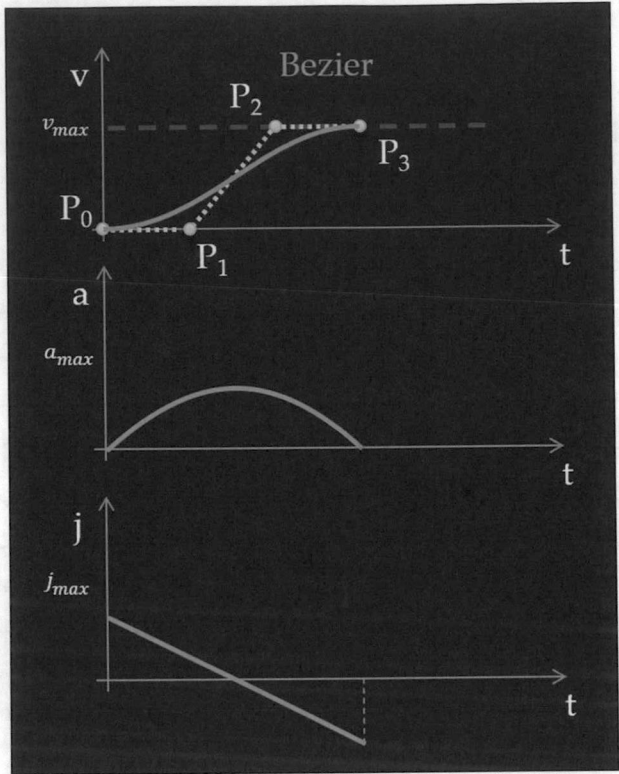

Figure 7-12. *Building a speed profile using a cubic Bezier curve*

The result is a very simple trajectory model, with fewer parameters to control than the S-curve. As a matter of fact, evaluating the speed over time does not even require the De Casteljau's algorithm we learned in Chapter 5, because the spline is uniform and depends only on the parameter *t*:

$$v_1 = \left(3t^2 - 2t^3\right)v \qquad (7\text{-}31)$$

By differentiating with respect to time, we can derive the equations for acceleration and jerk:

$$a_1 = 6(1-t)t\,v$$

(7-32)

$$j_1 = 6(1-2t)\,v$$

(7-33)

The maximum values of acceleration a and jerk j depend solely on the target velocity v and are directly related to each other, not independent as they were in the S-curve. The peak acceleration is reached at $t = 0.5$, while the peak jerk is reached at the beginning and end of the movement.

$$a = \frac{3}{2}v$$

(7-34)

$$j = 6v$$

(7-35)

As a consequence, we cannot choose limit values for acceleration and jerk individually. In practice, that is usually not an issue, because most robot operators only choose an acceleration limit and do not bother selecting a specific jerk. We can automatically set a jerk limit in the control program: a common reasonable value is ten times the acceleration limit.

Alternatively, we could use a higher order spline (i.e., a quartic Bezier curve), by adding extra control points to be able to set jerk and acceleration limits independently. It is a solution that would provide more flexibility but has the drawback of an increased number of parameters in the model.

The path distance covered during the acceleration phase of the cubic spline can be found integrating the speed profile in Equation (7-31) over the interval $t = [0...1]$:

$$d_1 = \int_0^1 \left(3t^2 - 2t^3\right)v\,dt = \frac{v}{2}$$

(7-36)

All the equations given above are normalized for the interval
$t = [0...1]$. In practice, however, every movement has a specific duration T
determined by the set limits of acceleration and jerk. Luckily, the Bezier
spline scales linearly, and we can use the factor T to stretch the profile
along the time axis.

If time moves along the interval $t = [0...T]$, the speed profile becomes
as follows:

$$v_1 = \left(3\left(\frac{t}{T}\right)^2 - 2\left(\frac{t}{T}\right)^3 \right) v$$

(7-37)

The final speed is still v, but it now takes time T to reach it, which can
be longer or shorter than 1.

By differentiating the new speed profile with respect to time, we can
derive the equations for acceleration and jerk and observe that their
maximum values are also scaled linearly:

$$a = \frac{3}{2T} v$$

(7-38)

$$j = \frac{6}{T^2} v$$

(7-39)

If $T > 1$, the acceleration phase will take longer to complete because the
peak values for acceleration and jerk are lower than before. Conversely, if
$T < 1$, the dynamic values are higher and the curve will be shorter.

In practice, the operator typically sets a maximum allowed
acceleration (because of the motor's peak torque or other mechanical
limitations) and possibly a limit jerk. If no limit jerk is set, then we impose
it automatically in the control program as described above.

We then use Equation (7-38) or Equation (7-39), whichever is longer, to calculate the time needed to accelerate from standstill to the programmed speed v.

Compared to the seven-zones S-curve, the cubic Bezier spline offers a simpler computational model and smoother acceleration profile, usually at the cost of a longer time to reach target speed.

Let's repropose the same numerical example of the previous section, to investigate the difference between the two profiles. We consider a movement from standstill to a target speed of $v = 1$ m/s, imposing the maximum limit for acceleration $a = 10$ m/s^2 and jerk $j = 100$ m/s^3.

Computing the total acceleration time for the polynomial S-curve results in $\Delta t_1 = 0.2$ s. Conversely, evaluating Equation (7-39), we find that the Bezier profile will take $\Delta t_1 = 0.24$ s. Notice, however, that the actual maximum acceleration is only about $a = 6$ m/s^2. The smother profile of the Bezier acceleration is gentler on the mechanics at the cost of a slightly longer time to execute.

In conclusion, the piecewise polynomial approach of the S-curve is superior to both the alternative options we investigated if the ultimate goal is to cover the same path distance in the shortest time within the given dynamic limits. The other approaches are superior when less computational complexity and smoother acceleration and jerk profiles are required.

Time-Optimal Movements

The speed profiles theory presented so far works well for individual axes (e.g., for jogging a single joint) and their direct interpolations (i.e., a PTP movement). We can easily set the speed, acceleration, and jerk limitations for an axis based on its motor and mechanics.

However, the same principle does not immediately apply to path interpolated movement. The relation between path and joint space is highly nonlinear, especially near singular points: setting a maximum speed value for the interpolated path trajectory does not impose any speed limitation on the individual joint axes.

Consider, for example, the movement shown in Figure 7-13: the robot moves along a line on the Y axis and crosses a wrist singularity. The orientation of the TCP stays constant along the line, which forces the wrist axes J_4 and J_6 to flip their angles by 180 degrees from start to end pose. The vast majority of that rotation only happens in a small section of path distance around the singularity.

Figure 7-13. *Path interpolated movement crossing a singularity*

If we were to generate a standard jerk-limited trajectory for that path s at constant speed $\dot{s}$ (see S-curve in Figure 7-14 left), J_4 and J_6 would have to move at extremely high speed around the singular point to hold the TCP orientation fixed. In practice, their motors cannot keep up with that requirement because they cannot realistically accelerate to such high speed in such a short time. The result is that the TCP will momentarily leave the programmed path, unless an error on the individual servo drives interrupts the movement altogether or the mechanical structure of the robot gets damaged.

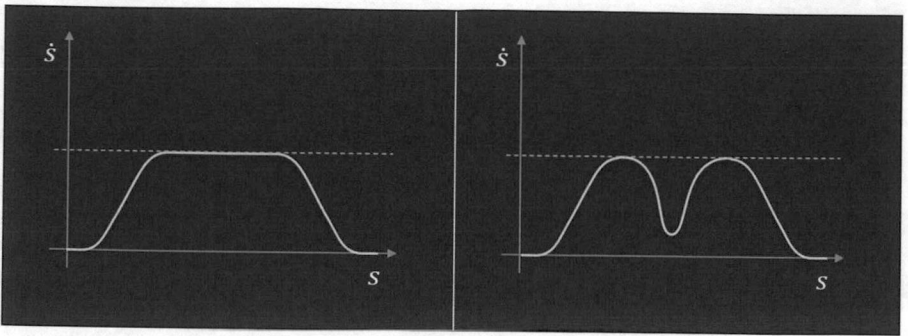

Figure 7-14. *Jerk-limited trajectory for a path interpolated movement without (left) and with (right) additional constraints imposed by joint axes dynamic limits*

The solution is to slow down the actual path speed in the critical area, thus allowing the individual joints J_4 and J_6 to move without violating their speed and acceleration limits (see modified speed profile in Figure 7-14 right).

The crucial question is when to slow down the path speed and by how much.

Let's analyze a generic speed profile along a planned path (see Figure 7-15). The path s can be described parametrically and covered in the range $t = [0...1]$. The point $t = 0$ is where the path begins and $t = 1$ where it ends, regardless of the kind of path it is (could be a line, circle, spline, etc.) and regardless of the time it takes to complete. The vertical axis represents the path speed $\dot{s} = ds / dt$.

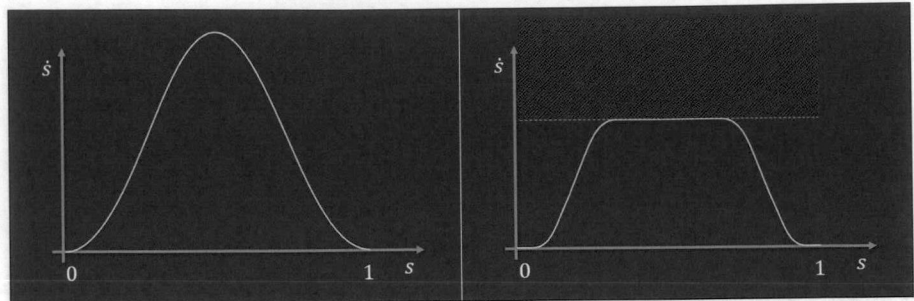

Figure 7-15. *Interpolated path speed profiles without (left) and with (right) constraints*

If no speed limit is programmed by the operator, we can simply accelerate and decelerate to reach the target. The trajectory is only limited by the maximum jerk and is the fastest way to cover the path (left profile in Figure 7-15).

In case a path speed limit is given, we impose a first constraint and reduce the maximum speed (right profile in Figure 7-15). That is typically the case of cutting or painting applications, where the process only allows a specific maximum path speed: a laser cannot travel too fast; otherwise, it does not cut through the metal sheet; a paintbrush cannot move too quickly; otherwise, it does not spray paint uniformly on the workpiece.

Notice that the covered path is the same in both cases: we always start and finish in the same positions. However, the time that it takes to complete the movement is different because the actual path speed has changed.

Let us now introduce a further limitation: maximum speed values for the joint axes. These constraints are not linear, and they might look like something shown in Figure 7-16. We will show later how to compute that curve; for now, let's assume it is already given.

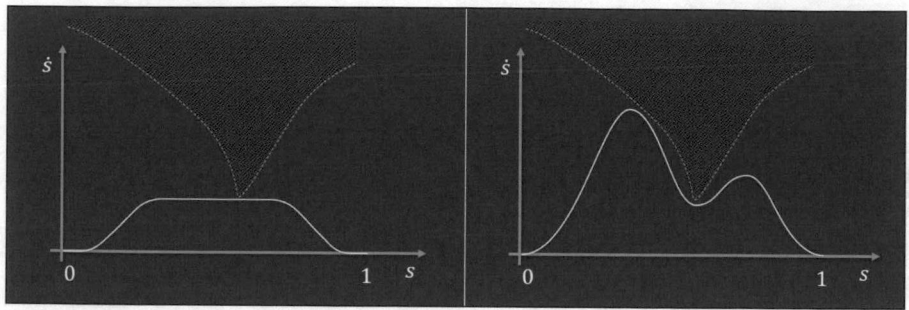

Figure 7-16. *Introducing joint speed limitations*

The easiest solution to execute the movement without violating the new constraints is to limit the maximum path speed, so to avoid reaching limits on the joints (left profile in Figure 7-16).

Such an approach is clearly very inefficient: it decreases the entire productivity of the robot. Worse even, in case the planned path passes near a singularity, the path speed needs to be lowered to almost 0, which cannot be forced to the entire movement; otherwise, it will never complete.

The solution is to generate what is called a **time-optimal movement**: a trajectory that traverses the path as fast as possible, still without violating the given dynamic constraints. The resulting path speed is not constant anymore: it is higher (up to the maximum path speed) when the joints move at sublimit dynamics; it is slower (almost down to zero speed around singularities) when at least one joint reaches its dynamic limits (right profile in Figure 7-16). In other words, at any point in time, we either reach the maximum path speed or we saturate at least one joint at its limit speed/acceleration. That is the fastest we can let the robot move without violating any constraint.

The equation for a time-optimal speed profile is found numerically: we start integrating the maximum acceleration forward and the maximum deceleration backward until either we hit the edge of the violation space or the two curves meet. This approach is called *bang-bang*, because we are always either accelerating or decelerating.

The procedure is rather simple, but it requires knowing the maximum allowed path speed at each point of the path.

Differential Kinematics

There are two ways of calculating the instantaneous maximum path speed that does not violate speed limits on the joint axes: a numerical approach and a closed-form solution. The first is very simple and works well in practice, despite being a rough approximation. The second is much more accurate but also very complex to implement. We briefly present both methods here.

We start with the numerical approximate solution. Imagine the robot's TCP is currently in the pose $X_{TCP} = [X, Y, Z, A, B, C]$ and is moving along a path. We can calculate the current joint axes angles θ_i by means of the inverse kinematics:

$$X_{TCP} \xrightarrow{IK} \theta_i \qquad (7\text{-}40)$$

The control software generates a new target position for the TCP at a fixed frequency tick (e.g., 1kHz). The offset of the new position from the previous cycle depends on the path speed profile. Using the inverse kinematics, we can then calculate the new corresponding positions for the joint axes:

$$X_{TCP} + \Delta X_{TCP} \xrightarrow{IK} \theta_i + \delta_{\theta_i} \qquad (7\text{-}41)$$

We now use an iterative procedure: check if the increment δ_{θ_i} in the cycle time interval Δt has violated any speed limit on the joints; reduce the path displacement ΔX_{TCP} until no violation occurs.

In other words, we dynamically adjust the position of the next sampled point along the planned path so that all joints' increments are small enough to avoid violating their speed limits.

Considering that the cycle time interval is usually small (typically around 1ms or less), the inverse kinematic functions can be linearized, and the required reduction of the path displacement can be approximately derived from the calculated violation of the joint axes.

This approach is clearly an approximation, but it is very easy to implement and works well in practice for systems that do not require very high accuracy. Using this technique, you can let a robot move through a singularity using a path interpolated movement: you will observe the robot slowing down while approaching the singularity and then speeding up again on the other side of the critical area once the joint axes have performed their required rotations without violating their speed limits.

A more sophisticated approach, typically used in high-end industrial systems, is finding a closed-form solution to the path speed given the individual joint speeds: we basically need to differentiate the kinematic model with respect to time.

The TCP pose $X_{TCP} = [X, Y, Z, A, B, C]$ is a function of the joint angles θ_i via direct transformations, which is always uniquely defined for serial kinematics:

$$X_{TCP} = f(\theta_i) \tag{7-42}$$

We now differentiate both sides with respect to time to express the TCP path speed as a function of the joints' speed. Since the right term is a composite function, we need to apply the chain rule for derivatives:

$$\frac{dX_{TCP}}{dt} = \frac{\partial f(\theta_i)}{\partial \theta_i} \frac{d\theta_i}{dt} \tag{7-43}$$

The expression represents the *differential kinematic model* of the robot and can be expanded into a matrix product, where all the dot symbols represent derivatives with respect to time:

$$
\begin{bmatrix} \dot{X} \\ \dot{Y} \\ \dot{Z} \\ \dot{A} \\ \dot{B} \\ \dot{C} \end{bmatrix} = \begin{bmatrix} \dfrac{\partial X}{\partial \theta_1} & \cdots & \dfrac{\partial X}{\partial \theta_6} \\ \vdots & \ddots & \vdots \\ \dfrac{\partial C}{\partial \theta_1} & \cdots & \dfrac{\partial C}{\partial \theta_6} \end{bmatrix} \begin{bmatrix} \dot{\theta}_1 \\ \dot{\theta}_2 \\ \dot{\theta}_3 \\ \dot{\theta}_4 \\ \dot{\theta}_5 \\ \dot{\theta}_6 \end{bmatrix}
\tag{7-44}
$$

The matrix of the partial derivatives of each path axes with respect to all joint axes is called the **Jacobian**, which we denote by J. Each element indicates how much a single joint axis movement will influence the TCP path movement along a single direction or orientation.

Equation (7-44) provides a *closed-form way to calculate the maximum possible speed of the TCP axes given the speed limits of the joints*. Notice that the relation is linear. Also, keep in mind that the numerical value of the Jacobian needs to be updated at each cycle time interval, because it depends on the current position of the joint axes.

In a similar way, we can also solve the inverse problem: given a target TCP path speed, find the required speeds on the joint axes. The solution is simply to invert the Jacobian matrix.

$$
\frac{d\theta_i}{dt} = J^{-1} \frac{dX_{TCP}}{dt}
\tag{7-45}
$$

One thing we immediately notice is that the inverse differential kinematics admit a unique solution: given a TCP path speed, we can calculate a unique speed for the joints. That is unlike the position inverse kinematics, where given the pose of the end effector we could find different possible configurations for the joints.

In practice, there is one potential issue we need to be well aware of: when the robot encounters a singularity during its motion, the Jacobian matrix loses rank and its determinant becomes 0. Mathematically, a zero determinant means that the matrix cannot be inverted. Physically, it means that even a very small speed along the path axes requires an extremely high speed on the joint axes. The situation is not new to us: we had faced it already while studying the inverse kinematic model in the Section on Singularities in Chapter 4. Now we have a deeper mathematical interpretation of the problem.

The Jacobian for a six-axes robot is a 6x6 matrix, because there are six joint axes and six TCP coordinates: three translational components along the X-Y-Z axes and three angular velocities around the X-Y-Z axes, i.e., along the rotating A-B-C axes. The combined vector of these six velocities $\dot{X}_{TCP}$ is called spatial velocity, or **twist**:

$$\textbf{\textit{Twist}} = Jacobian \times Joint\ Velocities \qquad (7\text{-}46)$$

The individual columns of the Jacobians represent the twist caused by one moving joint at unit speed when all other joints do not move: that is the influence of one individual joint over the TCP pose. The total TCP speed is then just the linear vector sum of the influences from each joint.

Using Equation (7-44), we can now draw the $(s,\dot{s})$ diagram of Figure 7-16 and visualize the actual path speed limit of a planned movement. The procedure, however, is not unique, because there are different ways of defining a path speed $\dot{s}$ given the path axes speeds $\dot{X}_{TCP}$.

Path Speed Definitions

When programming the movement for a robot, the operator usually needs to specify a target speed. Imagine we are executing a PTP movement, jogging the first axis J_1 from -30 to $+30$ degrees. It seems natural in that case to define the desired speed as degrees per seconds.

However, if the programmed movement is a linear path interpolated movement, say from $Y = -150$ to $+150$ mm, then it feels more natural to define the target speed in millimeters per second.

Finally, if the movement involves both a translation along a path and at the same time a change of tool orientation, we have the option to decouple the movement into translational and rotational components. Since the two components are typically required to complete the movement at the same time, we select the slower of the two definitions and let the other one follow along.

By providing all possible options in the control software, we allow the operator the freedom to program any kind of movement using any speed definition: for example, running a PTP movement with a speed in mm/s.

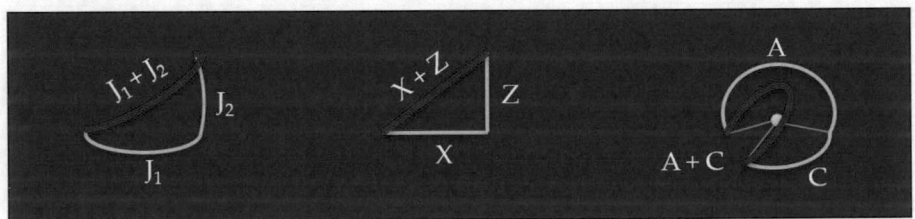

Figure 7-17. *Joint, Cartesian, and angular movements all allow for different speed definitions*

Let's see how to calculate the path speed according to different definitions. The time it takes to complete a movement (at constant speed) is the ratio between the length of the movement and the speed programmed by the operator.

For PTP movements (Figure 7-17 left), the joints are interpolated linearly, and the total joint path length L_J is measured in degrees. The most common definition of the speed in this case is degrees per seconds.

Because the interpolation is linear, the total *joint speed* v_J is decomposed quadratically as the sum of the speeds of each individual joint v_{J_i}.

$$v_J = \sqrt{\sum_i v_{J_i}^2} \quad (deg/s) \tag{7-47}$$

The movement will take $\Delta t = L_J/v_J$ to complete.

Note that interpolating the joint axes linearly is not necessarily the default behavior for PTP movements. There is no defined industry standard, and some manufacturers do not interpolate the joints, but let them run independently all at the same programmed speed. The result is that all joints arrive at destination at different times. You can implement your program in either way, typically depending on any specific application requirement.

The next movement we consider is a pure translation along the X-Y-Z axes (Figure 7-17 middle). In this case, the Cartesian path length L_C is in mm, so the typical speed definition is in mm/s and can be seen as a composition of the speeds along the X-Y-Z axes (*Cartesian speed*):

$$v_C = \sqrt{v_X^2 + v_Y^2 + v_Z^2} \quad (mm/s) \tag{7-48}$$

The movement will take $\Delta t = L_C/v_C$ to complete.

Note that the Cartesian movement length can be calculated for any kind of movement, not necessarily a line. We learned in the path-planning chapter how to calculate Cartesian lengths of PTPs or splines using a segmentation of the path. In other words, we linearize the path by decomposing it into small linear segments to approximate the original path.

For instance, if the operator programs a PTP movement with a speed defined in mm/s, we first need to calculate the Cartesian length of the movement and then find out the time it will take to complete. After that, we can generate the required trajectory.

Finally, we consider the *angular speed* of rotational movements (Figure 7-17 right). The TCP rotates around a fixed position, and the corresponding path length can be calculated in degrees. Recall that Euler angles do not add up linearly, so the correct way to calculate the path length is using quaternions (see the SLERP Section):

$$L_\alpha = cos^{-1} \frac{q_1 \cdot q_2}{\| q_1 \| \| q_2 \|} \tag{7-49}$$

The angular speed v_α is therefore typically programmed in deg/s. The time to complete the movement will be as follows:

$$\Delta t = \frac{L_\alpha}{v_\alpha} \tag{7-50}$$

In a practical software implementation, we let the operator decide what kind of speed definition to use for each programmed movement. We then calculate the length of the movement in that particular coordinate system: either in the joints space, or in the Cartesian system, or in the quaternion system. Once the movement's length is known, we can apply the trajectory generating algorithm to the specific block.

The movements execution times given here are calculated considering a constant path speed. In practice, we generate a complete speed profile taking into account all the other dynamic constraints: acceleration and jerk limitations in the path and joint space.

Optimal Motion in Practice

So far, we have considered trajectory limitations forced by constraints on the joints' speeds.

However, we normally also want to impose constrains on higher orders of derivatives: limiting acceleration and jerk makes the movement visibly smoother, gentler on the mechanics, and also more conservative on the motors' currents.

In order to introduce those additional constraints, we proceed mathematically with further differentiations of the Jacobian, so that the approximation of the dynamics in a local neighborhood of the current state includes acceleration and jerk of the joints. The calculations are a bit more involved, but the concept is the same.

For example, calculating the accelerations of the TCP along the six path axes given the joint dynamics requires a differentiation of the twist. Recall the expression of the twist as function of the Jacobian:

$$\frac{dX_{TCP}}{dt} = J\frac{d\theta_i}{dt} \qquad (7\text{-}51)$$

Differentiating with respect to time gives the following:

$$\frac{d^2X_{TCP}}{dt^2} = \frac{dJ}{dt}\frac{d\theta_i}{dt} + J\frac{d^2\theta_i}{dt^2} \qquad (7\text{-}52)$$

The expression can be rewritten in compact form using the dot notation for the time derivatives:

$$\ddot{X}_{TCP} = \dot{J}\dot{\theta}_i + J\ddot{\theta}_i \qquad (7\text{-}53)$$

The acceleration of the TCP along the path axes depends both on the speed $\dot{\theta}_i$ and acceleration $\ddot{\theta}_i$ of the individual joints. A similar expression can be derived for the jerk values. These formulas allow us to calculate the required constraints on the path speed profile in order to avoid any violation of the dynamic limits in the joint space.

Solving the inverse problem is also possible. Given the target path accelerations $\ddot{X}_{TCP}$, we can calculate the required joint axes accelerations by inverting the Jacobian matrix:

$$\ddot{\theta}_i = J^{-1}\left(\ddot{X}_{TCP} - \dot{J}\dot{\theta}_i\right) \tag{7-54}$$

The operation is only possible if the Jacobian is not singular, as we already explained in the case of the inverse differential kinematics for speed.

Imposing limits on the joints' speed, acceleration, and jerk is usually enough to guarantee the generation of a valid and realistic trajectory for the robot. Specific applications or high-end control systems also impose constraints on the torques required by the motors to allow the robot to follow the desired trajectory.

Electric motors cannot exceed a maximum torque before being permanently demagnetized and damaged. Being able to generate a path speed profile that also limits the individual joint torques can be very convenient and prevents the servo drives to go in error state during a movement. In Chapter 8, we will study how to calculate motor torques using the dynamic model or the robot, and we will briefly show how to use them to generate optimal trajectories.

One last practical tip regarding the optimization task is the careful planning of the corresponding path. In fact, the problem of optimizing a trajectory can be simplified enormously if the underlying path is well optimized in the first place.

Cleverly modifying the path can be often extremely beneficial to trajectory speed: for example, introducing round edges to smooth corners or using a PTP movement instead of a path interpolation wherever the actual path is not relevant. Also, trying to reroute the planned path to stay clear of singularities always helps.

Finally, when everything else fails, you can always cheat and filter out the generated trajectory in the time domain. We will explore this technique in the next section.

Optimizing a trajectory in the broadest sense is not limited to an increase of the path speed. While minimizing a movement's time duration is usually the priority in the vast majority of practical applications, it not necessarily the only goal that a trajectory planner might try to optimize.

For instance, there are situations where the execution time of a movement is not critical, but the energy required to perform the movement is far more important. In that case, the cost function to be optimized depends on the amount of power required by the motors along the trajectory. Besides reducing accelerations peaks, the optimal solution would require modifying the configuration of the joints in order to take advantage of the robot's mechanical structure to achieve better compensation for gravity, especially in the case of heavy loads. Depending on how the links are oriented in space, the motors' required contribution to hold a load in place can vary greatly. We will briefly talk about mechanical forces in the section about statics in Chapter 8.

Time Filtering

Optimizing a trajectory can be tricky. Sometimes, adding a simple filter to the generated set values can be a quick and cheap (but effective!) way of smoothing a sharp edge in the profile.

Imagine we have a rough geometrical transition between two consecutive programmed paths. Our control algorithm would detect a large angle between the two adjacent tangents and would plan a complete stop for the trajectory in that point (see red curves in Figure 7-18).

We could smooth out the path with a spline creating a round edge. In that case, the speed at transition will immediately increase: with a large radius of curvature, the required path acceleration will lower considerable (see yellow curves in Figure 7-18). That is typically the best solution, as long as the application allows it.

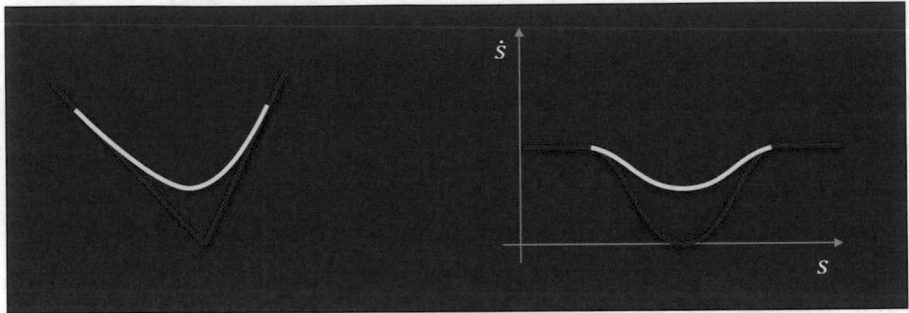

Figure 7-18. *Smoothing a path (left) can also smooth the trajectory (right)*

An alternative quick way to smooth out trajectories is to introduce a filter in the time domain. Once a trajectory is generated in the path space, the inverse kinematic function transforms the set points into angles in the joint space. These values are individually sent to each servo drive controlling the motor of the corresponding axis.

However, instead of immediately following the incoming stream of commanded angle values, each servo drive can add a small filter effect to the set values (see Figure 7-19), with the effect to reduce jerk and therefore soften the vibration on the mechanics. In most applications, a bit of time smoothing is actually always active.

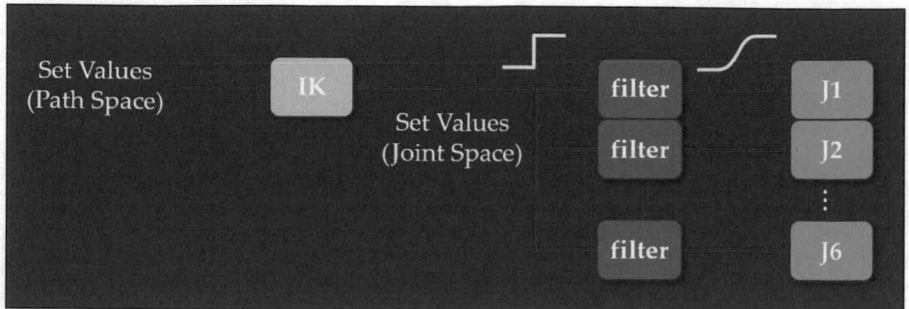

Figure 7-19. *Introducing time filters in the joint space*

This kind of filtering is totally independent from the shape of the planned path and from the corresponding generated trajectory. It is a fixed filter in the time domain only: imagine a black box reading an input stream of incoming numbers arriving at a fixed time interval (e.g., 1 ms) and generating an output filtered (smoothed) stream at the same rate. The filter does not know anything about the physical meaning of those numbers.

The advantages are simplicity and effectiveness. The drawback is a slight modification of the originally programmed path, because each axis individually manipulates its own set position independently from the other interpolated axes. The magnitude of the path deviation depends on how aggressive the filter is configured and on how fast the robot moves. A stronger filter allows the robot to increase its speed considerably without stopping at every corner, but it also visibly distorts the planned path. As usual, the optimal setting depends on the application requirements.

The most common type of filter used to smooth a curve is the **Gaussian filter**. It consists in modifying an incoming stream of data through the convolution of the original signal with a Gaussian curve.

Let's consider the case study of a linear movement close to a singularity, as shown in Figure 7-20. If the trajectory is generated with a constant path speed, the joint axis J_4 is forced to flip its orientation very quickly: its position jumps rapidly and its speed reaches very high values (see Figure 7-21). Higher derivatives, acceleration and jerk, are almost unbounded.

Figure 7-20. *Path linear movement that needs filtering*

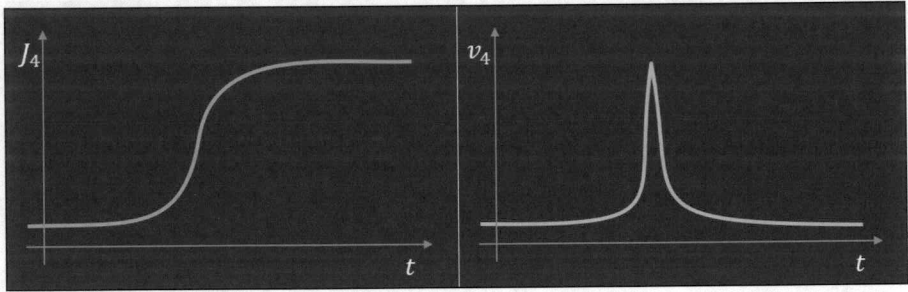

Figure 7-21. *Position (left) and speed (right) of J_4 during the path linear movement*

We now let a Gaussian curve of finite size (*filter window*) slide over the original speed profile along the time axis (see Figure 7-22). When multiplying the two curves with each other, everything outside the filter window results in a zero value, while the parts of the original signal that fall inside the filter window are averaged out. The effect is stronger in the middle of the window and weaker at the sides. The operation is similar to taking an average, but using variable weights. The immediate result of this approach is that sharp peaks in the speed profile are smoothed out.

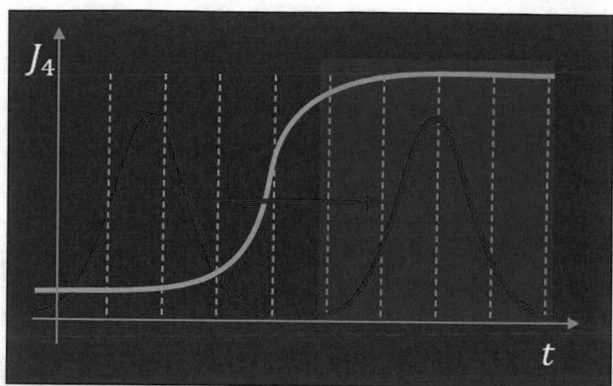

Figure 7-22. *Convoluting the original signal (yellow) with a Gaussian curve (red)*

Mathematically, the operation takes the original signal $f[n]$ and convolutes it with the Gaussian curve $g[n]$, defined over a window size N as follows:

$$g[n] = e^{-\frac{1}{2}\left(\frac{n-N/2}{\sigma N/2}\right)^2}$$

(7-55)

Note that we work with functions in the discrete variable n, not continuous time t, because the trajectory generator samples points at each time tick of the controller.

The convolution generates the new signal profile $z[n]$:

$$z[n] = (f * g)[n] = \sum_{i=0}^{N} f[N-i]g[i]$$

(7-56)

Since the sum of the Gaussian points does not add up to 1, the final signal is actually reduced in amplitude and needs to be normalized:

$$z[n] \leftarrow \frac{z[n]}{\sum_{i=0}^{N} g[i]}$$

(7-57)

The important parameter is N, which is the number of samples that we decide to include in the filter window. For example, if the control software runs with a cycle time of 1 ms and we want a 10 ms filter window, then we set $N = 10$.

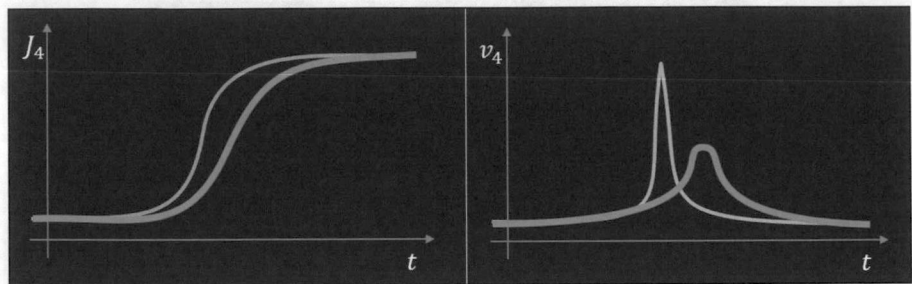

Figure 7-23. *The result of the convolution is a smoother speed profile*

The result of the convolution is shown with a blue curve in Figure 7-23. It rises less sharply than the original one and has a much lower peak speed. Similarly, you can infer that both acceleration and jerk are also much reduced. After filtering, the J_4 axis performs a smoother movement and likely causes no vibrations on its mechanical axis. However, the total movement turns out to be slower than initially programmed, and the actual TCP position will likely leave the original path before reaching its target destination.

In conclusion, adding a filter is typically a good idea, but filtering over a very large window of samples causes large position violations. The best compromise depends on the actual application requirements.

External Path Corrections

You might wonder why we should bother adding time filters in the joint space if we already have the ability to properly optimize the trajectory in the path space. One reason is that trajectory optimization is a complex task to implement and, unless you are developing a serious industrial system, you might as well add a simple filter to your hobby robotic project and still achieve great results.

The second, most important reason is that not all movements executed by the robot are processed by the trajectory generator. We introduced external path corrections in Chapter 5, describing them as changes from the originally planned path added from external sources: for example, a conveyor, a force controller, or any other signal that could influence the robot's behavior and that is not included in the originally planned path.

Consider, for example, the situation shown in Figure 7-24. The robot receives the target position to pick up a product from point A and plans a movement accordingly along the red line. Suddenly, the conveyor starts to move and the robot is forced to follow the yellow path to end up reaching the product in point B. The total execution time does not change, because the trajectory has been generated already. However, since the actual path turned out to be longer than planned, the joint speed limits might have been violated.

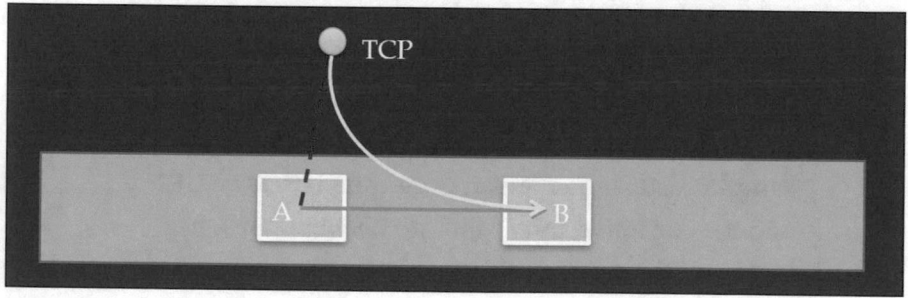

Figure 7-24. *Unplanned path corrections added by a conveyor-tracking function*

The original generated trajectory is based on the planned path and is guaranteed to respect all dynamic constraints of the individual axes. But if the actual path suddenly changes while the movement is being executed at runtime, the trajectory cannot be aware of that. There is no guarantee that the dynamic limits are complied with.

In this kind of situations, it is very useful to activate time filters on the individual axes, so that sudden jumps coming from external signals are smoothed out and the likelihood that a servo drive triggers an error is reduced.

When working with six-axes robots, it is wise to keep the configuration of the joints so that the robot will not have to run into a singularity while following the conveyor. Conveyors usually run at constant speed in the path space, which can easily cause large speeds in the joint space around critical areas. Since tracking movements are usually not planned, they cannot be optimized before starting, and they might result in execution errors, unless a large filter is applied.

Other practical examples where external corrections are added to the planned path and the addition of a filter helps with smoothing the actual movement are as follows:

- The laser head of a cutting robot, which must be kept at a constant distance from the material to be cut in order for the laser to focus correctly and cut efficiently. The movement's path is planned ahead of time by teaching a few fixed points, but the actual position of the workpiece might slightly vary between samples. A distance sensor is mounted on the TCP, and the controller forces the gap between the laser head and the surface of the workpiece to be constant by introducing small corrections to the robot's actual position.

- The tool tip of a stir-welding robot, which must exert a constant force on the workpieces to be welded together. The process is based on heat generated by friction and is only effective if the tool provides enough pressure to generate the required amount of heat and the required amount of force to forge the two parts with each other.

Again, the path is generated according to taught points and controlled in position, but external corrections are added dynamically according to the feedback from a force sensor mounted at the TCP.

Summary

Generating an optimal trajectory for a movement is probably the most complex task in industrial robotics. The S-curve profile is the best starting point to build a trajectory, but further optimization via differential kinematics is required to avoid violating a number of dynamic limits imposed by the physical constraints of the actuators and the mechanics. That is especially critical when the robot approaches singular configurations. Adding a bit of filtering action in the time domain is a cheap and convenient trick in practice.

If you were able to implement some of the movement profiles in your simulation environment, you should now be able to see your robot gracefully dancing around in space. That is definitely a great achievement!

CHAPTER 8

Statics and Dynamics

In this chapter, we introduce the concepts of statics and dynamics for robots. These are advanced topics and are not strictly necessary for most practical applications, but it is good to have at least an idea of what they are and what advantage they can add to a robotics control system.

Statics

Statics is the branch of physics that deals with forces. So far in this book, we have learned how to control the position and speed of a robot, because that is what the majority of typical industrial applications require: e.g., pick-and-place, palletize, cut, weld, paint, etc.

However, there are situations where we also need to *control the forces at the TCP*. For example, imagine a robotic arm deployed to clean a glass window: Not only we have to move the TCP along a predefined path along the window surface, but we are also required to keep the TCP in close contact with the window by applying a specific force F_N perpendicular (*normal*) to the glass (see Figure 8-1). The force control accuracy must be high: too much force would end up breaking the window; too little would not clean the surface effectively. Position control alone is not enough in this case: an additional force control system is also required. Combining position and force control together is called **hybrid** control.

© Fabrizio Frigeni 2023
F. Frigeni, *Industrial Robotics Control*, Maker Innovations Series,
https://doi.org/10.1007/978-1-4842-8989-1_8

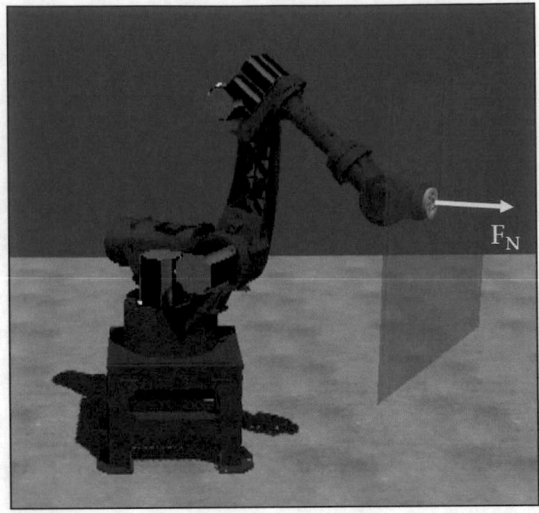

Figure 8-1. *Applying a force at the TCP normal to the workpiece*

Specifically, for a six-axes robot, besides the usual control of position and orientation of the TCP, we can also decide to control the forces and torques exerted at the TCP. The forces $[F_X, F_Y, F_Z]$ act linearly along the X-Y-Z axes, while the torques $[T_A, T_B, T_C]$ act rotationally around the A-B-C axes (see Figure 8-2).

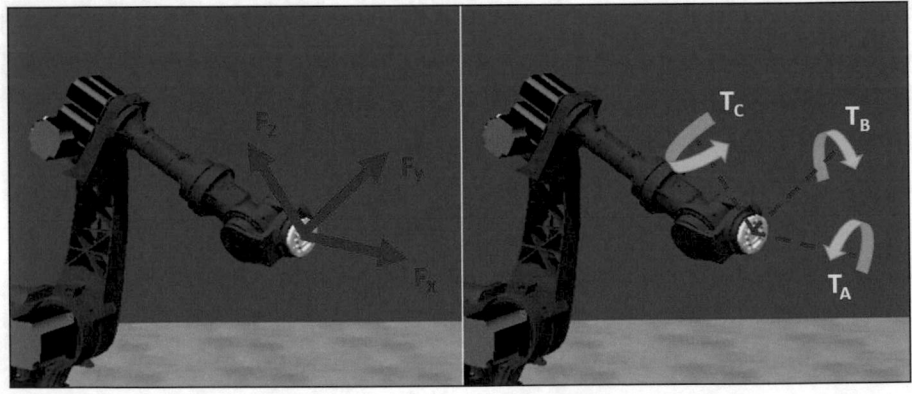

Figure 8-2. *Wrench components at the TCP*

The resulting six-dimensional vector of forces and torques acting at a point in space is called **wrench**.

$$F_{TCP} = \left[F_X, F_Y, F_Z, T_A, T_B, T_C \right]^T \tag{8-1}$$

In practice, the wrench at the TCP is generated by the torques of the motors driving the robot's joints. In order to perform torque control with our control software, we need to be able to calculate how much torque is required from each joint motor in order to achieve a specific target wrench at the TCP.

We say that the robot is in *static equilibrium* if the power generated by the motors at the joints is the same as the power experienced at the TCP:

$$W_J = W_{TCP} \tag{8-2}$$

We are assuming, for now, that all the power generated by the motors is used to generate a wrench vector at the end of the robot (see Figure 8-3). We will see later that this 100% efficiency is not always feasible, because often the motors have to generate additional torque to contrast other forces: typically, friction and gravity.

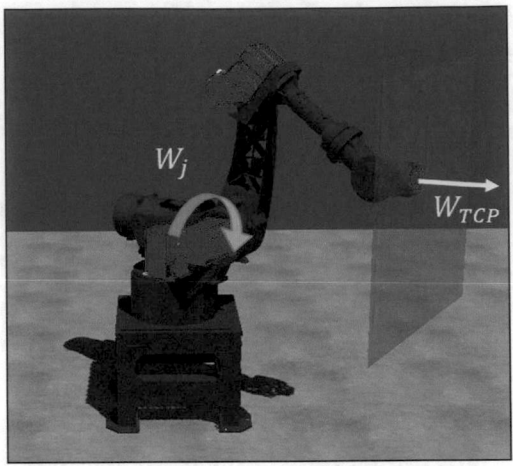

Figure 8-3. *Equivalence of power in the joint space and in the path space*

Mechanical power is the product of force and speed. Since we are dealing with rotational axes, the power generated at the joints is given by the product between the torques τ_i acting on them and their angular speeds ω_i. In vectorial form,

$$W_J = \tau^T \omega \tag{8-3}$$

The power experienced at the TCP is a combination of linear forces and speeds and rotational torques and angular speeds. For the forces and torques, we use the wrench vector from Equation (8-1); for the linear and rotational speeds, we use the twist vector from Equation (7-43).

$$W_{TCP} = F_{TCP}^T \dot{X}_{TCP} \tag{8-4}$$

When solving differential kinematics in the previous chapter, we showed that the twist is the product of the Jacobian and the vector of the joint angular speeds. Accordingly, we can rewrite Equation (8-4) as follows:

$$W_{TCP} = F_{TCP}^T J \omega \tag{8-5}$$

Applying the condition of static equilibrium (Equation (8-2)), we obtain the following:

$$\tau^T \, \omega = F_{TCP}^T \, J \, \omega \qquad (8\text{-}6)$$

After a final simplification, we find the relation between the TCP wrench and the corresponding joint motor torques:

$$\tau = J^T \, F_{TCP} \qquad (8\text{-}7)$$

Let's now go back to the initial example of applying a constant force normal to the glass window. Given the orientation of the window with respect to the TCP frame, we first compose the corresponding wrench vector, which in this case will only include linear forces, as no torques are required. Then we calculate the Jacobian for the specific TCP pose, transpose it, and pre-multiply it by the wrench vector. The result will be the vector of all the required joint torques.

Two additional observations are required here. The first is that the resulting calculated torques are considered to be at the joints. In practice, the robot's joints are driven by electric motors, but not necessarily with a direct connection, because there might be a gearbox in between them. The ratio of the gearbox must be included in the calculation to scale down the actual required torques from the motors.

The second issue is that, as mentioned earlier, additional torque components might be required just to hold the body of the robot in place and to overcome friction at the joints. Parts of these components can be compensated as static TCP forces that only depend on the position of the TCP (e.g., gravity), while others also depend on the speed and acceleration of the robot (e.g., centrifugal forces, dynamic friction) and will be analyzed in the Section on the Dynamic Model.

Singularities

As usual, we can also ask the opposite question: what TCP force can be generated by applying a given torque at the joint axes? The answer is found by simply inverting Equation (8-7):

$$F_{TCP} = \left(J^T\right)^{-1} \tau \qquad (8\text{-}8)$$

When we studied differential kinematics in the previous chapter, we noticed that the Jacobian cannot be inverted at singularities because its determinant goes to 0. The effect in the kinematic domain was that small TCP movements can cause extremely high joint speeds. A similar problem arises when we consider the domain of static forces: if the robot approaches a singularity, then a very small torque on the motors can cause a great force or torque at the TCP.

Such characteristic can actually be exploited as mechanical advantage and used as a lever: either to actively generate a high wrench or to passively oppose a high external force, with little torque from the motors.

In the extreme configuration of a singularity, the Jacobian completely loses rank, and one component of the wrench vector is not controllable by the joints anymore. The robot is in a peculiar condition: it experiences total loss of mobility along one of the TCP path axes, but at the same time it can oppose an infinite force (or torque) along that same axis without doing any work (clearly, up to the physical strength of the mechanical structure itself).

Consider, for example, the robot in a wrist singularity pose ($J_5 = 0$), as shown in Figure 8-4. No movement around the TCP local Z axis (along the C orientation) is physically possible, unless we first reorient the wrist to a different configuration.

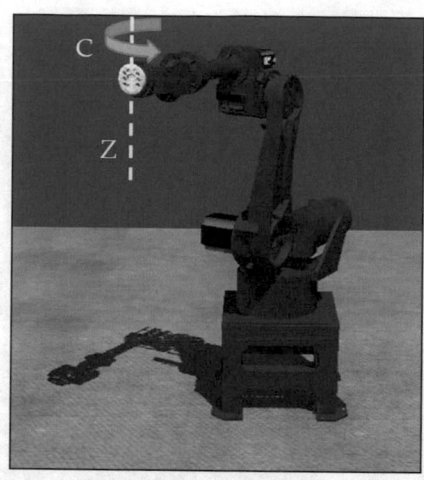

Figure 8-4. *A robot in this pose can oppose an infinite torque around the TCP local C axis without performing any actual work*

Applying an arbitrarily large external torque around the C axis will not cause any movement of the robot, because no joint is physically responsible for that orientation. The whole external power applied to the TCP around C will be resisted entirely by the mechanical structure itself.

In general, the maximum magnitude of possible spatial forces (wrench vector) and spatial velocities (twist vector) that we can generate at the TCP, in any given configuration of the robot, are inversely related to each other. In the pose of Figure 8-4, we can resist an infinitely large force around C, but at the same time we cannot generate any speed. Conversely, we can produce a high TCP speed along the Z axis, but the force exerted will be relatively small.

There is actually a way to display the maximum amount of twist and wrench that can be generated at the TCP. Using the magnitudes of the linear speeds along the X-Y-Z axes of the TCP frame, we can build a three-dimensional ellipsoid, called the *manipulability ellipsoid*, which shows how much mobility the TCP has in that particular configuration (see Figure 8-5 left). Similarly, we could build an ellipsoid for rotational speeds.

Figure 8-5. *Ellipsoids for linear speeds and forces*

According to the inverse relation between forces and speeds, we can also draw manipulability ellipsoids for the wrench vector at the TCP (see Figure 8-5 right). The force ellipsoid shows the linear forces $[F_X, F_Y, F_Z]$ along the three positional axes of the TCP frame. The forces are aligned exactly as the linear speeds but have an inverse relative magnitude: directions along which the robot can achieve high speed can only generate small forces. Vice-versa, low-speed directions can exert large forces. The same concept can be derived for rotational axes, considering the torques $[T_A, T_B, T_C]$ instead of the linear forces.

Critically, the dimensions of the ellipsoids strictly depend on the current pose of the robot: in different poses, the TCP can achieve different maximum speeds, even though the limit speeds of the joints are always constant. In particular, in a singular pose, one of the dimensions of the manipulability ellipsoid goes to 0, showing that the robot loses mobility along one axis.

There is also a way to numerically measure the degree of manipulability of any specific pose: the determinant of the Jacobian.

$$\lambda = \left| det\left(\boldsymbol{J} \right) \right|$$

(8-9)

Poses with higher values of λ offer large mobility to the TCP. On the other hand, lower values of λ show that the robot is approaching a singularity and will quickly lose mobility. This value can be calculated in real time by the control software and can be used as a warning to the operator to show that the robot is about to enter a region where high TCP speeds cannot be achieved.

Generally, it is convenient to work in subsets of the joint space that guarantee a high degree of manipulability to the TCP, so that the robot can quickly move around space without limitations. However, there are applications where the opposite is true. We might require high quality of control, therefore high forces and low speeds: in those cases, it is actually better to work with lower degrees of manipulability.

Let's consider, for example, two particular wrist configurations shown in Figure 8-6. The pose on the left side allows the TCP to move at high speed along the vertical Z axis, but the wrist is at mechanical disadvantage: holding a large payload requires a high torque (therefore power consumption) from the motor on J_5. This is a convenient choice if the robot is required to perform vertical painting: high speed and low load.

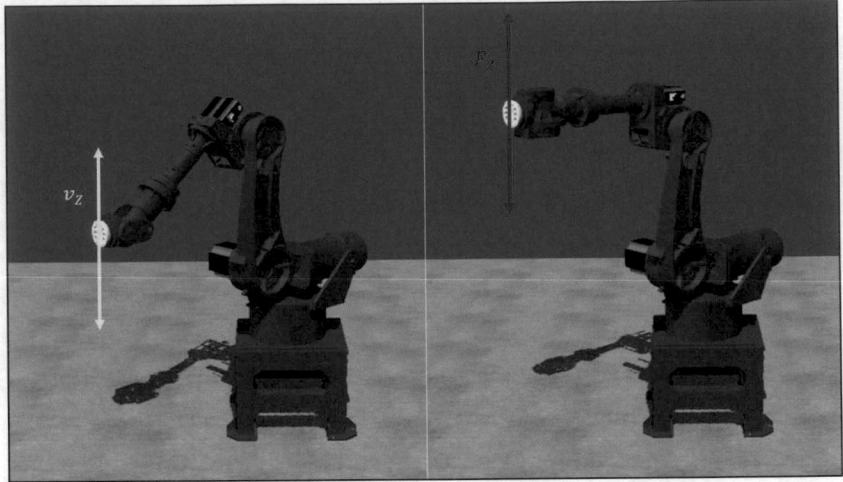

Figure 8-6. *Different wrist configurations allow for speed or force optimization*

On the other hand, if we turn J_4 by 90 degrees, the robot entirely loses speed mobility along the vertical Z axis. However, it can hold much heavier weights without any electric load on J_5. The actual load will be shared only by the lower axes J_2 and J_3, which are actuated by much larger motors. Such configuration is a better choice if the robot is required to transport a heavy payload across a horizontal path.

Dynamics

When talking about statics in the previous section, we derived the relation between joint torques and TCP forces assuming the robot was in static equilibrium: we did not introduce any motion in the equation yet. Let's now generalize the problem to a robot in any possible dynamic state, with all joints free to move at different speeds and accelerations.

The question now becomes finding how much torque is required from each motor to generate a certain motion profile at the joints (see Figure 8-7):

$$\tau = f\left(\theta,\dot{\theta},\ddot{\theta}\right) \tag{8-10}$$

Figure 8-7. *Finding the torque required to move a joint*

This function defines the *inverse dynamic* of the robot and is useful for a number of practical applications that we will describe later on.

The required torque τ depends not only on the speed $\dot{\theta}$ and acceleration $\ddot{\theta}$ of the joints but also on their actual positions θ. Typically, that is because of the gravitational effects on the structural links. Consider, for instance, the robot in Figure 8-8. If both the second and third axes J_2 and J_3 are in a vertical position (shown on the left side), the torque required to achieve a certain movement on J_2 is much lower than if the axes were in a horizontal position (shown on the right). The gravitational effects are much stronger in the second case and require considerably more torque from the actuating motor. That is why large robots mount a spring on the side to provide additional torque to support the motor on J_2.

The force exerted by the spring depends on the position of the axis: it is lower when the axis is in a vertical position, and larger when the axis moves down toward the ground.

Figure 8-8. *Different joint positions can require different torques to achieve the same motion profile*

Similarly, the torque required by the motor driving the first axis J_1 is different in the two cases. J_1 performs a horizontal movement, so it is not affected at all by gravity. However, accelerating a body in a rotational motion requires a torque directly proportional to its moment of inertia. The two poses shown in Figure 8-8 exhibit very different inertia values, increasing according to the extension of the arm away from the center of the robot.

As a consequence, the dynamic equations of the robot must consider all the links at the same time, because their positions and movements influence each other in quite complex ways.

For instance, we described how the positions of J_2 and J_3 have an effect on the torque of J_1 because of the varying moment of inertia. Interestingly though, a movement of J_1 also has an effect on the torques of J_2 and J_3. The reason is the effect of the centrifugal force that pulls the TCP load outward (shown in Figure 8-9), with an intensity that increases with the rotational speed of J_1. In practice, the J_2 axis requires a negative torque from its motor in order to hold its position still to counteract centrifugal effects, even if the axis itself is not required to move!

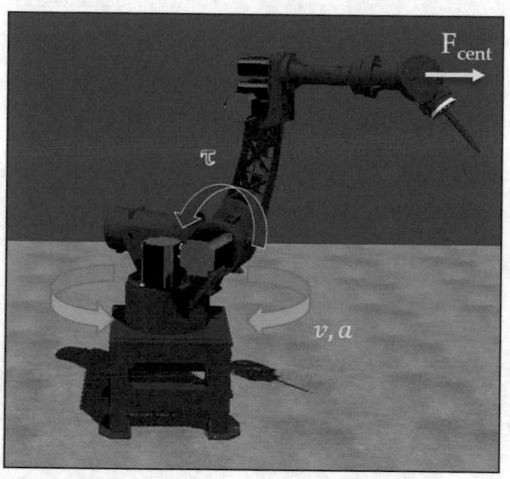

Figure 8-9. *A movement of J_1 increases torque requirements on J_2*

You probably start getting a feeling of how complex the dynamic model of the robot is and how many parameters are involved. We will briefly describe the procedure to solve the inverse kinematics in the next sections.

Inverting Equation (8-10) solves the *direct dynamics* problem: finding joints accelerations given their initial state of position and speed, plus the input torques from the actuators.

$$\ddot{\theta} = f^{-1}\left(\theta, \dot{\theta}, \tau\right) \tag{8-11}$$

This is less of practical importance and usually only done for simulation purposes.

Dynamic Model

When we built the kinematic model for the six-axes robot, we had to identify a number of mechanical parameters that had an influence on the final solution (see Mechanical Structure Section in Chapter 3). Similarly, we now want to build a dynamic model for the robot and need to determine what physical quantities are responsible for the torque requirements of a movement.

We start by analyzing a single individual link (see Figure 8-10). A six-axes robot is an open chain of six links; therefore, its dynamic model will require all the dynamic parameters for each link, plus those for the (optional) attached load.

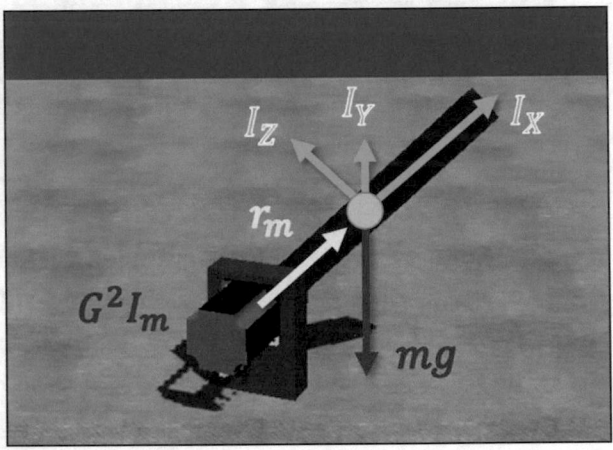

Figure 8-10. *Some dynamic parameters for an individual link*

Imagine the motor needs to move the link: what parameters define the relationship of the link's speed and acceleration with the applied motor's torque?

- **Mass** $[m]$: The gravitational force mg depends on the mass of the mechanical body and requires the motor to generate a torque even at standstill (otherwise the arm would fall down).

- **Center of mass** $[r_m]$: The way the mass is distributed in the link's volume also influences the magnitude of the required torque. For instance, a link whose center of mass is more offset toward the TCP will require more torque to accelerate because of the increased load inertia seen from the joint.

- **Inertia** $[I_X, I_Y, I_Z]$: The rotational acceleration of a link under applied torque depends on its moment of inertia. In general, the inertia of a rigid body is described by a generic 3x3 matrix. However, if we conveniently place the origin of the coordinate system at the link's center of mass and align it with the main axes of rotations X-Y-Z, the matrix reduces to a diagonal, with the three elements $[I_X, I_Y, I_Z]$ describing the inertia of rotation around the three main axes.

- **Motor inertia** $[I_m]$: The motor itself also has an inertia value, which needs to be taken into account in the dynamic model of the link.

- **Gear ratio** $[G]$: If the motor is attached to the link using a gearbox, then the motor's inertia value must be scaled by the square of the gear ratio. Higher gear ratios are useful to increase the apparent inertia of the motor and make the axis control easier, as we will learn in Chapter 11 when tuning servo loop parameters. On the other hand, the generated speed is reduced linearly by the gear ratio.

- **Friction** $[F_s, F_v]$: An additional factor to consider in the torque calculations is friction. When the motor starts turning, it needs an additional amount of torque only to overcome the static friction: the total required torque to move the axis will jump immediately at zero speed. After that, viscous friction comes into play, which increases slowly with the speed of the axis. Friction effects are much larger when using gearboxes.

- **Stiffness** $[\kappa, \delta]$: Finally, one last effect that influences the dynamics of a link is the flexibility of the joint: the mechanical coupling between the motor and the load. The connection acts like a spring, with a specific stiffness κ and damping δ, and introduces vibrations into the motion affecting the required torque. Additional complications are introduced if the link itself is not completely stiff and does not behave like a rigid body. We will not consider that case here.

These are all the parameters to consider for each link of the robot in order to parameterize the dynamic model of the robot. In addition, we need to include mass and inertia of the payload, which can itself bring a large impact on the dynamics of the robot.

Given the large number of parameters involved and the complex cross relations between all the six links, you can quickly guess that the final equations for the dynamic model are quite complicated. The derivation is a long and tedious process, and we will not show it here. We will only describe the general concept of two common derivation techniques: a Lagrangian formulation and a Newton-Euler approach.

Lagrangian Method

The Lagrangian method used to describe the robot's dynamic model is based on the principle of energy equilibrium. It is conceptually a more elegant solution, but it is also computationally very demanding, especially for robots with several degrees of freedom.

The total energy of a mechanical system (the robot's body, in our case) is the sum of its kinetic T and potential V components. On the other hand, the *Lagrangian* of a system is defined as the difference between the two energy components:

$$L = T - V \tag{8-12}$$

The kinetic energy depends on the position and speed of the joint axes: $T = T\left(\theta, \dot{\theta}\right)$. The potential energy only depends on their position: $V = V(\boldsymbol{\theta})$.

According to energy equilibrium constraints, a variation of the robot's Lagrangian can only be caused by externally applied torques: i.e., the motor torques (net of friction) and any possible wrench on the TCP.

Describing a generic variation of the Lagrangian requires taking its partial derivatives with respect to all its variables: the position and speed of the individual joint axes. For each joint $i = 1...6$, we can write the Euler-Lagrange equation:

$$\frac{d}{dt}\left(\frac{\partial L}{\partial \dot{\theta}_i}\right) - \frac{\partial L}{\partial \theta_i} = \tau_i \tag{8-13}$$

We now plug in Equation (8-12) and find the useful formulation:

$$\frac{d}{dt}\left(\frac{\partial T}{\partial \dot{\theta}_i}\right) - \frac{\partial T}{\partial \theta_i} + \frac{\partial V}{\partial \theta_i} = \tau_i \tag{8-14}$$

The Euler-Lagrange equation provides a physical meaningful way to calculate the torque τ that must be applied at a joint axis in position θ and rotating with speed $\dot{\theta}$, in order to make it accelerate at $\ddot{\theta}$. Such a torque depends on the three main components that can be identified on the left side of the equation:

- An **inertial** term $\dfrac{d}{dt}\left(\dfrac{\partial T}{\partial \dot{\theta}_i}\right)$, which depends on the *acceleration* of links

- **Centrifugal** and **Coriolis** effects $-\dfrac{\partial T}{\partial \theta_i}$, which depend on the *velocity* of links

- A **gravitational** term $\dfrac{\partial V}{\partial \theta_i}$, which only depends on the *position* of links

Deriving the actual energy components for each link as a function of all the other links is quite involved, especially for the kinetic part. Let's examine a very simple example for the dummy one-link robot of Figure 8-10.

The kinetic energy of a body depends on its linear and rotational speeds according to its mass and inertia. It can be shown that, for a rotational link with inertia I_l driven by a motor with inertia I_m, the total kinetic energy results in the following:

$$T = \frac{1}{2} I_l \dot{\theta}^2 + \frac{1}{2} I_m G^2 \dot{\theta}^2 \tag{8-15}$$

As we mentioned earlier, the motor's inertia is amplified by the square of the gear ratio, in case a gearbox is used in between.

The potential energy for the link is a function of how much of its mass is affected by gravity:

$$V = mg\, r_m \left(1 - \cos\theta\right) \tag{8-16}$$

If the link is completely vertical to ground ($\theta = 0$), then its potential component is 0, and the axis can stay in place with no power required from the motor. On the other hand, if the axis is parallel to ground ($\theta = \pm \pi/2$), then the motor needs to exert a large amount of torque to hold the link from falling down, even if no movement is required.

We can now plug the energy expressions (Equations (8-15) and (8-16)) into the Lagrangian and take the partial derivatives with respect to speed and position of the joint axis:

$$\left(I_l + I_m G^2\right)\ddot{\theta} + mg\, r_m \sin\theta = \tau - F_v\, \dot{\theta} \tag{8-17}$$

Notice how, on the right side, we also have to consider the torque to overcome viscous friction, which depends on the axis speed. The final motor's torque will have to compensate for that effect as well.

Equation (8-17) is the solution for one independent link. Adding more links in series requires looking at relative interactions between them: for example, how the inertia seen by the first axis depends on the position of the all the other axes. You can imagine how complex the equations become with the six links of the robot in place.

Newton-Euler Method

An alternative way to solving the dynamic model of a robotic manipulator is a simpler recursive algorithm, which can be efficiently programmed for open chains of any number of axes. Unlike the Lagrangian method, which is based on energy calculations, the Newton-Euler method solves the inverse dynamics by balancing forces and torques acting on each link.

The basic idea is to go through each link, analyze what forces act on it, and from there find out the required actuator's torque to balance them out.

This approach does not provide a closed-form solution: it is a numerical method built on a recursive algorithm. However, it has the advantage that it can be easily extended to systems with any number of links, whereas the dynamic equation using the Lagrangian formulation would get prohibitively complicated.

The procedure starts with a *forward pass* (Figure 8-11): given the input values $\left[\theta,\dot{\theta},\ddot{\theta}\right]$ for the ith joint and the spatial velocity and acceleration $\left[T_{i-1},\dot{T}_{i-1}\right]=\left[v_{i-1},\omega_{i-1},\dot{v}_{i-1},\dot{\omega}_{i-1}\right]$ of the previous $(i-1)^{th}$ link in the chain, we calculate the spatial velocity and acceleration $\left[T_{i},\dot{T}_{i}\right]=\left[v_{i},\omega_{i},\dot{v}_{i},\dot{\omega}_{i}\right]$ of the ith link, starting from the base all the way up to the TCP.

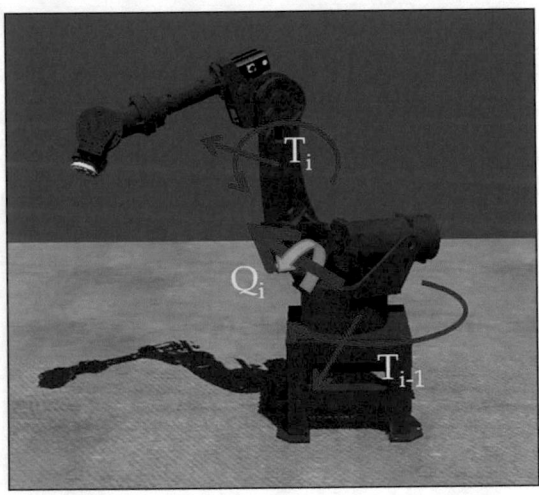

Figure 8-11. *Forward pass to calculate spatial motion of each link*

Then, we execute a *backward pass* (Figure 8-12): take the twist and its derivative $\left[T_{i},\dot{T}_{i}\right]=\left[v_{i},\omega_{i},\dot{v}_{i},\dot{\omega}_{i}\right]$ of the ith link that we calculated in the forward pass; add the wrench vector $w_{i+1}=[F_{i+1},\tau_{i+1}]$ caused by the following $(i+1)^{th}$ link in the chain, and calculate the wrench $w_{i}=[F_{i},\tau_{i}]$ of the ith link. Finally, from the wrench of each link we extract the required input torques at the corresponding joint.

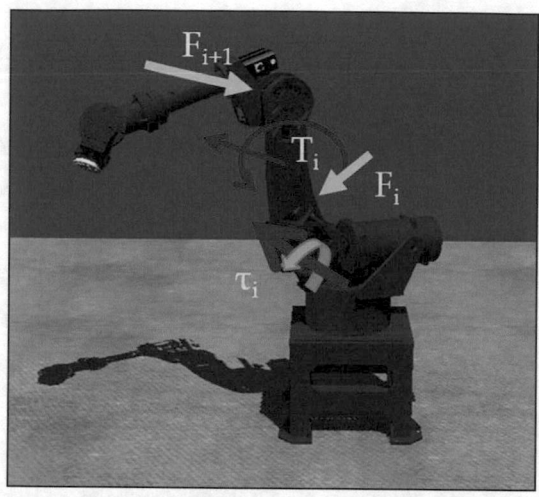

Figure 8-12. *Backward pass to calculate the torques of each joint*

The backward pass starts from the TCP and goes all the way down to the base of the robot. The wrench vector at the TCP is 0, unless external forces are applied.

The method is called Newton-Euler because it requires the calculations of linear components of velocity and accelerations (according to Newton's laws) and the rotational components (according to Euler's laws).

Parameters Identification

Solving the dynamic equations of motion, regardless of what method you decide to use, leads to the same result. However, in order to solve the calculations and derive a numerical value for the motor's torques, we need to plug in the values of all the dynamic parameters: mass, inertia, friction, and all the others we described in Dynamic Model Section.

Some of the parameters are relatively simple to measure: for instance, the mass of a link. Some could be derived from the mechanics CAD design: any 3D modeling software can provide the inertia matrix of a rigid body. However, in most cases, the majority of the required parameters are either not available or too difficult to measure in practice. That is where the procedure of parameters identification comes into rescue.

The idea is very simple, and it is based on the crucial observation that the equations of inverse dynamics can be written as a linear system in the parameters P:

$$\tau = P\theta \tag{8-18}$$

We normally use the equation to find the torques τ given the actual joint angle values and their time derivatives $\theta = \left[\theta, \dot{\theta}, \ddot{\theta}\right]$.

Here, we solve the inverse problem: we collect a large dataset of experimental values $[\theta, \tau]$, and from there we reconstruct the parameters vector P by least-square approximation.

Collecting data is simple: we let the robot move around at different speeds, different accelerations, and different spatial positions, and we record all the torque values used by the motors at the joints. Recording the torques requires reading the motors' currents and then multiplying by the motor's electric constant (see Current Sensing Section in Chapter 16).

The resulting dataset can be arranged in a very large non-square matrix shown in Equation (8-19), with as many columns as the number of recorded samples N and 19 rows for the number of recorded features (for a six-axes robot).

$$[\theta, \tau] = \begin{bmatrix} \theta_0 & \dot{\theta}_0 & \ddot{\theta}_0 & \tau_0 \\ \theta_1 & \dot{\theta}_1 & \ddot{\theta}_1 & \tau_1 \\ \vdots & \vdots & \vdots & \vdots \\ \theta_N & \dot{\theta}_N & \ddot{\theta}_N & \tau_N \end{bmatrix} \tag{8-19}$$

Solving the linear system in Equation (8-18), given the dataset in Equation (8-19), is a standard linear regression problem: you can plug the values into your favorite least-square solving algorithm to find the best fit result for the dynamic parameters. We will describe one of these algorithms when talking about tool calibration in Chapter 10.

The simplicity of this solution is possible because the dynamic model is linear in the parameters.

As with any other data-based problem, the quality of the dataset is critical for best identification results:

- The more data the better: collect a lot of samples.

- The more various data the better: use different speed profiles to stimulate different frequencies on the mechanical structure. Do not use constant speed profiles only.

- Try to avoid generating response behavior that is not modeled by the equations. For instance, if rigid links and joints are assumed in the model, do not excite a flexible behavior by using too high-frequency signals (add filters to the speed profile if necessary).

- Focus on the behavior that is expected during the actual application of the robot. If a manipulator will be used exclusively to perform palletizing applications, then you can identify the dynamic parameters performing that movement, as the predictions will be more accurate. Conversely, if the robot will perform generic applications, then it is better to identify the parameters while running random movements.

The resulting parametrized model should be the best fit for the specific robot that was used for the identification procedure. Different robots, even if sharing the same kind of mechanics, all have slightly different

249

characteristics, especially in terms of friction and elasticity (if modeled) at the joints. Ideally, we should identify a unique set of parameters for each specific robot.

In fact, even the same individual robot will probably drift over time because of temperature, wear, and age. For most applications, however, that is not an issue. Dynamic models are typically used to improve overall control quality, but they are not required or expected to be absolutely accurate.

Whenever accuracy requirements are stricter, or large variations over time are expected, a valid approach is adaptive tuning via online learning. There are a number of different possible ways to do that. One approach is to first approximate the model into a neural network using supervised learning and then apply reinforced learning to improve the policy over time using batches of online data. Any learning algorithm that can handle deterministic policies will work fine. The reward to be maximized is typically a function of the position control accuracy over all the robot axes.

In general, the dynamic model finds use in several practical applications: we will present some of them in the next sections and describe what advantages it brings to the robot control software.

Torque Feed-Forward

The most common application for the dynamic model is to use its calculated torques as feed-forward set values in the servo drive control. We will study the details of the servo drives software architecture in the next chapter, but we can anticipate the general idea here.

The robot control software, which includes the trajectory generator, runs on the main controller, which cyclically executes the calculations, and generates a target position vector θ for all the joints of the robot. The actual task of controlling each motor is left to the individual servo drives (see Figure 8-13).

250

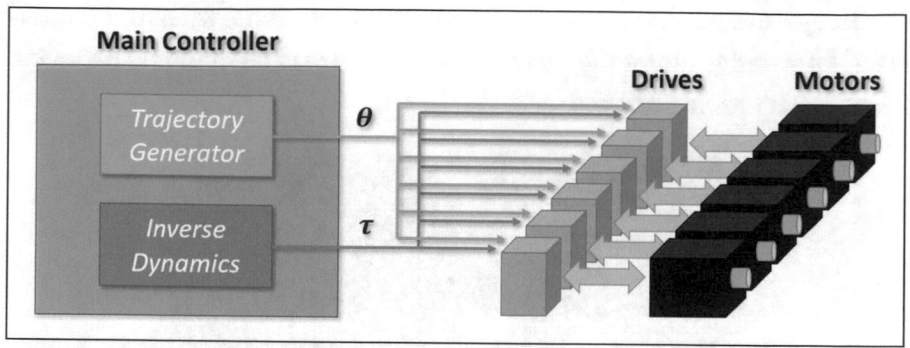

Figure 8-13. *Sending torque set values along position values increases control quality*

For reasons of simplicity and robustness, most of the industrial robots are only controlled in position. Each servo drive receives a position command every controller clock tick (e.g., 1ms) and makes sure that the motor follows the set position by providing the correct current value. This control technique works fine for most applications, but it is not particularly accurate, because the drive itself has no knowledge about the system under control.

Given the complex kinematic structure of the robot, the amount of torque required by each axis to achieve a certain speed profile depends strongly on the position and dynamic state of all other joints. Therefore, decentralizing the control and ignoring the cross-correlations between the axes are not an ideal approach.

The solution is to let the main controller calculate the set motor torques τ using the inverse dynamic equations and cyclically send them as feed-forward values to the drives along with the target positions. By adding that extra information from the entire robot status, a much better control quality can be achieved.

In practice, the result is that the positioning error of each joint is reduced (see Figure 8-14), and the individual controller of each axis can be softened, with the convenient side effect of decreasing vibrations. The advantages are particularly visible during accelerations, when the model-based knowledge about the mechanical system and its dynamic behavior bring considerable benefits.

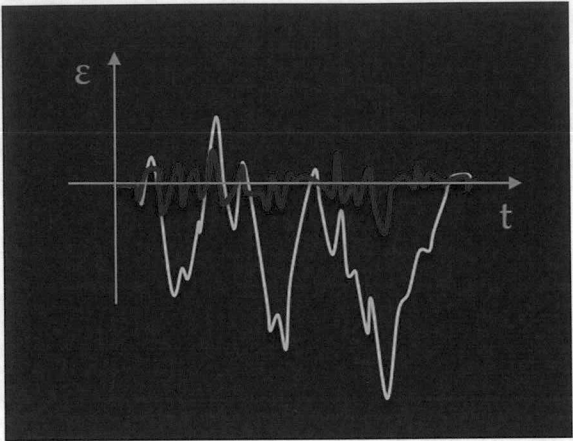

Figure 8-14. *Axis positioning error before (yellow) and after (red) adding model-based torque feed-forward*

One drawback to consider in practice is the complexity of the calculations, which significantly increases the computational load of the main controller.

Trajectory Optimization

We learned in Chapter 7 how to limit the path speed of a robot in order to avoid violating the joint dynamic constraints of speed, acceleration, and jerk. We can now introduce an additional constraint to the optimization problem: the joints maximum torques. In other words, we want to avoid a trajectory that requires too much output torque from a motor, so that we can prevent damage to the mechanics and its actuators.

You might wonder whether limiting the acceleration of the joints would be enough to reduce peak torque values, since torque and acceleration are proportionally related to each other by the inertia value of the load. That would work for a single axis machine. However, in case of serial kinematic structures, the inertia seen by an axis's motor is not constant over time: it depends on the actual load mounted at the TCP and, more importantly, on the real-time state of all the other axes.

Consider, for example, the robot shown in Figure 8-15, which is lifting the same load at different distances from its base. The inertia seen by the motor of J_2 in the two cases changes significantly. Consequently, the trajectory generator can use two different acceleration values and complete the movement with different timings using the same amount of peak torque from the motor. In particular, the case on the right exhibits lower inertia, and its movement can be performed much faster.

Figure 8-15. *A change in the apparent inertia allows the movement to be planned at different speeds without changing the torque requirements from the motors*

Besides inertia, the motor torque requirements are also affected by velocity-dependent terms, such as centripetal and Coriolis forces. Therefore, in general, limiting the differential kinematics when generating a trajectory is not a sufficient condition to limit the actual motor torques exerted during a movement.

A more complete solution requires to consider the robot's dynamic model and use it to predict the torques required at each instant of the programmed path. That information can then be exploited to generate a valid trajectory that never violates the motor's maximum torque values.

In a practical software implementation, we need to add the torque values as additional constraints to the optimization algorithm described in Chapter 7. First, we estimate the actual torque values at each point of the trajectory expressed as a function of the joints' positions (gravitational term G), speeds (centripetal and Coriolis terms C), and accelerations (inertia term I):

$$\tau = G(\theta) + C(\theta)\dot{\theta}^2 + I(\theta)\ddot{\theta} \qquad (8\text{-}20)$$

Then, we can proceed recursively: we reduce the path speed linearly until the torque values are not violated along the trajectory. Often, the trajectories optimized using additional torque constraints achieve lower-speed performance than those considering only acceleration limits (see red vs. yellow curve in Figure 8-16).

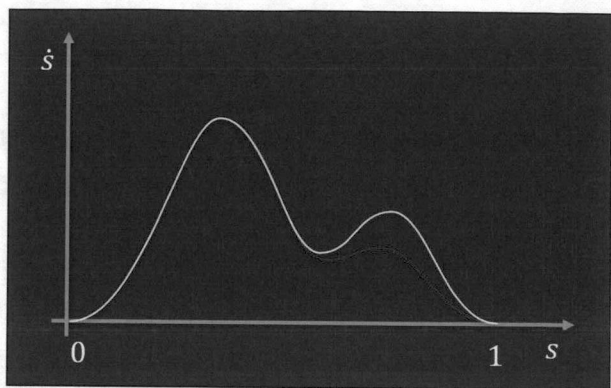

Figure 8-16. *Adding torque constraints (red curve) can further reduce speed limits in trajectories*

One important thing to remember when dealing with torque limits is to always keep some safety margin. The dynamic model is always an approximation both in its mathematical formulation and in its empirically identified parameters, and it might include inaccuracies. For example, the actual joint friction might be higher in practice than what was initially modeled or identified.

Allowing some torque margin avoids complete saturation of the electric motors and leaves them a chance to recover and correct unplanned positioning errors.

Teach by Hand

We now explore a nice application where the concepts of statics and dynamics come to work together: the method of *teaching by hand*.

The word "teaching" in industrial robotics describes the action of moving the TCP to a target pose and saving it (either in the path space or in the joint space) in the application program. After several points have been taught by the operator, the control software can automatically generate a trajectory that passes through them, so that the robot can perform a specific movement without the need of manually entering the target coordinates.

Traditionally, moving the robot to a specific position was done with jogging commands, either one axis at the time in the joint space or one direction at the time in the path space.

More recently, a more convenient teaching by hand method has been introduced. A six-axes force sensor is mounted on the TCP, and the operator can literally grab the robot with a hand and move it around in space (see Figure 8-17).

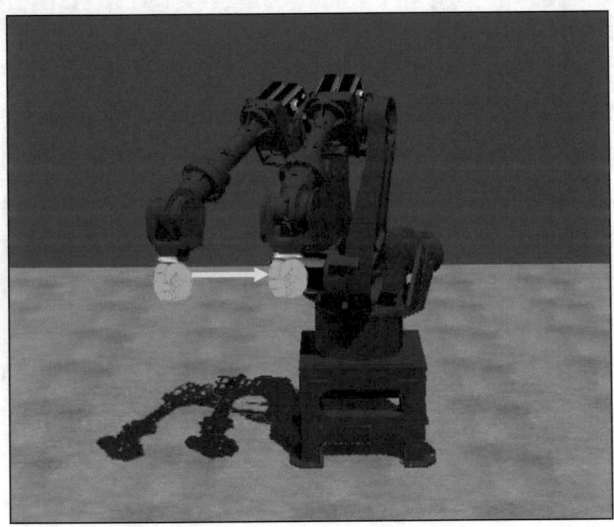

Figure 8-17. *Teaching by hand with a force sensor*

However, a robot in standard operating state would normally oppose the external force and keep the TCP stationary, because the servo drives have active position controllers and are programmed to hold their position.

There are a couple of tricks we need for this to work. First, we need to disable the position controllers so that the motors do not oppose the external force applied by the operator's hand. That would possibly cause the robot to collapse to the ground under its own weight, unless we actively compensate for it. Here is where the dynamic model comes into rescue: the calculated torques at zero speed keep the robot up against gravity.

If we call θ_{act} the current position of the joint axes, plugging the tuple $\theta = \left[\theta = \theta_{act}, \dot{\theta} = 0, \ddot{\theta} = 0 \right]$ in the inverse dynamic equations returns the torque vector τ_0 required to hold the robot stationary.

In this situation, the motors are not position controlled anymore, but they are controlled with an open-loop set torque. That is sometimes called *zero-gravity mode* and allows us to switch off the position controllers safely, while the robot literally floats in the air and at the same time can be moved around with the touch of a hand. The torque feed-forward compensates for gravity forces at the joints, but does not oppose the force coming from the operator's hand.

Typically, a force sensor at the TCP is used to amplify the applied wrench signal, which can be a combination of translational forces and rotational torques. The conversion from TCP wrench to joint torques is performed using the statics calculations we have learned in earlier in this Chapter:

$$\tau_{ext} = J^T F_{TCP} \tag{8-21}$$

The generic Euler-Lagrange equation, Equation (8-13), can be rewritten to include the contribution of the external contact forces at the TCP, which generate additional torques at the motors and push the robot in the direction (or orientation) desired by the operator during the teaching process.

$$\frac{d}{dt} \left(\frac{\partial L}{\partial \dot{\theta}} \right) - \frac{\partial L}{\partial \theta} = \tau - J^T F_{TCP} \tag{8-22}$$

In theory, one could sense the external forces directly at the motors by monitoring their current levels. However, that is only possible if the motors are not coupled through large gear ratios, which would render the feedback signals too small and noisy to detect.

Motor Sizing

One last application we present in this chapter, which also takes advantages of the robot' dynamic model computations, is motor sizing. Imagine you build or purchase a mechanical arm and need to pick suitable motors to drive it.

Motors are expensive and heavy, so you want to avoid selecting them too large (J_2 in Figure 8-18 middle). However, if you use very small motors (J_2 in Figure 8-18 right), then the robot might not be able to accelerate as fast as you would like it to, because small motors do not generate enough torque and/or do not have enough inertia to balance the load.

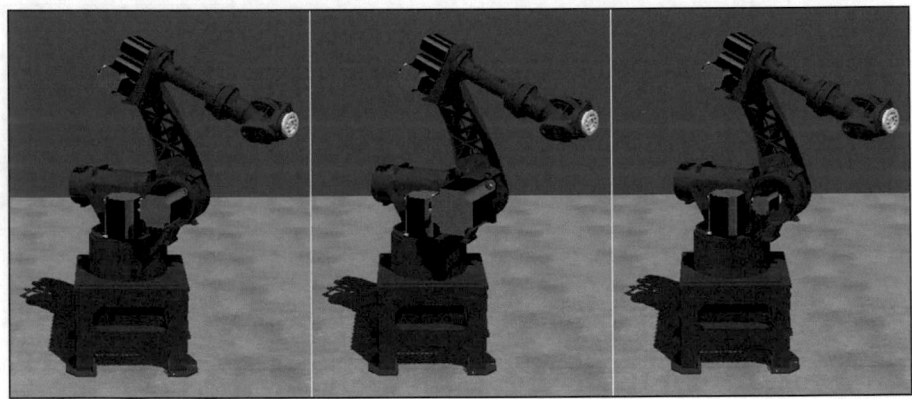

Figure 8-18. *Selecting the right motor for J_2*

Calculating the required actuator's torques given a specific motion trajectory is exactly what the dynamic model can do, specifically, using the inverse dynamic equations. As long as we know the planned trajectory of the robot, then we can find out the torques and choose sensible motors for the application.

We work in simulation and generate a speed profile for the robot. Then we run the trajectory through the dynamic model and monitor the output torques of each axis, as shown in Figure 8-19. In particular, we observe two parameters: the maximum torque, which provides an estimation for the peak torque of the motor, and the RMS torque, which suggests a minimum value for the motor's nominal torque (see the Section on Motor Sizing in Chapter 14 for more details on motor's parameters).

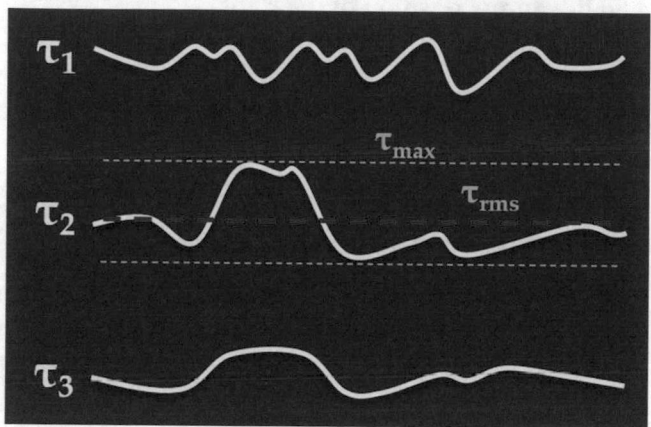

Figure 8-19. *Monitored motor torques during simulated movements*

The applied trajectory should be as close as possible to real application of the robot. If no specific application is given, then we should solve for many possible variations and pick the highest recorded values, so that the motors are never undersized in any circumstance. Generic speed profiles help to make robust predictions.

Note that running the dynamic model requires a valid parameterization of the robot. Most parameter values can be provided by the mechanical manufacturer; friction, on the other hand, cannot be estimated in simulation and is typically left at 0. Therefore, when making the final sizing decision, we should always keep a margin of at least 10 % ~15% to account for friction and other possible disturbances not considered in the theoretical model.

Finally, some large robots have a mechanical spring on the second axis to help reduce torque requirements of the corresponding actuator. In that case, we need to account for the spring effect in the model's calculations and also keep in mind that the force exerted by the spring is not constant but depends on the actual position of the joint J_2.

Summary

Generating a trajectory in space and moving the robot with simple position control techniques are enough for most applications. However, in some cases, we face more advanced requirements. In this chapter, we described the statics equations of the manipulator, which we can use to introduce force control at the TCP, in addition to standard position control. The combination of the two is called hybrid control.

We then presented the basic concept of the dynamic model, which complements the kinematic model by adding calculations for the actual joint torques required to move the robot. Practical applications range from torque feed-forward control used to improve movement accuracy to zero-gravity mode used when teaching by hand. The solutions to the equations of the dynamic model are complex and lengthy, so we only outlined the procedure here.

PART III

Robot Software

We now start the second half of the book, where it is finally time to put in practice all the math we learned in the previous chapters and build a control system for the robot. We divide the task in two main parts: the software architecture and the hardware electronics design. We talk about the software in this section and leave the hardware to the last part of the book.

Software is a generic word that describes all the programs we use to control and monitor the robot. In particular, we will analyze the details of the following components:

- **Firmware**: This is the core control system of the robot. It is where all the complex calculations for kinematics and dynamics take place and where the operator's commands are transformed into actual robot's movements. Writing the firmware is the most challenging task for a robotics developer.

- **Calibration**: The firmware is developed as a generic piece of software and must always be parameterized and tuned to the specific robot it is controlling. This operation is called calibration and is usually performed by the robot's manufacturer before shipping the robot to the end user.

- **Commissioning**: Robots are versatile machines that can perform a number of different tasks. As a consequence, there are quite a few parameters and functions that need to be adjusted according to the concrete application the robot is going to work on. This operation is typically performed by the robot's operator.

- **Simulation**: A digital twin is a helpful companion to each deployed robot for a number of reasons: mechanical and electrical sizing during the design phase; rough pre-tuning during the commissioning phase; and offline programming, remote monitoring, and control during the application phase. All robot manufacturers develop and distribute digital twins for their robots.

- **Machine vision:** While vision is not technically required to drive an industrial robot, the number of applications where vision-related functions find place is increasing all the time. The addition of an electronic eye gives the robot much more flexibility and awareness of its surroundings, with significant improvements in output quality, productivity, and safety.

The number of people and companies involved in the software development for a robot can vary greatly. Large robot manufacturers employ entire teams of engineers to develop their own control systems (though more often firmware than hardware), provide commissioning service, and typically also use their own robots in their production facilities.

On the other hand, small robot manufacturers tend to outsource most of the work: they build the mechanics, but they purchase control systems from automation providers (both firmware and hardware) and use system integrators to commission their robots to end users.

Ideally, you should learn to do it all by yourself: build a robot, design its electronics controller, and write the driving software. That will give you a more interdisciplinary understanding of the entire system and will make your skills much more marketable.

CHAPTER 9

Firmware

The quality of a robot greatly depends on its firmware: it should provide a rich set of functionalities; it should be reliable and robust; it should be easy to configure and intuitive to use.

All too often we find robots on the market that advertise a large number of functions and features but are much too complicated to use and set up for the average operator. User-friendliness is a must in the consumer electronic world, but the same concept has had a hard time penetrating the industrial world.

Flexibility is also an important requirement. The firmware you develop should be able to adapt to run under a large variety of circumstances and applications. A control system that is great at palletizing but performs poorly in arc-welding does not sell well in the robotics market.

The core firmware that controls a robot typically performs a number of different functions:

- Reads commands and parameters from the operator. This is the interface between the outside human world and the internal machine world: in short, **HMI** (human-machine interface).

- Translates the input commands into motion instructions. This task is performed by the **interpreter**.

© Fabrizio Frigeni 2023
F. Frigeni, *Industrial Robotics Control*, Maker Innovations Series,
https://doi.org/10.1007/978-1-4842-8989-1_9

- Plans a path and generates a trajectory. In practice, translates motion instructions into target position and torque values, using all the heavy theory we studied in the first half of the book. These libraries run on the **main controller**.

- Drives the motors to move the robot. This is the software part that runs on the **servo drives** and controls the behavior of the electric motors according to the set values commanded by the main controller.

- Outputs status and diagnostic information; also performed by the HMI.

We will now analyze all these sections in detail from top to bottom: from the input commands on the touch screen down to the motor electric signals that make the robot move.

Human-Machine Interface

The HMI (or visualization) is the main interface between the robot and its human operator. It is a piece of software used to configure the robot and tell it what to do. While this sounds easy enough to build in a few lines of code, you should keep in mind that the look and feel of a visual interface plays a major rule in the market acceptance of a product.

The recent trend in the industrial world is to simplify the user interface layer, which has been lacking attention for years in favor of more complex core functionalities. The idea is to make everything more graphical and user-friendly, so that the operator is not required to have an advanced mathematics degree in order to safely jog the robot.

Some practical solutions implement completely *codeless automation*: all the functions and movements of the robot can be simply configured instead of being programmed. The advantages are faster time to market, higher productivity, and a reduction in application errors.

The functions typically provided by a standard HMI can be divided in three main areas:

- **Configuration**: This is where the user can modify the values of the robot's parameters: mechanical dimensions, joint and path limits, zero position offsets, workspace zones definitions, dynamic profiles, motors and encoders settings, servo loop gains, and many more (see Figure 9-1 for a basic example).

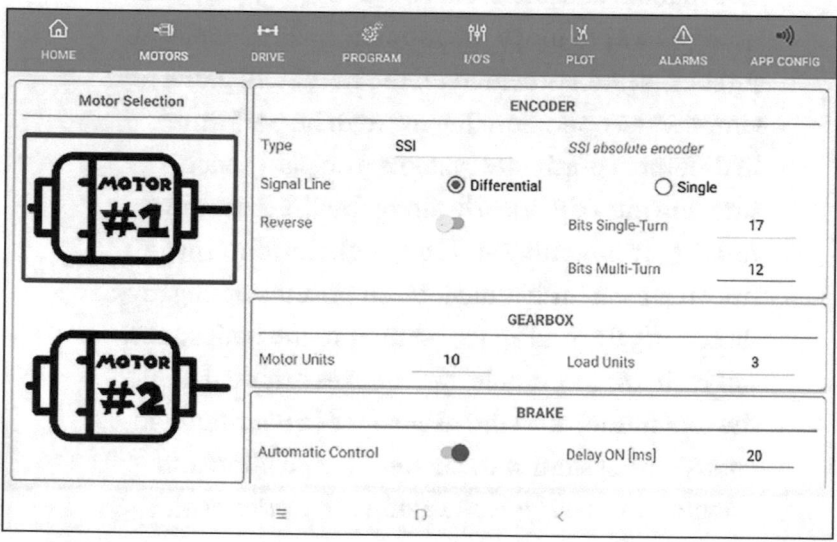

Figure 9-1. *Motor parameters configuration*

Some of these parameters are very critical for a safe functioning of the robot, and different levels of access are usually required to access and modify their values. For example, the manufacturer's service engineer would be granted a higher level of access than the end-user application engineer. This approach is called *user management*. As the developer of the control software, you should grant yourself the highest rights, so that you can access all the possible fields.

- **Movement programming**: This is the section of the HMI where the operator can directly trigger robot's movements: it is usually split in manual and automatic control.

  The manual control, often called *jogging*, provides a simple way to move the robot in space. A small joystick is often provided on the handheld panel to simplify the operation. Jogging can be performed in different coordinate systems: the path space (moving the TCP linearly along the X-Y-Z axes and rotating around the A-B-C angles), the joint space (moving each individual joint in positive or negative direction), the tool space (similar to the path space, but with the coordinate system axes oriented along the tool frame), and the user space (a user-defined coordinate system with customized position and orientation axes, typically along the workpiece).

  The action of teaching means manually repositioning the robot to a specific pose and saving the corresponding coordinates in a list of points, which can be accessed when programming automatic movements.

  The automatic control functionality requires the operator to program an entire sequence of operations for the robot that will later be executed automatically by the main controller. Programs can be provided either as text files (as shown in

Figure 9-2) or as graphical diagrams. They can either
be executed in step-by-step mode (normally for
testing purposes) or in continuous mode without
interruption (for standard production work).

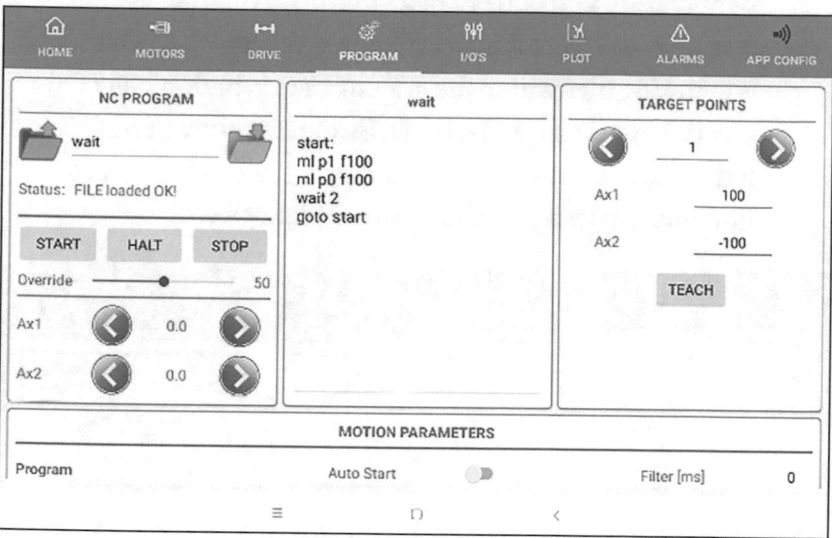

Figure 9-2. *Automatic movement programming*

Programming a robot through its HMI interface
using the teaching and testing functions is normally
referred to as *online programming*. More often,
though, the robots' programs are automatically
generated, tested, and tuned using simulation
software for convenience and safety reasons. The
resulting program is then uploaded to the real robot,
which can immediately start running. This approach
is known as *offline programming*.

- **Diagnostic**: Besides reading input commands, the
  HMI also needs to provide feedback to the operator
  about the current status of the robot: the position of
  all joints and TCP axes in different coordinate systems,
  temperature and current of all motors, eventual
  warnings or errors that are triggered during operation,
  and the log files where the system state is periodically
  saved. Most data are shown as instantaneous values,
  but some should also be plotted over time to offer more
  in-depth analysis possibilities (see Figure 9-3).

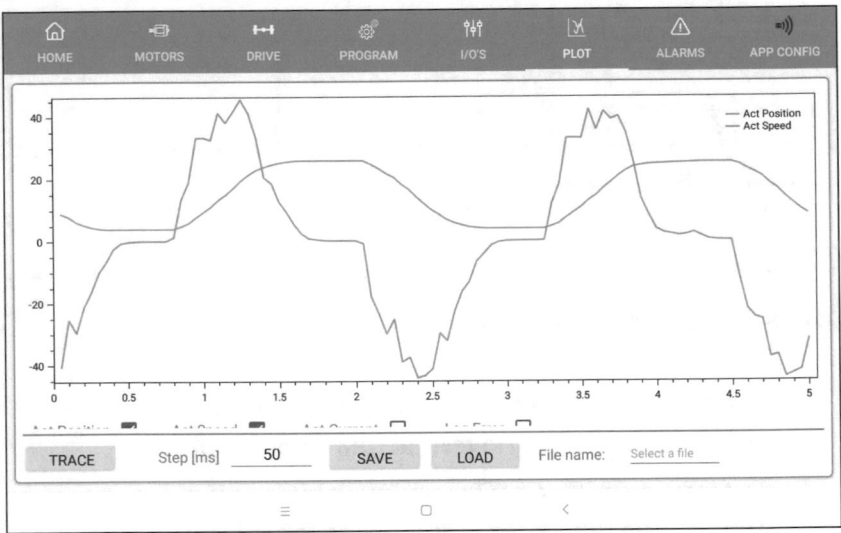

Figure 9-3. *Plotting data over time*

A monitoring function must also be provided for all
the I/Os (inputs/outputs) of the robot. The operator
is typically allowed to manually force certain output
values for testing purposes.

Unlike the rest of the control firmware, the HMI application does
not need to run on a real-time OS and does not communicate in a
deterministic way with the main controller. That is because changes to

configuration parameters and human commands are not as time-critical as sub-millisecond motion calculations. The only exception where a high-priority response is required is the E-Stop (emergency stop) button, which must be hard wired to the IO modules and processed in real time by the main controller program.

HMI software runs either on a dedicated low-speed device (e.g., a handheld panel with an Arm processor and a Windows IoT OS) or, more commonly, on a decentralized generic device (e.g., a consumer tablet running an Android application). See the Display Section in Chapter 18 for more details on the hardware.

As for the development suite, there are a number of different options to choose from. If your hardware is microcontroller-based, then a lightweight graphic library such as LVGL offers a great deal of features to build your HMI pages. Their Squareline Studio is a very convenient tool to quickly drag and drop widgets on a screen without writing any line of code. On the other hand, if your hardware is more powerful, then you have access to a wider range of graphic tools. Personally, I have used Xamarin for quite a few projects I worked on: it is part of the .NET framework and builds applications that run across multiple platforms (Android, Windows, and iOS), so you can quickly reconfigure your HMI to target whatever system the customer requires. In addition to standard libraries, Xamarin also provides access to NuGet packages, which are created and shared by users in the .NET community.

Interpreter

The HMI provides an interface for the user to input a sequence of actions that the robot will then execute in automatic mode. This process is called robot programming. Regardless of how the actions are described (text lines or graphical icons), they need to be parsed and translated into low-level motion instructions to be processed by the motion libraries. The program responsible for that operation is called interpreter.

Every robot on the market comes with its own programming language: usually, a very intuitive, high-level language that the robot's manufacturer defines. There is no standard across the industry, although most languages are very similar to each other, and many instructions overlap. We show here a simple example of robotics programming language that you could implement for your own controller.

Each instruction is a combination of a command and some characterizing parameters. For example, a command to drive the TCP along a line would take as parameters the target pose and the desired path speed. The values of the parameters could be programmed explicitly (i.e., writing out all the coordinates for the target pose), but that would make the syntax very heavy and repetitive. In practice, parameters are saved in memory tables, and the motion commands only need to specify the index of the required object.

For example, we could name the linear motion command ML (move line) and write an instruction out explicitly as follows:

```
ML X=1755 Y=0 Z=2660 A=0 B=0 C=0
```

This command would move the robot described in the Numerical Test Section in Chapter 3 to its home position. The syntax is not user-friendly though and is prone to input errors. It is much more convenient to define a table of target points P_i in memory, save the home position at P_0, and write the same command simply as follows:

```
ML P0
```

Other parameters that are typically saved in tables are workpiece frames (Z) and tools (T). Therefore, moving the robot along a line to reach point P_1 defined in the frame Z_1, and using the tool T_1, can be written as follows:

```
ML P1 Z1 T1
```

The interpreter needs to break this instruction down into parts: the type of motion command requested (an interpolated line in this case), the target point, the reference frame for the point's coordinates, and the size of the tool currently mounted on the robot. The resulting numerical values are then passed onto the path-planner for geometrical calculations. If no frame or tool is specified, they are assumed to be 0, i.e., using the robot's own local frame, and no tool mounted on the MP.

A movement also needs a speed to run. In numerical machine control, movement speeds are traditionally referred to as *feed rate* and symbolized with the *F* letter. In robotics, path speed definitions are not unique, and their exact meaning and measurement depends on the programmed movement (as we described in the Section on Path-Speed Definitions in Chapter 7). For example, a path-interpolated movement would typically expect a path speed in mm/s, while a PTP movement would assume a speed given in deg/s. However, the operator can easily override the standard settings and use the symbols *FC* and *FA* to specify a Cartesian or angular speed, regardless of the movement kind. The following movements are likely to execute all at different speeds:

```
ML  P3  FC100
ML  P3  FA100
ML  P3  F100
```

The first command only considers the Cartesian length of the movement; the second command only considers the angular length, while the last command is the most generic and includes both Cartesian and angular calculations. If the specified speed type does not apply to a movement, then the generic default speed calculation is used: for example, if the target point has the same orientation of the starting point, the angular length of the movement is 0, and the Cartesian coordinates are used for the speed calculations instead. See the Path Length Section in Chapter 5 for details on how to calculate the length of different movements.

Finally, when two or more movements are programmed in series and they are all to be executed at the same speed, the parameter F can simply be omitted, and the speed definition and value will be carried on to all subsequent movements until a new one is specified.

Given a series of movements, the operator can also decide to add a round edge to smooth out the intermediate transitions. The parameter R can be used in this case, along with a specified radius of curvature:

```
ML  P5  R10
ML  P6
```

Note that the rounding radius cannot exceed half of the segment's length and will be automatically clamped otherwise. The user can also program a zero-radius command ($R0$), which forces the movement to stop at the transition point, no matter if the actual transition is tangential or not (see Transitions Section in Chapter 5).

Unlike linear movements, circular movements require two points to be specified: a target point P and an additional point Q on the circle along the path. Both their indexes refer to the same table of points taught by the operator and saved in memory. We could name the circle motion command MC (move circle):

```
MC  P2  Q1  F10
```

Critically, the two points cannot be collinear; otherwise, the calculations will result in an error, and the program execution will abort.

Here is a complete example with linear and circular movements (see Figure 9-4):

```
ML  P1  R100  FC100
ML  P2  R50
ML  P3  R100
```

```
MC  P5  Q4  R100
MC  P7  Q6
```

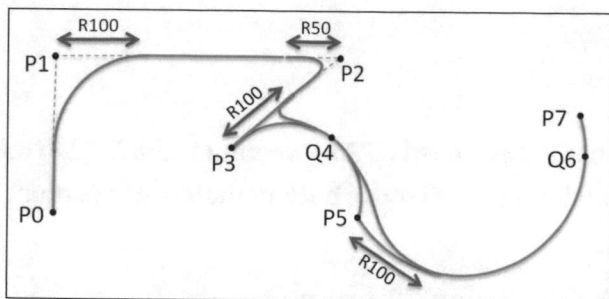

Figure 9-4. *Example of programmed movement with lines, circles, and rounded edges*

The length of a circular movement can also be forced to be equal to an angle H, which can be either positive or negative and can range across several revolutions:

```
MC  P2  Q1  H-720
```

The (optionally) specified H value has priority over the target point P, and the end of the movement will possibly not correspond to P.

The next movement we consider is the spline, and we call its corresponding command *MS* (move spline):

```
MS  P1  F100
MS  P2
MS  P3
```

Splines are geometrically built tangentially to each other by definition, so the movement between a series of splines is always continuous and does not stop at each transition point. For the same reason, round edges programmed between splines are ignored.

So far, we have looked at commands for interpolated movements in the path space. If we want the robot to interpolate the joint space (see Section on PTP Movements in Chapter 5), we can call the command *MJ* (move joints):

```
MJ P1 F100
```

The most commonly used PTP movement is the *HOME* command, which brings all joint axes to move back to their zero position:

```
HOME F100
```

When programming a movement, either in the path or joint space, the operator needs to define a target point. The coordinates of that point can be taught either in the path space [X, Y, Z, A, B, C] or in the joint space [$J_1...J_6$] and saved accordingly. When executing a movement in the path space (i.e., *ML, MC, MS*), the control program first transforms all target points in path space coordinates and then plans the movements. Since the direct transformations are always uniquely solved, the procedure does not present complications.

On the other hand, if a PTP movement is programmed with the target point given in path space coordinates, then the control program needs to transform the point into joint space coordinates. However, since the inverse transformations are not always uniquely solved, a choice must be made between all the possible joint configurations.

A default rule could be to select the configuration closest to the pose where the movement starts, measuring the distance in the joint space. However, there are situations where the default choice leads to unwanted poses or even to errors when the solution violates the workspace or exceeds mechanical limits.

Therefore, the user must be given the possibility to override the internal pose selection by specifying a desired reference position for each of the joints. The robot will then select the joint configuration closest to the programmed reference values. Figure 9-5 shows the robot running a PTP movement reaching for the same pose (programmed in path space coordinates) but choosing two different joint configurations.

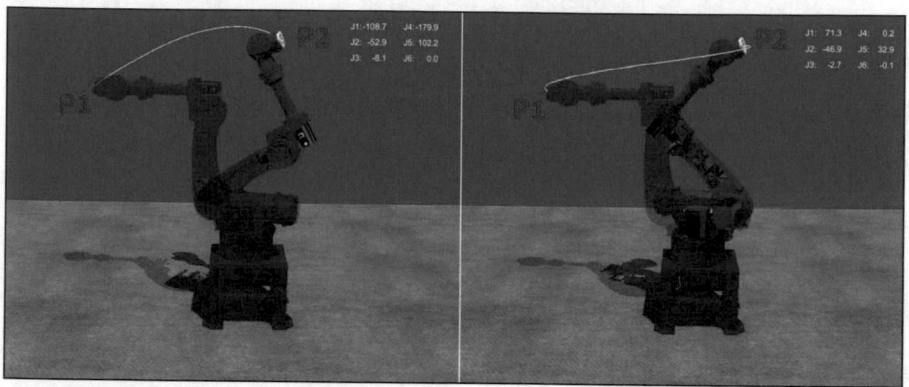

Figure 9-5. *Provide joints reference positions to force a specific target configuration for PTPs*

If the starting point P_1 has coordinates $J_1 = -100$ and $J_2...J_6 = 0$, then the two movements can be programmed, respectively, as follows:

```
MJ P2
MJ P2 J1=0
```

The first command leads the control program to choose the configuration on the left (with a target $J_1 = -108.7$, which is closer to the initial value of $J_1 = -100$), while the second command results in the configuration on the right (with a target $J_1 = 71.3$, which is closer to the programmed reference value of $J_1 = 0$).

We can summarize all the motion commands introduced so far in Table 9-1 and their respective parameters in Table 9-2. The parameters in bold are mandatory; all others are optional.

Table 9-1. *Motion commands*

Command	Description	Parameters
ML	Linear interpolation of the path axes	**P**, F, Z, T, R
MJ	Linear interpolation of the joint axes	**P**, F, Z, T, R, Ji
MC	Circular interpolation of the path axes	**P**, **Q**, F, Z, T, H
MS	Spline interpolation of the path axes	**P**, F, Z, T
HOME	Move joints to their home (zero) position	F

Table 9-2. *Parameters for motion commands*

Parameter	Description	Notes
P	Target position	
F/FC/FA	Path speed Generic/Cartesian/angular	Units depend on speed definition
Z	Frame index	0 is base frame
T	Tool index	0 is no tool
Q	A point on the circle path	Must not be collinear with P
H	Rotation angle	
R	Round edge radius	Must be a nonnegative value
Ji=	Optional reference value for joint axes	Only used for MJ with target point specified in path axes

Next, we could add other motion-related commands more targeted to specific applications. For example, the TRK command is used to (de) activate a conveyor-tracking synchronization:

```
MJ P0 F100
TRK 1 //synchronize with Conveyor #1
ML P1 F100
ML P2 F100
TRK 0 //desynchronize from Conveyor #1
ML P3 F100
```

All the movements within the *TRK 1 … TRK 0* section are performed, while the robot is also following the direction of the moving conveyor. The orientation frame of each conveyor must be defined in the control program, so that the path correction is added in the right direction. The actual speed of the conveyor is read from an external encoder and added to the planned movement's path at execution time.

Another example of application-related command is the (de)activation of a tangential axis. Typically used with an asymmetric tool for cutting or grinding operations, a tangential command forces the orientation of the TCP to be always aligned with the path, when projected over a surface (usually, the X-Y plane). The *TANG 1* command activates the tangential axis, while the *TANG 0* command switches back to normal mode (see Figure 9-6).

```
MJ P0 F100
ML P1 F500
TANG 1 //activates tangential mode
MS P2
MS P3
MS P4
MS P5
MS P6
MS P7
TANG 0 //deactivates tangential mode
ML P8 F6000
```

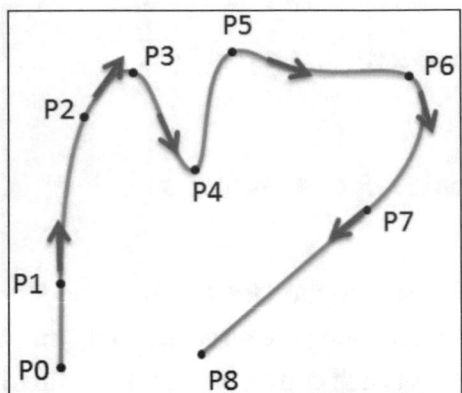

Figure 9-6. *The activation of tangential mode automatically modifies the tool orientation during movements*

Note that tangential mode only affects interpolated movements in the path space, not PTP movements. Also, the path speed might be affected if the curvature of a path section is very high, because the rotation speed required to keep the tool aligned cannot violate axes speed limits. An extreme case is a non-tangential transition between consecutive segments, where the path has to halt its movement and will only resume after the tool has realigned itself with the next path segment (see Figure 9-7).

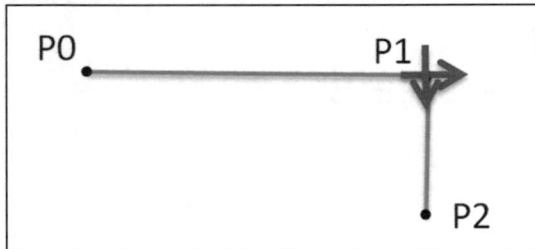

Figure 9-7. *The path movements are halted at sharp corners to let the tool realign*

An optional offset angle between the actual path and the tangential axis can be programmed with the parameter *H*. For example, the command *TANG 1 H45* forces the tool to align itself at 45 degrees with respect to the true tangent. This feature is useful to compensate for fixed mounting angles of the tool on the robot.

Besides commands that trigger movements, we also need to offer a variety of auxiliary commands to perform other functions. For instance, the robot requires commands to interact with the rest of the working cell using IOs and shared variables and also needs to make decisions according to the resulting state of logic operations.

We define the command *WAIT* to delay the next movement step by a specific time and the command *WAIT DI* to halt further program execution until an external digital input signal is received. That is typically used for hardware synchronization with other machines.

We define *SET DO* and *RESET DO* to, respectively, set and reset a digital output. For example, the program could trigger a laser head on and off between movements while working on cutting applications; or it could activate a mechanical gripper after reaching a target position while palletizing some loads.

On request, we could also add *M* functions, which originate from the CNC industry and are an old-fashioned way to synchronize the robot program with the main control program running on the PLC. *M* commands can be defined as synchronous or asynchronous. In the first case, the program execution halts until it is reset by the application: that is used in situations where a movement must be synchronized with a specific action. In the second case, the program only sends a notice flag but continues operation without any speed dip.

M commands can also be used as a simple way to synchronize several robots being controlled by the same software system. For example, if two robots are programmed to work on the same workpiece at the same time, their programs will both require a call to the same M function and halt their execution. Only when the application sees that both robots have

281

activated their M commands (i.e., they have reached the same starting point), then the M flags can be reset and the two programs can continue synchronously.

Finally, we need to add a few instructions to add flexibility to the programming style. We can add a *GOTO* command to jump to a target label, specified as a string. While such statements are typically discouraged in high-level programming languages, they can indeed be very useful and intuitive when programming short blocks of robotics instructions in our own-defined simple language, where variables and constructs are very limited. Robots normally perform the same operation over and over, so a typical program would run an infinite loop:

```
HOME F300
MJ P1 F100 //move to starting point
LOOP:
ML P2 T1 Z1
ML P3 T1 Z1
ML P4 T1 Z1
WAIT 2
MC P2 Q5 T1 Z1
GOTO LOOP //jump back to label "LOOP"
```

If the loop only needs to be repeated a finite number of times, then we could add an index to the *GOTO* command and move on to the rest of the program after the loop has been repeated enough times:

```
HOME F300
MJ P1 F100 //move to starting point
LOOP:
ML P2 T1 Z1
ML P3 T1 Z1
```

```
GOTO LOOP 5 //repeat this loop 5 times
MJ P1 F300
END
```

This approach simplifies the operator's programming style avoiding using variables for counters.

We could further add a *SUB* command to execute a specific subroutine, which has to be repeated many times in several parts of the program:

```
HOME F300
MJ P1 F100 //move to starting point
SUB CUT //execute the SUB program
ML P2 F1000
SUB CUT 3 //execute the SUB program 3 times
END
//subroutine defined here
CUT:
  ML P5 T2 Z2
  ML P6 T2 Z2
END
```

Finally, we should also provide the classical IF...ELSE construct to allow for branches based on the status of an IO or of a variable:

```
IF DI3
  ML P2
ELSE
  ML P3
ENDIF
```

Table 9-3 shows a complete list of auxiliary commands we suggest implementing in a basic robotics programming language.

Table 9-3. *Auxiliary commands and statements*

Command	Description	Parameters
TRK	Start tracking conveyor	[0, index]
TANG	(De)activate tangential tool	[0,1], H
T	Modify active tool (without movement)	[index]
WAIT	Delay time	[seconds]
WAIT DI	Wait for a digital input signal to be TRUE	[index]
SET DO	Set a digital output signal	[index]
RESET DO	Reset a digital output signal	[index]
M	M-function to communicate with main controller program	[index]
IF…ELSE	Conditional statement	
GOTO	Jump to specified label	Label, Loop count
SUB	Execute local subroutine at specified label	Label, Loop count
LABEL:	Any string followed by ":"	
END	Abort main program execution or return from subroutine	
//	Comment	

The list could go on, and you could include increasingly complex commands and syntax to offer more functionalities based on the application requirements. Robot programming is much simpler than computer programming, and developing a suitable interpreter to parse the motion commands is fairly easy.

The typical automatic execution of a program starts from the first line and ends at the last one. However, for testing purposes, the operator could decide to execute only a part of the program. To fulfill this requirement, we can offer the option to specify a starting and ending line number. The program will start executing from the requested line and abort once the end line number is reached.

A common feature often used during tests is the *stepping mode*: the program executes one step at the time instead of in continuous mode. The movement is interrupted at the end of each line, so that the operator can check the correct axes positioning and then manually trigger the next step.

Program execution can be also halted and continued at any time by the operator. Also, the execution speed can be manually controlled with a *speed override* parameter, which usually ranges from 0 to 100%. Some systems even allow negative override values, which force the program to be executed in reverse until the starting point.

Main Controller

The main controller is the brain of the robot: it reads input commands (from the HMI, the IOs, the actuators' encoders), performs all the necessary motion calculations, and generates output instructions for the servo drives.

The most important software components of the main control program are interpreter, path-planner, set point generator, and the motion buffer.

We have already described the interpreter module in detail in the previous section; it is basically a text parsing tool that reads the user program line by line and extracts all the meaningful information to be translated into a motion instruction: the type of movement (PTP, line, circle, spline, etc.), the target position, the desired speed of the movement, the required tool, the user frame, the round edge radius, and many more.

All motion instructions are inserted into a *motion buffer*, and the interpreter keeps an index counter so that it always knows where to insert the new parsed information. The motion buffer has a circular structure of finite length, which forces the interpreter to halt its translation process either when it reaches the end of the user program or when the buffer is full. In the latter case, it waits until some space frees up and then continues its operation.

Not all the commands in the user program are strictly motion-related. For example, auxiliary commands can trigger instructions to set or reset IOs, wait for a time delay, or jump to a label in the program. These commands are still parsed by the interpreter and inserted in the buffer along with the motion instructions.

Figure 9-8 shows a simplified schematic of the interpreter operation: parsing each line from the user program and filling the motion buffer accordingly.

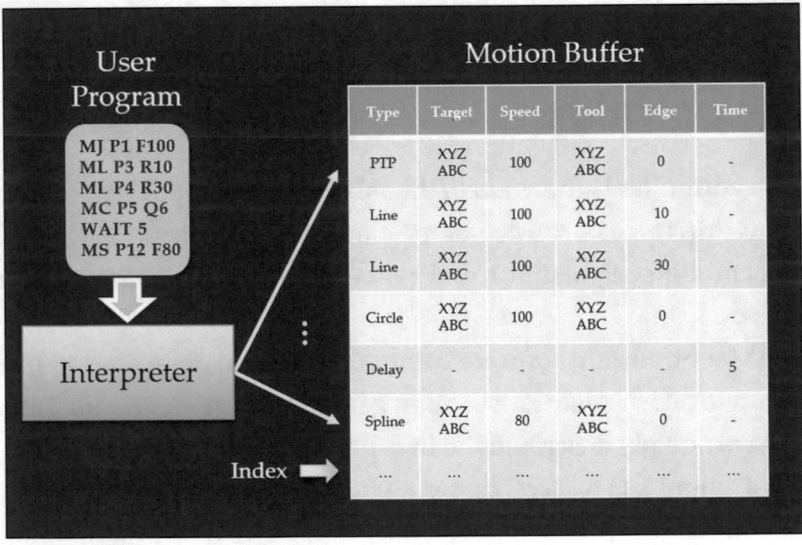

Figure 9-8. *The interpreter fills the motion buffer with information parsed from the user program*

The second module is the *path-planner*, which is responsible for all the geometrical calculations of the path. Each entry in the motion buffer is analyzed and expanded with new calculated values when needed, for example, the length of the movement, the quaternion orientation of the target point, all the circle parameters (center, rotation angles), the control points for splines, and many more. Instructions that are not motion-related do not require path-planning calculations and are left untouched.

Another task performed by the path-planner is the workspace monitoring (see Chapter 6 for details). After calculating all the geometrical path of a movement, the module checks that the entire movement does not violate the predefined workspace constraints and triggers an error in case of violations.

The path-planner also keeps its own index counter, which follows the interpreter index, without ever getting ahead of it. Figure 9-9 shows a simplified schematic of the path-planner module operation of filling the geometrical details of each motion instruction in the buffer.

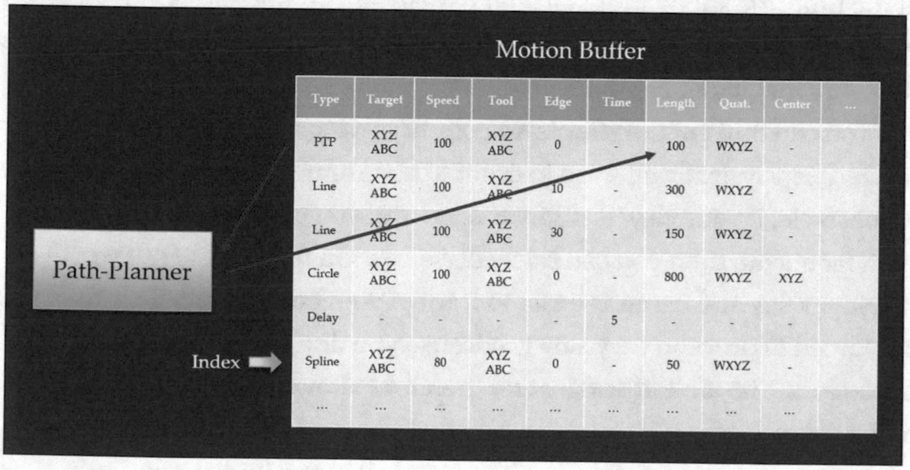

Figure 9-9. *The path-planner adds geometrical calculations to each motion instruction*

The last module is the *trajectory generator*, which reads the planned path from the buffer and sample points in time to generate the actual set position for the axes. The inverse transformations are called when needed, to calculate the joint values that are sent to the drives.

The trajectory generator module works in real time, synchronous to the movement of the robot: it generates a new set of position values at every tick of the controller's clock. The interpreter and path-planner modules, on the other hand, are always running ahead of the actual movement of the robot and fill in the motion buffer much before the instructions are going to be executed.

This approach is called *look-ahead*: it looks into the future and analyzes the next few programmed movements before starting to execute the current line. The reason is that the trajectory generator needs to know in advance the length of the total movement, in order to generate a high-speed motion profile, and still be able to stop the robot safely if needed. Conversely, if the control program were only allowed to look at one block at the time, the speed profile would have to stop at the end of each motion instruction, because it would not know what to do next. That would clearly be a very inefficient solution.

You can think of the simple analogy of driving a car: you only drive as fast as it is physically allowed in order to have enough time to brake before an obstacle. If your view is unobstructed and you can see the upcoming road far ahead, then you can safely drive fast. On the contrary, if you drive in a foggy day and cannot see far in front of your current location, then you are forced to proceed very slowly because you don't know where the next obstacle is. The look-ahead feature therefore allows for higher path speed of the robot.

One consequence of this architecture is that the interpreter finds possible errors in programmed blocks (e.g., a syntax error in the user program) much before the actual movement reaches that specific line. Similarly, the path-planner might find a workspace violation a few blocks ahead of the currently executed movement. In all these cases, the robot has enough time to output a warning message and slow down to a stop along the planned path.

However, there are other situations when both the interpreter and the path-planner module go through the whole user program without detecting anomalies, but then the trajectory generator suddenly finds an error, causing the robot to arrest its movement immediately. A typical example happens when the robot follows an external conveyor, which adds unplanned additional movements to the programmed motion instructions. Critical situations could arise: e.g., the conveyor could drive the robot's TCP to move too fast and violate the joint speed limits, or worse, it could drive it too far, out of reach for the arm. In both cases, the trajectory generator is unable to generate a new set position, and the robot is forced to a sudden violent stop, which could be harmful for the mechanics.

In general, the more and the better we plan a movement ahead of time, the fewer problems we should encounter in the actual execution at real time. Simulation here clearly helps, in order to run and visualize a complete user program without actually having to move the real robot. The robotics control system we design should have the ability to execute a program, read input correction values (e.g., a conveyor encoder), generate the set axes positions and reroute them to the virtual environment of our choice, while the servo drives stay disconnected. This approach provides a safe and realistic way to detect errors before running a program: most commonly, singularities along the path and workspace violations.

A similar execution mode is called *dry run* and does actually move the robot, but without activating the working tool. The reasoning behind this procedure is that an active tool (e.g., a laser cutting head or a welding gun) would irremediably damage the workpiece if errors in the user program are found during execution. Running the program for a first pass while keeping the tool turned off allows the operator to monitor the behavior of the robot without risking damage to the workpiece.

A further significant consequence of the look-ahead architecture is that the values of all parameters and variables referenced in the user program are read and evaluated by the interpreter much before the actual movement reaches that location in the program. In other words, parameters evaluation is usually not path-synchronous. This behavior poses critical problems and misunderstanding when not correctly accounted for. Let's explore two typical situations that often happen in practice.

Consider the following user program:

```
MJ P1 F100
IF DI1
  ML P2
ELSE
  ML P3
ENDIF
```

The interpreter reads through each line of the program and completes the parsing very quickly, reaching the end when the axes have barely started executing the movement specified in the first line of code. In particular, the line *IF DI1* was parsed well before the robot had reached the point P_1. It is possible that the value of the digital input DI1 changes, while the robot executes the movement to reach P_1, and the operator is left wondering why the robot has taken the wrong path at the IF branch.

A similar situation arises when the positions of the target points are dynamically assigned while the robot is executing a program. Consider the following user program:

```
MJ P1 F100
ML P2
```

If the value of P_2 is modified by the application while the robot is running the *MJ P1* line, the interpreter and path-planner have no way to know that since they have already parsed and evaluated the *ML P2* line. Consequently, the trajectory generator will reach the old target value, not the newly assigned one.

The solution to such cases is to introduce commands to block the execution of the interpreter and path-planner (either one or both, depending on the internal implementation) and wait until the path has reached the desired position or wait until the application says it is safe to continue parsing the next lines. M-commands are a common way to synchronize the program parsing operation with the rest of the application. The movement is stopped when reaching the line with the M-command and only restarted after the corresponding M flag has been reset in the application.

IF branches are typically implemented as path-synchronous commands and do not normally require additional synchronization flags. However, that configuration could be overridden by the operator, in case a continuous movement with no intermediate stops is more desirable and functionally acceptable.

Besides controlling the main robot axes, the main controller also needs to plan movements for external *auxiliary axes*. Typical examples are conveyors, electric tools, and turntables. Movements of auxiliary axes can be programmed and synchronized with the movements of the robot, so that they both start and end at the same time. Figure 9-10 shows an exhibition application where a six-axes robot transfers objects from a rotating roundtable to a moving linear conveyor. The auxiliary axes (roundtable and conveyor) are controlled by the same main software driving the robot and are required to move synchronously to the manipulator.

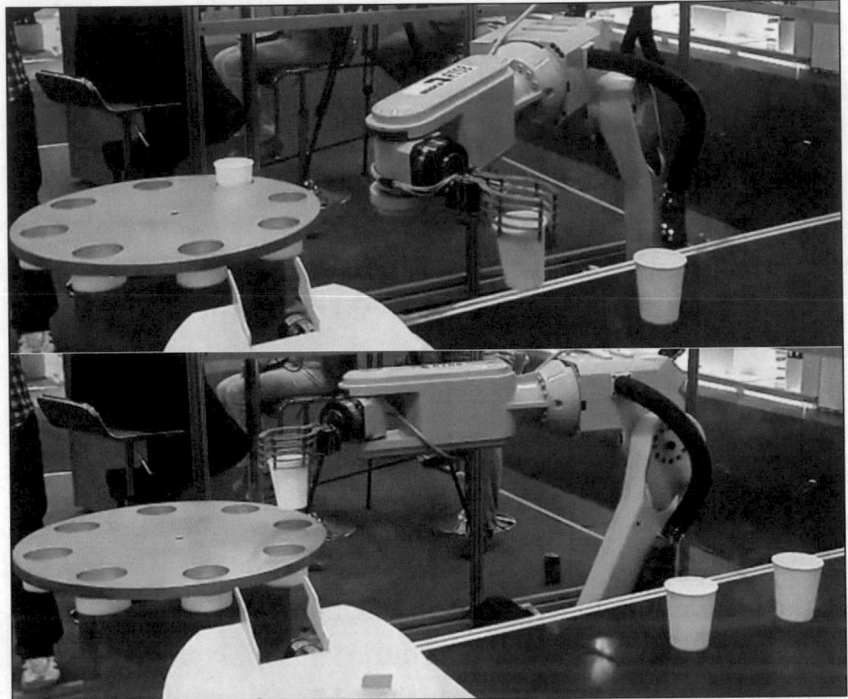

Figure 9-10. *A six-axes manipulator transferring parts from a*
rotating roundtable to a linear conveyor

In some applications, the movement of an auxiliary axis can directly
influence the actual position of the robot. Figure 9-11 shows the model of
a six-axes manipulator mounted on a linear actuator: the complete cell
turns into a seven-axes kinematic chain, which is actually a redundant
system, but with a much larger workspace than the single individual robot.
The assembly is typically driven either with simple PTP movements or by
programming path-interpolated movements in the moving base frame.

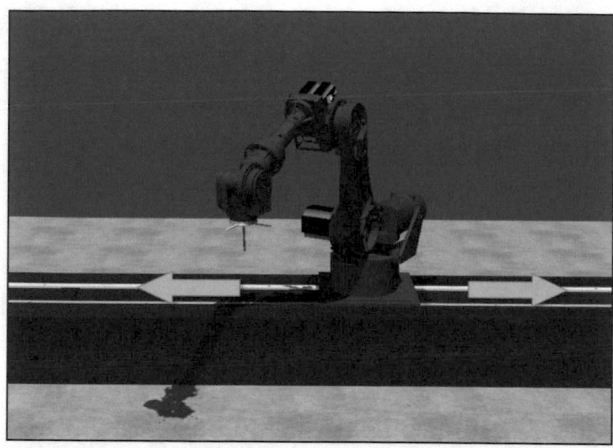

Figure 9-11. *A six-axes manipulator mounted on a linear actuator*

The calculations performed by the main controller are time-critical, and they must run on a real-time operative system. That requires a clock ticking at a specific frequency: a new set position value is generated by the controller at each time tick and synchronously sent to all the actuators.

The frequency of the controller depends on the required accuracy of the movements. Low-end systems run with a cycle time of $2\sim10ms$ and are typically deployed to control robots performing mostly slow PTP movements (e.g., large palletizing arms, spot welding applications). High-end systems run at a cycle time of $100\mu s\sim1ms$ and are used for fast and accurate requirements, where positioning precision along the whole path is critical (e.g., laser cutting applications).

The main controller program repeats all the time-critical calculations (trajectory generation, inverse transformations, workspace monitoring, etc.) once every cycle-time period. Therefore, it needs a quite powerful CPU to perform that task. The shorter the clock tick, the faster the CPU needs to be.

In fact, with the availability of more powerful CPUs relying on multiple cores, modern controllers are now often programmed to control the movements of several robots at the same time. The entire software control structure is distributed into an array of individual components, each of them responsible for one single robot. The advantage of such an architecture is that the entire software resources (points, frames, tools, workspace cells) are directly shared among all robots, removing all complex communication issues and making synchronizations between them much simpler to achieve.

Kernel Interface

All the main controller functionalities are typically encapsulated into a single software library, the core of robotics control, which can be accessed by the rest of the system via a well-defined interface. A common way of structuring the interface to a robotics library (and any motion library in general) is dividing it into three parts:

- **Commands**: The actions called by the application software to control the robot

- **Parameters**: The configuration of the robot and the commands

- **Monitor**: A detailed view of the current status of the robot

We show here a simplified example of the main components in a typical interface with a robotics control library: the commands are collected in Table 9-4.

Table 9-4. *Robotics library: commands structure*

Command	Description
Run program [*auto mode*]	Starts the execution of a user program, from the selected line number until the end line or end of file
Jog axis [*manual mode*]	Jogs an axis along a specific direction, either in the path space or in the joint space
Stop	Stops the current movement (both in automatic and manual mode)
Halt	Halts the current movement (only in automatic mode)
Continue	Continues a halted movement
Reference	Sets the position of the joints and auxiliary axes to the actual values read from the encoders
Reset	Resets all active errors
Calibrate	Runs the calculations for tool or frame calibrations
Identify	Runs the calculations for dynamic model identification

The reference (homing) command is typically executed when starting up the robot to let the motion functions know where the physical axes are positioned. Modifying the values of the joint axes causes a sudden jump on the path axes as well, so the application can only execute this command while the robot is not moving and also with the servo drives either switched off or disconnected.

All commands should only be executed after the robot interface has been configured with the correct parameters (shown in Table 9-5).

Table 9-5. *Robotics library: parameters structure*

Parameter	Description
Mechanics	Type of kinematic model (e.g., SCARA, Delta, six-axes Manipulator, etc.), links size, joints couplings, zero frame
Joint axes limits	Limit values for the position, speed, acceleration, and jerk of each joint axis
Auxiliary axes limits	Limit values for the position, speed, acceleration, and jerk of each auxiliary axis
Path space limits	Limit values for the path speed (linear and angular), acceleration, and jerk
Motor units	Conversion factor between the joint units and the actual motor units: includes gear ratio, encoder scaling, and direction
Workspace	Configuration of all the safe and forbidden zones, the orientation cone, the links capsules radii for self-collision and inter-collision detection
Calibration	An interface to the tool and frame calibration functions: accepts input values for recorded measurements and returns estimated frames positions and orientations
Dynamics	An interface to the dynamic model parameters and their identification function
Actual joint positions	Read directly from the encoders and used when referencing the axes at start-up
User program	Provides the interface to the user program that must be parsed by the interpreter. It can be input either as a text file or in tabular format

(*continued*)

Table 9-5. (*continued*)

Parameter	Description
Start/stop lines	The starting and ending lines for automatic program execution. By default, the whole program is normally executed
Single step mode	Specifies whether the program execution should be continuous or step-by-step
Simulation	Specifies whether the calculated axes positions should be sent to the servo drives or not
Override	A percentage speed reduction for testing purposes
Points	A list of target points taught by the operator and used during program execution. Each point specifies the target position of the robot axes, either in the joint space or in the path space. Auxiliary axes values can also be provided. Points given in the path space are valid in their corresponding workpiece frame
Frames	List of local workpiece frames taught using the frame calibration procedure and used during program execution
Tools	List of tools to be mounted on the robot and used during program execution. Their size represents the geometrical distance between the MP and the TCP and can be identified using the tool calibration procedure
Jogging configuration	An interface for the manual movement command: axis to be jogged (joint axis, path axis in base frame or tool frame, auxiliary axis), direction, speed, optional target position
Conveyor	Frame configurations of the conveyors to be tracked and real-time position to be followed by the robot

(*continued*)

Table 9-5. (*continued*)

Parameter	Description
Path corrections	External offset corrections provided by the application program and applied to the path axes. These values are not included in the path-planning phase, and care should be taken to avoid uncontrolled axes movements
Filter time	Specifies the filter time for the output position values sent to the servo drives. Larger values smooth out the movement dynamics but also cause larger deviations from the planned path

Finally, the results of the internal calculations are available in the monitor structure shown in Table 9-6.

Table 9-6. *Robotics library: monitor structure*

Variable	Description
Robot status	Current status of the robot (e.g., standstill, moving, halted, error, etc.)
Path status	Current TCP speed and position [X, Y, Z, A, B, C]
Joint axes status	Current speed and position of the joint axes
Auxiliary axes status	Current speed and position of the auxiliary axes
Set positions	Calculated motor positions to be sent to the servo drives They already include compensations for the gear ratios, encoder scaling and direction, homing offsets
Set torques	Calculated motor torque values to be sent to the servo drives
Line number	Shows the line number of the user program currently being executed in automatic mode

(*continued*)

Table 9-6. (*continued*)

Variable	Description
Block length	Shows the total length and already completed length of the currently executed movement
Target point	Shows the coordinates of the target point for the currently executed movement
Tool/frame	Shows the currently active tool/frame
Error	Shows the number and detailed information of any active error. Also, shows the line number that triggered the error in the user program

Servo Drives

Unless we work in simulation mode, all the set positions and torques generated by the core calculations in the main controller must be sent to the servo drives, which are then responsible for the actual movements of the motors.

While a few different actuators systems can be deployed to control robot's movements, we focus here on servo drives, because the overwhelming majority of industrial robots is actuated by servo systems. Possible alternatives could be brushed DC or stepper motors (for cheap and simple solutions) or hydraulic actuators (where very high forces are required).

Servo drives are essentially power inverters, which transform a constant DC bus voltage into a time-varying modulated voltage, with the goal of electronically commutating three-phase permanent-magnet synchronous (brushless) motors. The drives typically receive set values from the main controller and then use a closed-loop control system (servo loop) to calculate the voltage level that must be applied to each phase of the motor in order for it to generate the required torque and reach the desired position.

We will analyze the closed-loop structure and the commutation algorithms in the rest of this section, while the hardware components necessary to build a servo drive will be presented in Chapter 16.

The typical framework of a servo loop is a cascaded PID, as shown in Figure 9-12. The input control variables received from the main controller are the set position x_{set} and the (optional) set torque τ_{set}. The feedback variables received from the motor are the actual position x_{act}, the actual speed v_{act}, and the actual current i_{act}.

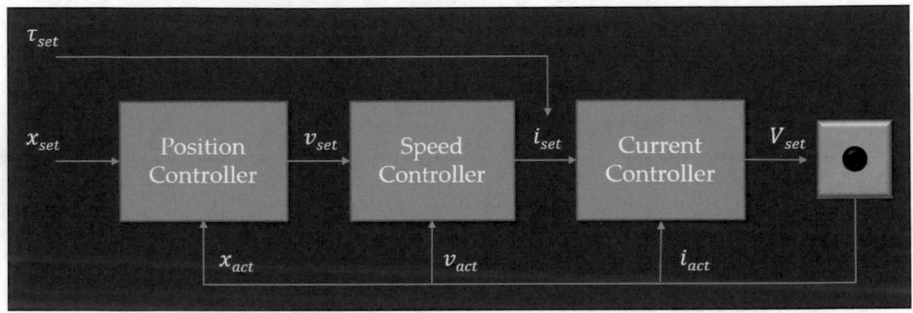

Figure 9-12. *Cascaded PID structure used for servo loop control*

The three individual controllers for position, speed, and current are usually implemented as PI controllers: They calculate the error e between the set and the actual values, and they generate an output correction y *proportional* to the error (P part) and to the *integration* of the error over time (I part):

$$y = k_p e + k_i \int e\, dt \qquad (9\text{-}1)$$

The constants k_p and k_i are called *gains* and determine how strongly the controller reacts to an input error. The goal of a controller is to keep the actual value as close as possible to the set value. As soon as the difference between them gets too large, the controller output increases accordingly and influences the motor, so that its actual state moves closer to the desired one.

Tuning the gains is not a simple task. A low gain causes the controller to respond too slowly and does not allow the motor to accurately follow the planned trajectory. The result is a large position error of the individual axis (also known as following or tracking error) and, consequently, a large error of the robot's TCP position. On the other hand, a high gain causes the controller to react too quickly and to possibly overshoot the set value, leading to an oscillatory behavior and, in extreme cases, to instabilities. Figure 9-13 shows the response of a proportional controller to a change in the set current value from 0 to 1A (purple curve), using either low or high gains (yellow and red curves, respectively).

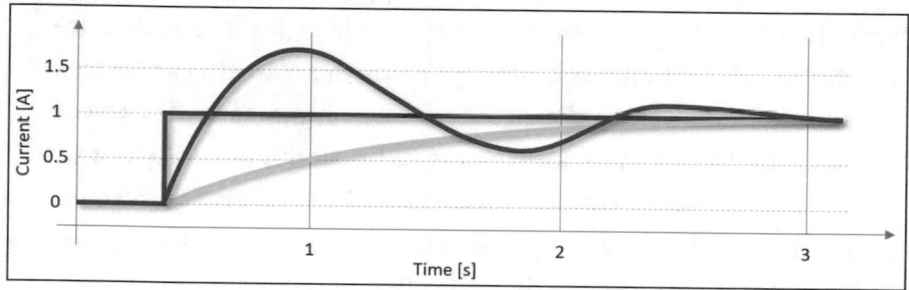

Figure 9-13. *Response of a P controller to a change in the set value according to gain settings*

Most modern servo drives provide autotuning functions to optimize the choice of gain values. We will discuss tuning procedures more in detail in Chapter 11 when describing the commissioning activities.

One drawback of a linear proportional correction is that the system often settles to a small steady error when constant external disturbances are applied. A hanging load on the motor is a typical example. In order to get rid of the residual error, we need to add an additional correction, which accounts not only for the instantaneous error but also for the history of the error over time. Such a function is called *integrator* because it integrates the error along time and generates a correction proportional to it. Even a small steady error will generate a large correction after some time and will eventually be eliminated.

Care should be taken when adding an integrating term, because its response can grow continuously over time and can generate too large corrections, which often lead to oscillations of the load. We either need to keep the integral gain k_i small or add a limitation to how large the total integration can be. The limiting function is called *anti-windup* and forces the integral correction to saturate at a predefined threshold to avoid instabilities.

PI controllers are not guaranteed to be optimal and actually not even to be stable. However, they are very simple to implement and are computationally very efficient, so they are the mainstream choice for motor control. One of their disadvantages is that they cannot deal well with delays between actions and feedback, but that is normally not an issue in motor control, because as soon as we send current to the coils, the motor will immediately react, and the encoder will follow along providing real-time position feedback to the controller.

There are a few ways to improve and optimize the PI controller. Firstly, using a series of cascaded controllers increases stability with respect to an individual one. That is why the common servo loop structure is based on three cascaded layers of corrections: position, speed, and current.

Then, the calculations should be executed very quickly in order for each controller to react as soon as possible to errors in the motor's status. Typical loop cycle times are in the order of $100\mu s$~$1ms$: at each cycle, the commanded set position is compared to the actual position of the motor, and a correction signal is generated and sent to the power stage.

A critical issue in practical implementations of servo loops is that the feedback signals should be clean of electric noise in order to avoid sudden jumps in the correction outputs and prevent unstable behaviors. Both position and current sensors typically employ hardware and software filters to guarantee smooth readings. Current sensing, in particular, needs to be of the highest possible accuracy to allow for high-quality control of the motor. Excessive filtering, however, removes high-frequency signal components and slows down the controller response.

Finally, the major drawback of a PID controller is that, by definition, it only generates an output correction when an input error is present. In other words, *an error must build up on the motor's actual status before the PID controller can start doing any work at all*: a PID alone cannot be the perfect motion control solution.

Let's consider a practical example: when a robotic axis is commanded to accelerate, its actual position starts lagging behind the set value, and the PID controller starts correcting the error by forcing more current through the motor. However, if we already know that a motor needs more current to accelerate, why not already use that knowledge and apply more current in advance, instead of waiting for the position error to build up?

The answer is that we can actually decide to apply a correction signal in advance, before waiting from the feedback from the encoder. The way to do that is use a **feed-forward** (open-loop) control signal in parallel to the standard closed-loop correction. The feed-forward value is essentially based on a prediction of how the system behaves when a set value is commanded. The more accurate that prediction, the better we can control the system before any error starts appearing.

Figure 9-12 shows a feed-forward signal τ_{set} being applied to the servo loop: the prediction is the torque value required by the motor to follow the commanded trajectory and calculated by the dynamic model based on the position, speed, and acceleration of all the robot's axes. The desired torque is transformed into an additive current value with the help of the motor's torque constant as in Equation (9-2) and then added to the output correction of the speed loop.

$$i = \tau / k_T \tag{9-2}$$

Being able to predict how much torque is needed by the motor to follow the input trajectory profile provides great help to the controller. If the dynamic model and its parameters are accurate enough, the PID will have little work to do to keep the loop errors small. The robot's movements will be more accurate and responsive. More importantly, the PID gains can be tuned softer, so that less oscillations and instabilities occur and the control loop will be more robust. In fact, since feed-forward control is an open-loop correction, it cannot add instabilities into the system.

In practical applications, the largest improvements after adding a torque feed-forward to the robot's controller can be noticed during accelerations, where set torque peak values are generated. Figure 9-14 shows a real-life example of angular error reduction in a robotic axis before and after adding torque feed-forward (blue and red curves, respectively).

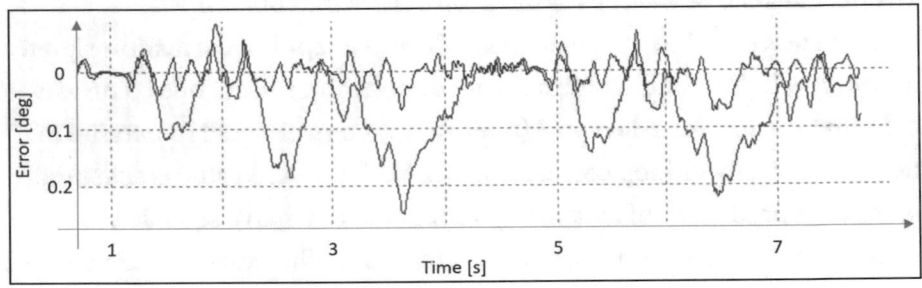

Figure 9-14. *Reduction of the position error of an axis after adding torque feed-forward control*

Torque feed-forward also plays a major role when the robot's axes are driven in torque control mode. Imagine you want to grab the robot with your hand and move it around, maybe to teach a trajectory. The position controllers of the axes must be switched off to avoid opposing your movements. However, the current control cannot be switched off; otherwise, the robot would collapse to the ground under its own weight: we still need torques from the motors to keep the robot standing up against gravity. The robot operates in a so-called zero-gravity mode, where the set torques calculated from the dynamic model are enough to fight off gravity, but do not stop the external additive force exerted by our hand. This procedure was already described with some more mathematical details in the Teach-By-Hand Section in Chapter 8.

The last stage of the servo loop is the current controller: its output is a voltage value, typically a percentage of the total available DC bus voltage, and is applied to the windings of the motor. In the next section, we are going to study in detail how that set voltage value is calculated.

Electronic Commutation

Commutation is the process of applying the correct voltage to each phase of the motor at the right time to make it move in the desired way. A brushed DC motor is mechanically commutated by means of brushes and only requires a simple DC voltage to run. Conversely, a brushless motor requires electronic commutation: the controller needs to dynamically calculate the voltage values for each phase depending on the set and actual states of the motor. (See Chapter 14 for more details on motors.)

We will study the hardware structure of electric motors in detail in Chapter 14. For now, we focus on the control algorithm to drive them. We first need to find a way to modulate the intensity of electric voltage via software, and then we need to calculate how much voltage each winding of the motor needs.

Let's start from the basics and increasingly build up from there. Imagine we have a constant source of DC voltage, either a battery or a power supply: Figure 9-15 shows how applying that constant DC voltage to a load can transform electrical power into heat (left), light (middle), and motion (right).

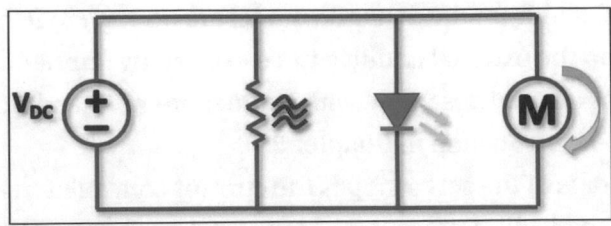

***Figure 9-15.** Applying a constant DC voltage to a resistor, an LED, and a motor generates heat, light, and motion*

Next, a more sophisticated application would require the generated output to be electronically controllable, so that we could tune the amount of heat, light, and speed produced by the load. The solution is to modulate the voltage applied by the DC source: for example, reducing the voltage across the LED dims its output luminosity.

The technique used to modulate voltage from a DC source is called **PWM** (pulse width modulation). Basically, instead of connecting the source to the load the entire time, we connect and disconnect it quickly in a series of ON/OFF pulses. If the frequency of the pulses is high enough (e.g., 20kHz), the ripples will average out generating the equivalent of a smooth DC voltage, whose actual value depends on the ratio between the ON and OFF pulse widths (hence the name pulse width modulation). For example, switching the signal 50% of the time ON and 50% OFF will generate an apparent voltage equal to half of the actual source voltage. The ON time duration as a percentage of the waveform period is called *duty cycle.* Figure 9-16 shows the effect of different PWM duty cycles: the real switched signal is shown in purple and the resulting apparent average signal in green.

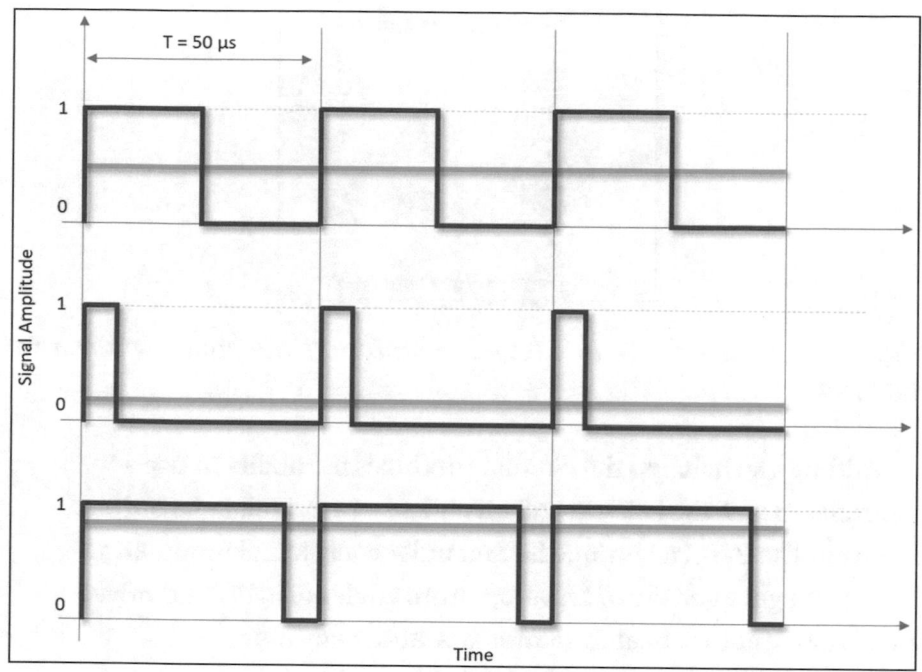

Figure 9-16. *Pulse width modulation is used to reduce voltage applied to the load*

In practice, when driving high inductive loads such as motor windings, the PWM voltage waveform generates relatively smooth current profiles because of the filtering effect of the coil inductance.

The high-speed ON/OFF switching is accomplished using transistors. We will study these components more in detail in Chapter 16, but for now you can think of them simply as electronically controlled switches, as shown in Figure 9-17.

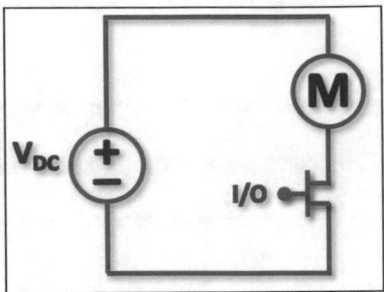

Figure 9-17. *Switches can be used to modulate the power intensity at the load*

Adding a switch into the circuit introduces the ability to use a controller to turn the load on and off and also to modulate its output intensity. The PWM technique is used universally for all modulating purposes: light intensity of LEDs, aperture angles of hydraulic valves, speed of DC motors, heat from resistors, and many more.

Opening the switch turns the load electrically off: in the case of a motor, the switch cuts it off from the power supply, not allowing any more speed to be actively generated. However, that does not mean that the motor immediately stops: the mechanical inertia of the driven link would keep the motor rotating until friction slowly stops it. Such an uncontrolled behavior of a mechanical axis is called *coasting* and in most applications is not desirable and often not even safe.

A better way to quickly stop a DC motor is shorting its terminals. The operation is called *dynamic braking* and forces a rapid deceleration of the motor, for reasons we will study later in Chapter 14. The energy stored in the motor then dissipates as the current slowly decays burning into heat. In order to add this braking feature, we need to introduce a second switch in the circuit, as shown in Figure 9-18.

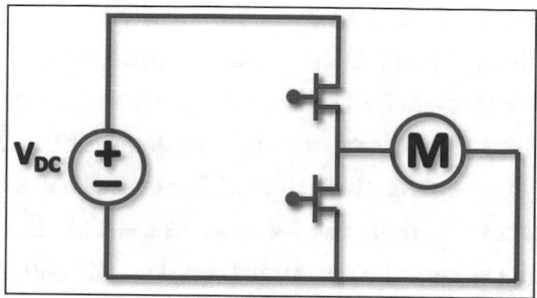

Figure 9-18. *Two switches can drive and brake a DC motor*

Although we are now able to quickly brake the motor, we are still missing a major operation requirement: inverting the direction of rotation. Applications like fans, blenders, and motorized pumps only need to rotate in one direction, and the two-switch circuit would suffice. But robot's axes definitely require a two-direction operation, which forces us to introduce two extra switches in the circuit and build a so-called **H-bridge** (because its shape looks like the letter "H"). Figure 9-19 shows an H-bridge: we call one leg A (switches A1 and A2) and the other leg B (switches B1 and B2).

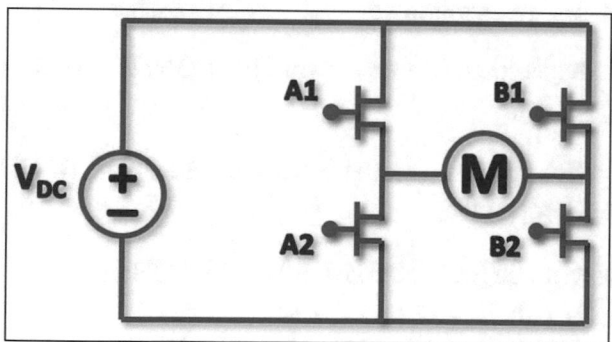

Figure 9-19. *An H-bridge can be used to control speed and direction of a DC motor*

Using different combinations of the four switches, we can now control both speed and direction of the DC motor. Figure 9-20 shows the most important states of operation of the bridge: positive voltage applied to the motor (left), negative voltage applied to the motor (middle), and motor terminals shorted together (right). The task of a motor driver is to activate the correct switches at the correct time, so that the motor receives the desired voltage value at its terminal, based on the calculations of the servo loop.

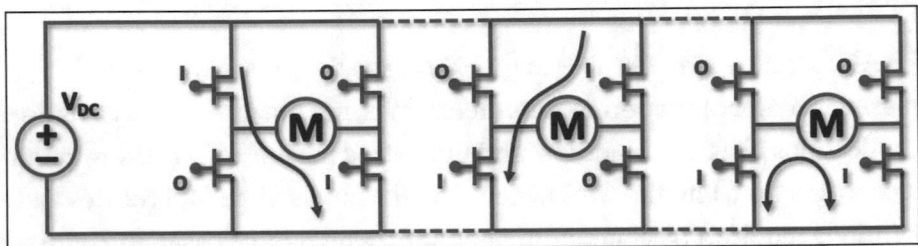

Figure 9-20. *Some states of operation of an H-bridge*

If you decide to use brushed DC motors in your robot, then you can program an H-bridge as follows:

- Positive voltage (rotate forward): A1 ON; B2 PWM; A2 and B1 OFF

- Negative voltage (rotate backward): A2 PWM; B1 ON; A1 and B2 OFF

- Zero voltage (quick brake): A2 and B2 ON; A1 and B1 OFF

Make sure you never activate the two switches in the same leg of the bridge at the same time (e.g., A1 and A2 both ON), because that would short the DC power supply to ground with unpleasant consequences. In practical cases, such an operation is usually overridden by the motor driver chip (see Gate Driver Section in Chapter 16).

The dynamic braking operation of shorting the motor terminals has the effect of burning off the energy stored in the motor into heat. A more efficient braking technique, called *fast decay*, consists in turning the bridge on in reverse, against the actual motor rotation. The effect is that the stored energy is regenerated back into the supply. Sure enough, that is only possible if the supply allows it, as in the case of a rechargeable battery. Using a fixed power supply will usually not work and would actually lead to damaged equipment, because power supplies do not typically allow reverse current flow, unless they are specifically built to regenerate energy into the power grid.

While brushed DC motors are convenient to drive, their mechanical commutation reduces their lifespan, makes them less efficient, and produces an output torque that is not very smooth (an effect described as *torque ripples*). Brushless motors avoid those problems using electronic commutation: the voltage applied to each phase is continually modulated by the motor driver in order to maximize efficiency and optimize the output torque profile.

Since this kind of electric motor has three phases (UVW), we need to expand the driving bridge by adding an extra leg: a total of six switches are now needed to drive the motor (see Figure 9-21).

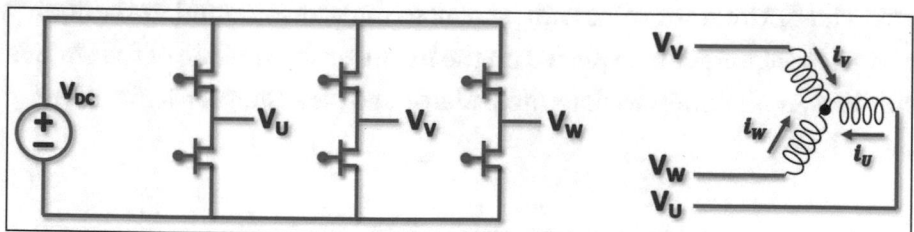

Figure 9-21. *A three-leg bridge can drive a three-phase brushless motor*

Each leg of the bridge generates an independent voltage value with respect to ground (V_U, V_V, V_W) for each phase of the motor. The minimum possible phase voltage is 0 (high switch open and low switch closed), while the maximum possible phase voltage is the supply voltage V_{DC} (high switch closed and low switch open). Any voltage in between can be generated by PWM switching.

The phase-to-phase voltage difference determines the amount of electric current flowing through the windings. For example, if all three-phase voltages are held at the same value (anything between $0V$ and V_{DC}), the three-phase currents are all 0.

Since all the windings are connected to each other and are electrically isolated from the rest of the system, the sum of the three-phase currents must always be 0 (*Kirchhoff's current law*):

$$i_u + i_v + i_w = 0 \qquad (9\text{-}3)$$

For example, if phase U is held high at V_{DC}, while phases V and W are held low at ground, then i_U is positive, while both i_V and i_W are negative and their absolute magnitude is equal to half of i_U.

Brushless motors are typically built using phase windings on the stator and permanent magnets on the rotor (see Figure 9-22 for a simplified cross view). The rotor spins with respect to the stator around the central axis. The magnets can be placed on the inside or the outside of the stator's windings; both configurations are widely used (see Chapter 14 for more details).

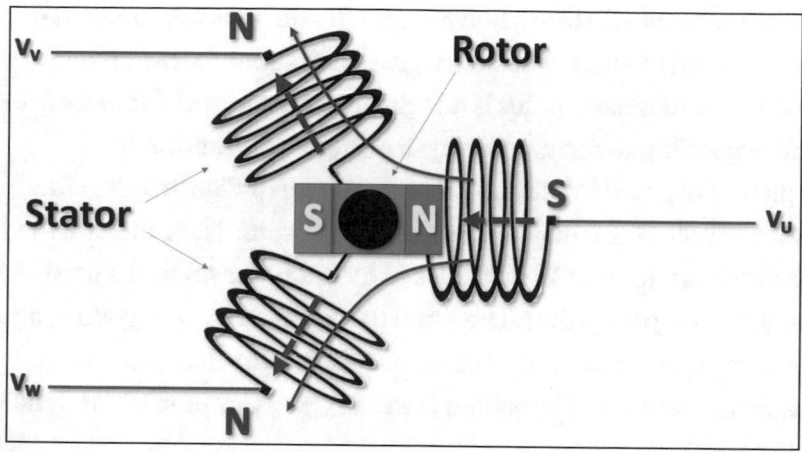

Figure 9-22. *Simplified cross section of a permanent-magnet synchronous motor showing the stator windings and the rotor magnets*

The three windings are oriented at 120 degrees from each other. A current flowing through a winding generates a magnetic flux in the radial direction (shown in red in the drawing). The combined effect of the three windings generates a total magnetic flux equal to the vectorial sum of the three individual components and causes the permanent magnets to rotate and orient themselves along the induced magnetic field. For example, applying a constant voltage at phase U while keeping phases V and W at ground causes a positive current to flow from U to V and W (shown in purple in the drawing) and forces the magnets to align in the same direction of the U winding. This procedure is used as a simple way to find the zero magnetic angle of the motor, which is typically chosen as the direction of the magnetic axis for phase U.

It should be now clear that the problem of positioning the rotor at a certain desired angle requires adjusting the individual phase currents so that the correct magnetic flux vector is generated.

In practice, motors have a large number of UVW windings connected in parallel to each other and a large number of magnetic poles, in order to reduce the actual distance between them and generate smoother movements and torques. The 360 degrees of rotation between two consecutive N magnetic poles is effectively a measure of the *electric angle*, not corresponding to the actual physical angle of rotation.

A picture of a real brushless motor is shown in Figure 9-23, where the stator windings are in the inside, while the rotor is on the outside. This motor has a total of 18 windings (6 for each phase) and a total of 20 magnets (10 N-S pole pairs). The three thick cables (green, yellow, and blue) are the UVW wirings that carry the phase currents. The bunch of thin cables is used for the position feedback (see Chapter 15 for details on encoders).

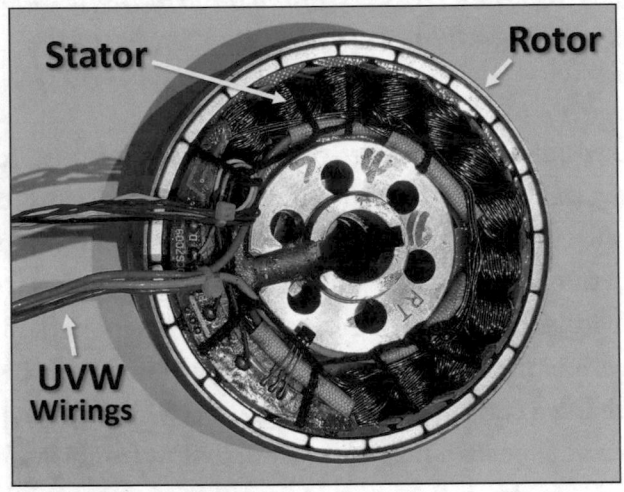

Figure 9-23. *Picture of a brushless motor showing the stator windings and rotor magnets*

The magnetic flux generated by each winding is proportional to the current flowing through it. The problem of determining the total generated flux is therefore equivalent to calculating the total current flowing through the motor.

We start by defining the current vector of each phase as a geometrical quantity, whose amplitude corresponds to the magnitude of the electric current flowing into the winding, while its orientation is the electric angle of the phase itself. In the example shown in Figure 9-24, the current flows into phase W and out of phases U and V.

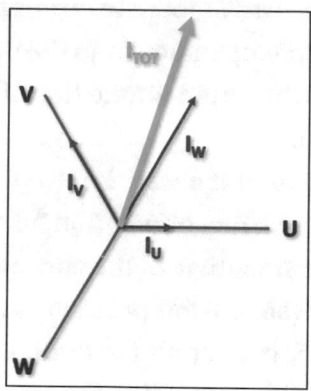

Figure 9-24. *Stator current state vector*

The sum of the three current vectors is the stator *current state vector*:

$$I_{tot} = I_u + I_v + I_w \qquad (9\text{-}4)$$

By modulating the magnitude of the individual current vectors, we can modulate the amplitude and direction of the total current vector and consequently that of the induced magnetic field. The question is that what current vector should we apply to the stator windings in order to let the permanent magnets of the rotor move with the commanded torque? In particular, we need to determine the required orientation and amplitude of the current vector given the actual position of the rotor and given the desired output torque.

We know from Lorentz's law that the force *F* acting between an external magnetic field (*B*, from the rotor's permanent magnets) and an electrically induced magnetic field (from the current *I* through the stator's windings of length L) is maximum when the two are perpendicular to each other ($\theta = 90$):

$$F = BIL \sin \theta \qquad\qquad (9\text{-}5)$$

In other words, to maximize the motor's efficiency and generate maximum torque with the least amount of current, we should always keep the induced magnetic field perpendicular to the permanent magnets. That will also guarantee a smooth output torque free of ripples, regardless of the rotor's position and speed.

The orientation of the rotor is always known from the motor's encoder feedback. That provides the target orientation for the current vector. On the other hand, the target amplitude of the current vector simply depends on how much torque we want the motor to generate and is dictated by the output of the speed controller plus the optional torque feed-forward component (see Figure 9-12). Remember that torque and current are directly proportional to each other.

The motor driving technique of modulating the total stator current vector is called *vector control* or *field-oriented control* (**FOC**) and is the control algorithm implemented in all standard industrial servo drives, because it is the most efficient way to produce the highest and smoothest torque out of a motor.

Taking the rotor coordinate system as the reference frame, the total stator current vector generated by the three stator windings can be decomposed into two components: the direct and quadrature components. The same stator current vector generates different components on the rotor according to the relative orientation between the two (see Figure 9-25).

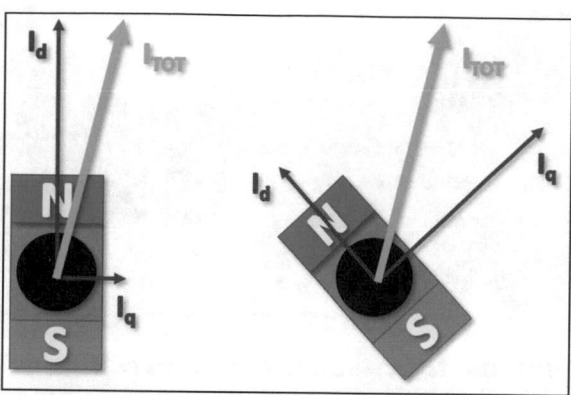

Figure 9-25. *Direct and quadrature components of the same current vector for different orientations of the rotor*

The direct current component (I_d) is parallel to the rotor and does not generate any torque: it only generates useless compression axial forces, reduces motor efficiency, generates heat, and, on the long run, wears out the motor's shaft. Needless to say, it must be minimized.

On the other hand, the quadrature current component (I_q) is perpendicular to the rotor and generates the torque that makes the motor move. Its value should be equal to the set current commanded to the current controller.

The internal structure of the current controller can now be reshaped using two parallel controllers: one for the quadrature component and one for the direct component, as shown in Figure 9-26. The set value for the direct current is always 0.

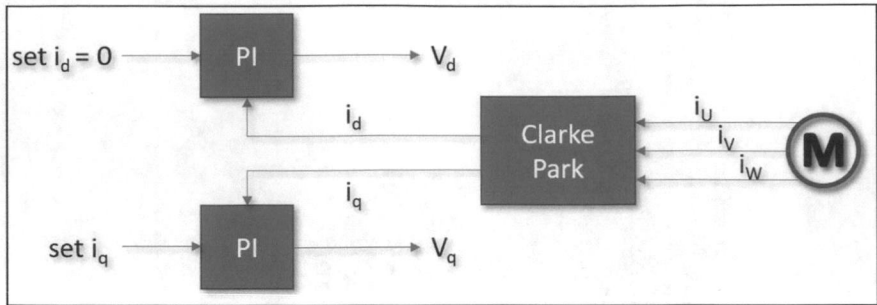

Figure 9-26. *Internal structure of the current controller stage of the servo loop*

Instead of working with the phase currents directly, the control algorithm focuses on their effects on the rotor. Therefore, the key idea of the FOC is to transform the problem from a time-variant control (the UVW currents are always changing) into a time-invariant control (the d-q current vectors are static DC quantities). That is achieved by switching the controller's reference frame from the static stator frame to the spinning rotor frame: from the point of view of the rotor's frame, the current's components are stationary, regardless of the actual rotational speed of the motor. The advantage is that the required bandwidth (refresh speed) of the PI controller can be much lower even for high-speed motors. We turned a difficult high-frequency AC control problem into a much simpler stationary DC control problem.

The translation of the current values from the static frame into the rotating frame is performed using a set of transformations called **Clarke-Park** (see Figure 9-27).

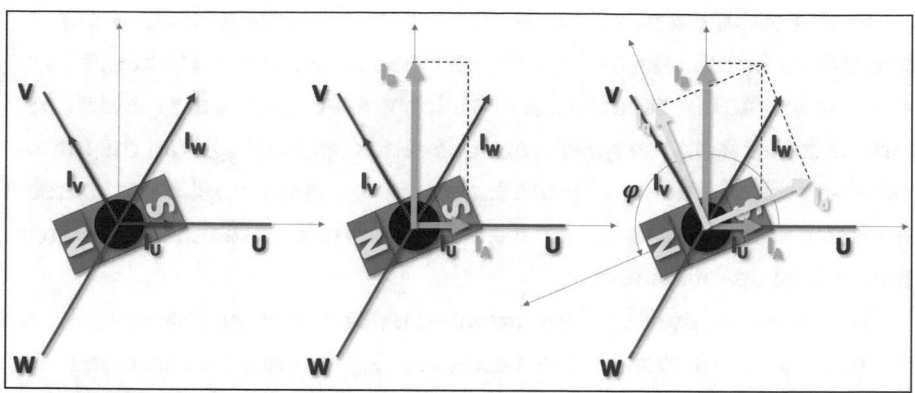

Figure 9-27. *The Clarke-Park transformations: from the fixed stator frame to the moving rotor's frame*

The first step consists in expressing the total current vector formed by the three windings currents (i_u, i_v, i_w) with only two components (i_a, i_b). The operation is possible because the three-phase currents are mutually dependent on each other as shown in Equation (9-3). Proportionality factors can be safely left out and will be compensated by the controller gains.

$$\begin{cases} i_a = i_u \\ i_b = \dfrac{1}{\sqrt{3}}(i_u + 2\,i_w) \end{cases} \qquad (9\text{-}6)$$

The next step is a simple frame rotation, which depends on the actual rotor orientation. The angle ϕ is measured by the encoder and must be transformed into electric angle units (i.e., the encoder angle times the number of magnetic pole pairs per motor revolution).

$$\begin{cases} i_d = i_a \cos\varphi + i_b \sin\varphi \\ i_q = -i_a \sin\varphi + i_b \cos\varphi \end{cases} \qquad (9\text{-}7)$$

At this point, the two PI controllers can start working, as shown in Figure 9-26, by reacting to the error between set and actual values. The set value for the direct current is 0, while the set value for the quadrature current comes from the speed controller and (optionally) from the torque feed-forward. The actual *d-q* current components can be slightly filtered if needed, before being fed to the respective controllers, to make the system more robust against noise.

The output of the PI controllers are the direct and quadrature components of the total voltage vector (V_d, V_q) that must be applied to the motor. This voltage vector is often called the reference voltage V_{ref}. The amplitude and angle of V_{ref} in the rotor frame are as follows:

$$
\begin{cases}
\left| V_{ref} \right| = \sqrt{V_d^2 + V_q^2} \\
\angle V_{ref} = \text{atan2}\left(V_q, V_d \right)
\end{cases}
\tag{9-8}
$$

Since the rotor frame is rotated by ϕ from the stator frame, we can calculate the orientation of the voltage vector in the stator frame as follows:

$$
\angle V_{ref} = \text{atan2}\left(V_q, V_d \right) + \varphi
\tag{9-9}
$$

The last step of the FOC consists in calculating the correct PWM values for the six switches of the driving bridge, in order to apply the correct amplitude and angle of the total voltage vector to the motor.

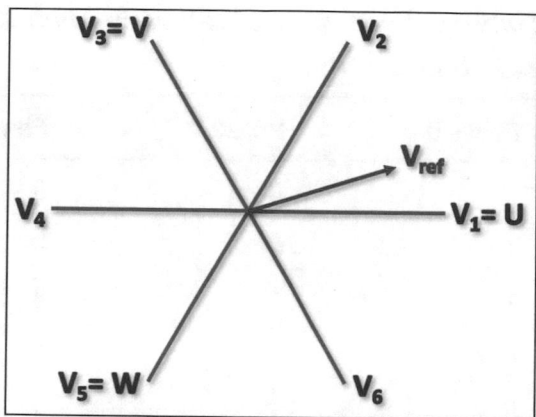

Figure 9-28. *Generating a voltage vector using the six available switches of the bridge*

Figure 9-28 shows how to apply a generic voltage vector to the motor using the six available switches of the bridge. Let's start from a simple case: if the required (electric) angle of the voltage vector is 0 ($\angle V_{ref} = 0$), then the voltage direction that we must apply to the motor corresponds to that of phase U (vector V_1 in the drawing). We switch phase U high (using PWM to set its amplitude equal to the desired voltage), while phases V and W are kept low (tied to ground).

Another simple situation is when the required electric angle is 60 degrees (vector V_2 in the drawing): in that case, both U and V phases are held high (with equal amplitude), while phase W is kept low. The pattern should be clear now: using a linear combination of UVW voltages, we can generate a generic voltage vector with any kind of amplitude and orientation.

The six main voltage vectors of the hexagon ($V_1 ... V_6$) can be generated by combinations of the individual phase voltages, as shown in Table 9-7, where 1 means that the phase is pulled high to the power supply voltage (high-side switch turned on), while 0 means that the phase is pulled low to ground (low-side switch turned on).

Table 9-7. *The main voltage vectors built as combinations of individual phase voltages*

Vector	Phase U	Phase V	Phase W
V_0	0	0	0
V_1	1	0	0
V_2	1	1	0
V_3	0	1	0
V_4	0	1	1
V_5	0	0	1
V_6	1	0	1
V_7	1	1	1

Switching all three phases low (or, equivalently, all three phases high) causes no current to flow through the windings because all phase-to-phase voltage differences are 0. These two vectors (V_0 and V_7) are used to reduce the amplitude of the total voltage vector.

In conclusion, any required V_{ref} can be generated using a linear combination of $V_0...V_7$: we first find the sector ($s = 1...6$) in which V_{ref} lies; then we calculate the UVW duty cycles using the two adjacent voltage vectors, plus V_0 (or V_7) if needed. Since the PWM duty cycles must be between 0 and 100%, the voltage values should be normalized with respect to the DC bus voltage level.

$$\begin{cases} T_1 = \sin\left(s * 60 - \angle V_{ref}\right)\left|V_{ref}\right| / V_{DC} \\ T_2 = \sin\left(\angle V_{ref} - (s-1) * 60\right)\left|V_{ref}\right| / V_{DC} \end{cases} \tag{9-10}$$

T_1 is the duty cycle of the vector preceding V_{ref}, while T_1 is the duty cycle of the vector coming after V_{ref}. For example, if V_{ref} lies in the first sector, T_1 and T_2 are the duty cycles of V_1 and V_2, respectively. Since V_1 is built using phase U, and V_2 is built using phases U and V together, we can apply the following PWM duty cycles to the three motor windings:

$$\begin{cases} U = T_1 + T_2 \\ \quad V = T_2 \\ \quad W = 0 \end{cases} \tag{9-11}$$

Another example: If V_{ref} lies in the third sector, then we need to combine vectors V_3 and V_4, which means a higher voltage on the V phase, some voltage on the W phase, and the U phase always grounded:

$$\begin{cases} \quad U = 0 \\ V = T_1 + T_2 \\ \quad W = T_2 \end{cases} \tag{9-12}$$

There are further tricks to optimize switching losses by using clever alternations of V_0 and V_7 for the deadtimes and using centered-based PWM signals. However, the general concept of the FOC works just fine applying the duty cycles in Equation (9-10) to the three phases of the motor according to the required orientation of the reference voltage vector. The phase voltages result is an induced magnetic field perpendicular to the rotor, which generates a total motor torque equal to the required value.

Summary

Writing a complete firmware package for a robot is a complex task that calls for a variety of skills: from front-end HMI, which needs to be captivating and user-friendly, to back-end motion control, which requires high accuracy and precise synchronization between all tasks.

In this chapter, we went through the main structure of a typical commercial robotics suite, showing what functions are commonly offered by the manufacturers and expected by the end user. Further customization is then required according to the specific industry for which the robot is targeted (e.g., adding a palletizing function or an interface for a welding gun).

We also described in detail how to transform motion trajectories into actual motor's movements, with the help of PID controllers at the core of a servo drive, down to electronic commutation to generate the voltage levels for the motor's phases.

In the next chapters, we will explore what additional tweaks are required in practice before the robot can actually be used in a real application.

CHAPTER 10

Calibration

We have now completed the development of our robotics control software, and we are ready to sell it. In theory, the customer could quickly insert the right parameters in the configuration pages, and the robot should work correctly. In practice, however, the process is not that straightforward.

There are a large number of parameters needed for the internal calculations: mechanical dimensions for the kinematic model, tool size, base coordinate system, etc. No matter how realistic our mathematical models are, if the inserted parameters are not entirely correct, then the results of the calculations will not be accurate, and the robot will not move to the required pose. We need to introduce a procedure to fine-tune the parameters in order to increase the robot's movement accuracy.

It is, of course, still possible to work with an uncalibrated robot, but there are limitations to its practical use: it cannot be programmed offline; the absolute positioning of the TCP will not be correct; and, more critically, movements will be distorted, as lines will not be straight anymore.

The reasons for inaccuracies are twofold: on the one hand, the operator does not always have the exact values for all parameters and sometimes provides only a rough estimate. A typical example is when mounting a third-party tool, which does not come with exact mechanical dimensions. On the other hand, it is also likely that the available parameters lack accuracy because of natural deviations in the production process and in the assembly phase, all of which introduce small errors within different degrees of tolerance. In other words, the real parameters are not exactly equal to the ideal (nominal) values.

© Fabrizio Frigeni 2023
F. Frigeni, *Industrial Robotics Control*, Maker Innovations Series,
https://doi.org/10.1007/978-1-4842-8989-1_10

Another common inaccuracy issue is the zero position of the encoders. Encoders are mounted at random angles on the motors, and their output value needs to be zeroed at the home position of the robot. This operation is normally performed manually by the operator and introduces errors.

Since the robot is a serial kinematic chain, the errors of each axis add up and are amplified from the first joint all the way up to the TCP. We need to provide a calibration procedure to compensate for these errors and identify a better set of parameters to improve the positioning accuracy of the robot.

There are typically three sets of parameters we can calibrate: the body of the robot, the mounted tool, and the workspace cell. Let's explore each of them in the next sections.

Robot Calibration

The first calibration step concerns the robot's body itself: we want to fine-tune the mechanical sizes of links, the angles between joints, and the encoder zero offsets. These parameters are used by the kinematic model and might be inaccurate because of building and assembling tolerances. Importantly, they are independent of the application for which the robot will be deployed and are typically calibrated by the manufacturer before the robot is shipped to the end user.

No further adjustments are then needed during normal operation because the mechanical system will not change anymore, unless very high accuracy is required (e.g., compensating for real-time temperature variations) or unless the robot suffers a major collision that causes mechanical misalignment.

The procedure to calibrate the mechanical parameters is shown in Figure 10-1. We first generate the nominal TCP position (TCP_{ideal}) using the kinematic model and the initial set of parameters. Then we utilize an external high-precision measuring device, normally a laser tracker with a special target mounted on the robot, and we measure the actual TCP position (TCP_{real}). The error $\Delta = TCP_{ideal} - TCP_{real}$ provides feedback to modify the parameters of the model and improve the output. For best results, several measurements in different poses are required. The optimal solution is usually found through an iterative least-square approach. We will describe the mathematical details of this method in the next section when dealing with tool calibration.

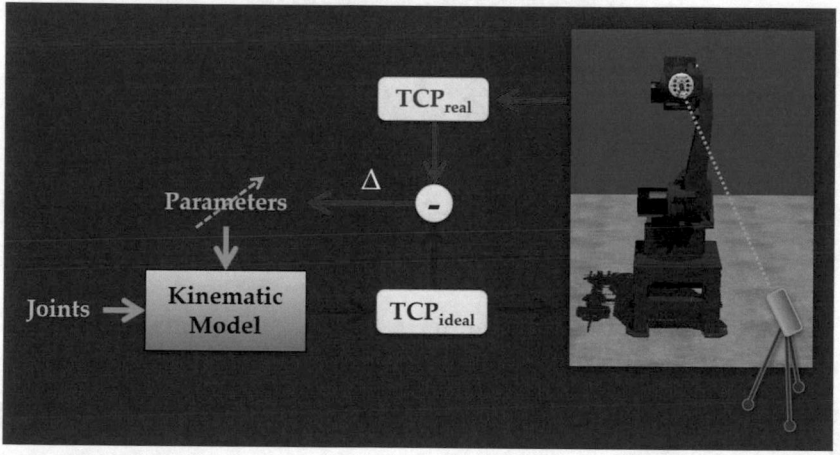

Figure 10-1. *Fine-tuning the kinematic parameters according to a high-precision measuring feedback*

There is, however, a subtle problem. Modifying the parameters alone might never lead to the most accurate solution. The reason is that the mathematical model itself might not be completely correct. Besides fine-tuning the parameters, we also need to calibrate the model.

Imagine, for instance, that the second joint of the robot is mounted with respect to the first joint at a nonzero angle around the X axis, because of assembly errors. The condition is shown (purposely exaggerated) in Figure 10-2.

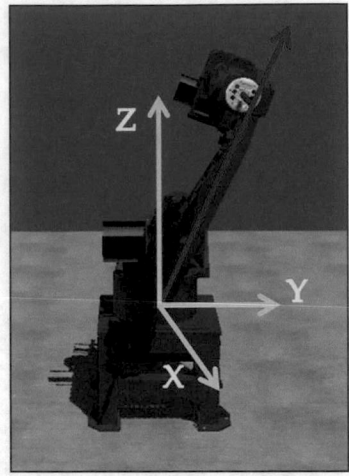

Figure 10-2. *Calibrating parameters does not help if the underlying model is wrong*

In that case the underlying model is wrong, and, no matter how well we tune the existing parameters, we cannot reach a good positioning accuracy. The original model simply did not consider the presence of an angle at that position.

We could build a mathematical model contemplating all possible compensations, but the math would quickly get extremely complicated, and the inverse transformations will likely not have a closed-form solution. A better approach is to keep using the original model and add an external compensation function in parallel to it, where all possible errors adjustments are accounted for (see Figure 10-3). The sum of the two outputs from the kinematic model and compensation model will result in a more accurate positioning value.

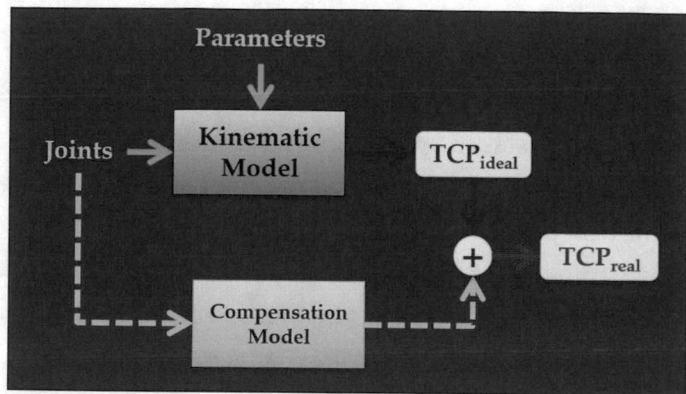

Figure 10-3. *Adding a compensation function in parallel to the original kinematic model*

The compensation model is typically highly nonlinear and cannot always be expressed in the form of analytical equations. To solve and calibrate it numerically, we turn it into a machine learning problem using a simple neural network as a function approximator.

This solution is very practical: by converting a model identification problem into a simple supervised learning problem, the network can be tuned with standard machine learning techniques. We collect a large dataset, where the training labels are given by the difference between the measured and the predicted TCP values, and then we run a gradient descent optimizer to tune the network's weights. The result will compensate for all possible mechanical variations from the standard kinematic model that would have been much harder to derive analytically.

In practice, this procedure is actually applied to the inverse kinematic model: the trajectory generator provides target TCP values, which are then transformed into target joint values and corrected with the addition of the error compensation, before being sent to the servo drives.

Tool Calibration

Once the robot's body is calibrated, it can be shipped to the customer and begin operations. In practical applications, the operator needs to mount different tools for different purposes and parameterize the tool dimensions in the configuration page of the control software. However, the actual size of the tool could be unknown or only available with an approximate accuracy. We describe here a simple and commonly used approach to numerically identify the size of the tool using only manual jogging operations.

Consider Figure 10-4, which shows the robot with an attached tool of unknown size. We first manually jog the TCP to a random fixed position, reaching it from several different orientations. Then we observe that, as we continue modifying the configuration of the robot, the mounting point will always be located on a spherical surface, as long as the final TCP position is not changed.

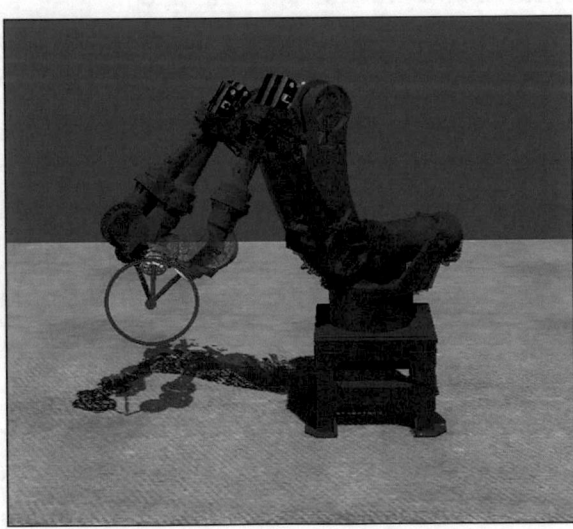

Figure 10-4. *Procedure to identify the tool size*

Now, if we collect enough points from the surface of the sphere, we can automatically derive its center point and its radius. Mathematically, at least five points are required to solve the problem. In practice, the more points we measure, the more accurate the final estimation will be. Taking too many points will also increase computational load, but that is not something to worry about, since tool calibration is normally done only once for each new tool mounted on the robot.

The complete procedure works according to the following steps:

- Move the robot to reach the same TCP position from several different orientations.

- Record the corresponding positions of the mounting point.

- Build a system of equations from all the recorded positions.

- Solve the system using least-square estimation and find the TCP position.

- Calculate the tool size using the equivalence
  $T = TCP - MP$.

The first step is easy: we need to position the TCP always in the same point, but with different orientations. Since we perform manual jogging movements, the accuracy will not be very high; therefore, we should collect a large number of measurement points to average out errors. Five points is the minimum requirement.

We report a numerical example here, using the simulated robot shown in Figure 10-4, which has a mounted tool measuring 380mm along the X axis. The collected measurements of the MP positions are reported in Table 10-1.

Table 10-1. *Example of recorded measurements of the MP position*

	P1	P2	P3	P4	P5
X	1577.256	1410.702	1659.391	2042.116	1873.821
Y	1243.340	1078.277	749.691	988.606	1234.896
Z	1192.628	1067.784	1067.892	1125.491	1221.280
A	-70.356	-41.706	81.321	86.419	155.174
B	50.789	26.505	26.524	36.737	58.236
C	-43.760	-0.183	74.411	163.088	-127.965

After collecting the measurements, we can use them to build a system of equations. Recall that all the positions of the mounting point are located on a sphere or radius R, whose center is the TCP position. That translates into the following equation:

$$R^2 = (MP - TCP)^2 \tag{10-1}$$

The radius R is the geometrical distance between the MP and the TCP, but it does not say much about the exact tool dimensions, which are generally distributed along the three axes [X, Y, Z] of the tool frame.

All the recorded measurements will satisfy Equation (10-1). Since we recorded five points, we have five of these equations.

$$\left(MP_{X_i} - TCP_X\right)^2 + \left(MP_{Y_i} - TCP_Y\right)^2 + \left(MP_{Z_i} - TCP_Z\right)^2 = R^2 \tag{10-2}$$

All the MP positions $[MP_X, MP_Y, MP_Z]$ are known, while the TCP positions $[TCP_X, TCP_Y, TCP_Z]$ and the radius R are unknown. We let the index i range from 0 to m, which means we have a total of $(m + 1)$ collected measurements.

The equation is based on geometrical properties and does not depend on the chosen TCP position, which means we can select any convenient position in space to perform the calibration procedure.

Next, we build a simplified system of linear equations: we subtract each equation with index $i = 1 .. m$ from the first one ($i = 0$) so that the R^2 element on the right side of the equations disappears:

$$TCP_X a_{1_i} + TCP_Y a_{2_i} + TCP_Z a_{3_i} = b_i \tag{10-3}$$

We end up with m equations, where all coefficients a_i and b_i are known:

$$a_{1_i} = \left[2\left(MP_{X_0} - MP_{X_i} \right) \right] \tag{10-4}$$

$$a_{2_i} = \left[2\left(MP_{Y_0} - MP_{Y_i} \right) \right] \tag{10-5}$$

$$a_{3_i} = \left[2\left(MP_{Z_0} - MP_{Z_i} \right) \right] \tag{10-6}$$

$$b_i = MP_{X_0}{}^2 - MP_{X_i}{}^2 + MP_{Y_0}{}^2 - MP_{Y_i}{}^2 + MP_{Z_0}{}^2 - MP_{Z_i}{}^2 \tag{10-7}$$

We are left with $n = 3$ unknowns: $[TCP_X, TCP_Y, TCP_Z]$, i.e., the position of the TCP. In theory, solving a system of three unknowns requires three equations. That would give us a unique solution, but likely not a very accurate one. Collecting more points generates more equations and creates an overdetermined system, which actually does not have a unique exact mathematical solution but can provide a better estimation of the real geometrical solution. That is the reason why we need at least $m \geq 4$ equations and therefore at least five measurement points.

The system in Equation (10-3) can be rewritten in compact form using a matrix A and a vector b.

$$Ax = b \tag{10-8}$$

The standard way to solve this system in mathematical terms is to turn it into a least-square minimization problem.

Let's rewrite the system as follows:

$$b - Ax = 0 \tag{10-9}$$

We mentioned that there is no unique exact solution, but we want to find the tuple $[TCP_X, TCP_Y, TCP_Z]$ for which the expression on the left side of Equation (10-9) is minimum, i.e., as close as possible to 0. That is equivalent to minimizing the cost function defined as the Euclidean norm of the residual vector:

$$\|b - Ax\|^2 = \sum_m \left| b_i - \sum_n a_{ij} x_i \right|^2 \tag{10-10}$$

That is a typical least-square problem, which actually has an analytical closed-form solution:

$$\hat{x} = min \|b - Ax\|^2 = \left(A^T A \right)^{-1} A^T b \tag{10-11}$$

The resulting $\hat{x}$ is the optimal value that minimizes the cost function. Unfortunately, despite being directly solvable in closed-form, this approach involves a matrix inversion, which is computationally expensive and, most critically, unstable.

There are a few alternative ways to solve the linear system, which usually involve some tricks to rewrite the system so that a matrix inversion in the calculations can be avoided. One powerful way of doing it is called the *QR factorization method.*

We decompose A as the product of two other matrices, Q and R, and rewrite the system in Equation (10-8) as follows:

$$QRx = b \tag{10-12}$$

Q is an orthogonal matrix, while R is an upper triangular one. Taking advantage of the property of orthogonal matrices that their inverse is equal to their transpose, the equation simplifies to the following:

$$Rx = Q^T b \qquad (10\text{-}13)$$

Transposing, instead of inverting a matrix, makes computations faster and much safer. Also, since R is triangular, the solution to the system in Equation (10-13) is immediate by back substitution, from the last equation up to the first one.

The remaining question is how to factorize A into Q and R. Remember that we are trying to solve the system $Ax = b$. There is a nice way to obtain R and Q at the same time, by combining the matrix A and the vector b into a single matrix $\hat{A}$:

$$\hat{A} = [A, b] \qquad (10\text{-}14)$$

If you are following along with the calculations of our numeric example from Table 10-1, you can verify that the matrix $\hat{A}$ is equal to the following:

$$\hat{A} = \begin{bmatrix} 333.109 & 330.125 & 249.688 & 1163070.427 \\ -164.268 & 987.296 & 249.471 & 999984.760 \\ -929.718 & 509.467 & 134.274 & -958315.072 \\ -593.128 & 16.887 & -57.303 & -1071703.361 \end{bmatrix} \qquad (10\text{-}15)$$

Finally, after triangulating $\hat{A}$ into a new matrix $\hat{R}$, it can be shown that we can quickly extract the matrix R and the vector $Q^T b$ from $\hat{R}$, as shown in Figure 10-5.

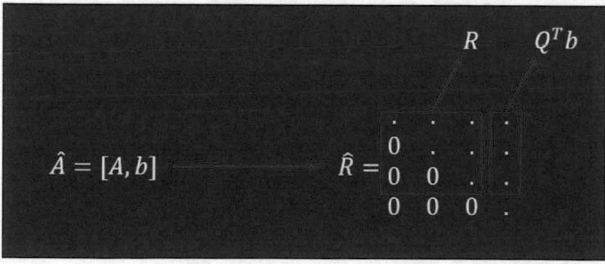

Figure 10-5. *Extracting R and Q from a triangulated form of A*

The standard method to triangulate a matrix is based on the *Householder transformation.* The idea is to proceed column by column and transform the matrix into an equivalent one with zeros under the main diagonal. The algorithm is shown in pseudo-code in Figure 10-6.

$$
\begin{aligned}
&\text{for } k = 1 \text{ to } n: \\
&\quad v = A_{k:m,k} \\
&\quad u_k = v - \|v\| e_1 \\
&\quad u_k = u_k / \|u_k\| \\
&\quad A_{k:m,k:n} = A_{k:m,k:n} - 2u_k \left(u_k^T A_{k:m,k:n} \right)
\end{aligned}
$$

Figure 10-6. *Pseudo-code for Householder transformations*

Let's proceed one step at a time. The loop goes through each column of the matrix: we first take the column vector v and then create the error vector e (of the same size of v), with all elements equal to zeros except for the first one, which is ± 1, taking the opposite sign of the first element of v. In our numerical example, the first element of v is 333.109, so the first element of e will be -1:

$$
v = \begin{bmatrix} 333.109 \\ -164.268 \\ -929.718 \\ -593.128 \end{bmatrix} \quad e = \begin{bmatrix} -1 \\ 0 \\ 0 \\ 0 \end{bmatrix} \tag{10-16}
$$

We then proceed with the transformation by building the new vector u_k:

$$
u_k = v - \|v\| e_1 \tag{10-17}
$$

After normalizing $\boldsymbol{u}_k$, we finally start generating a new matrix in the last line of the loop. Note that we always start from index k, so the further we go ahead in the loop, the smaller the submatrices will become. In other words, we will only modify an increasingly smaller subset of the entire matrix. After going through all the columns (in our case three, because we have three unknowns), we find the resulting triangulated matrix $\hat{R}$.

$$\hat{R} = \begin{bmatrix} -1163.668 & 460.519 & 41.813 & -1503679.490 \\ 0 & -1063.720 & -354.337 & -1464093.049 \\ 0 & 0 & -136.345 & -122465.284 \\ 0 & 0 & 0 & 1.956e-06 \end{bmatrix} \quad \text{(10-18)}$$

From $\hat{R}$ we can now extract the submatrix R (also triangular) and vector Q^Tb. Notice that we didn't even need to find the actual Q matrix: we already conveniently have the vector Q^Tb, which is all we need to solve the original system, as was shown in Equation (10-13).

$$\begin{bmatrix} -1163.668 & 460.519 & 41.813 \\ 0 & -1063.720 & -354.337 \\ 0 & 0 & -136.345 \end{bmatrix} \begin{bmatrix} TCP_X \\ TCP_Y \\ TCP_Z \end{bmatrix} = \begin{bmatrix} -1503679.490 \\ -1464093.049 \\ -122465.284 \end{bmatrix} \quad \text{(10-19)}$$

We solve for the three unknowns and find the TCP position, which is highlighted with a red dot at the end of the tool in Figure 10-7.

$$\begin{bmatrix} TCP_X \\ TCP_Y \\ TCP_Z \end{bmatrix} = \begin{bmatrix} 1750.758 \\ 1077.189 \\ 898.194 \end{bmatrix} \quad \text{(10-20)}$$

Since we now have the positions of both the mounting point (actually many of them; just pick whichever one you want) and of the tool center point, we can quickly find the tool size as the difference between them:

$$\begin{bmatrix} T_X \\ T_Y \\ T_Z \end{bmatrix} = \begin{bmatrix} 380 \\ 0 \\ 0 \end{bmatrix} \tag{10-21}$$

The result matches the tool size of 380mm along the X axis.

Figure 10-7. *Close-up view of TCP and MP*

Cell Calibration

A group of robots and machines working together on a workpiece is called a work cell. Calibrating the cell essentially means finding out the exact locations of all the components working together in that system, so that a unique reference frame can be shared among all the programs running on each robot.

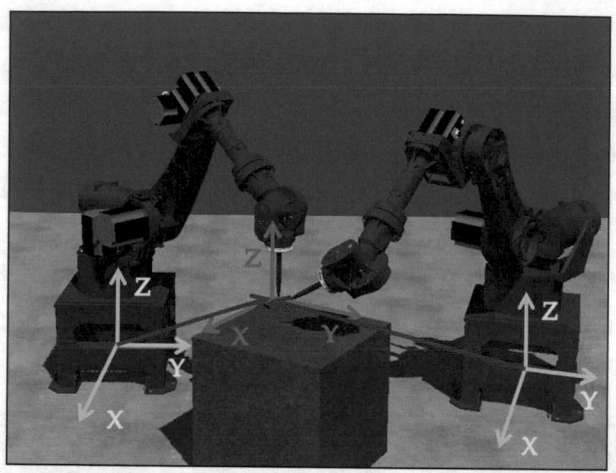

Figure 10-8. *Two robots sharing the same workpiece in a single cell*

Consider, for example, the two robots in Figure 10-8. If each robot would be programmed in its own independent frame, the coordinates of the workpiece seen from the robot on the left would be different from those seen from the robot on the right. That would generate confusion and would make offline programming unnecessarily complicated, hence the requirement for a uniform coordinate system for the entire cell. Only then we can be sure that all the components related to the cell live in the same exact coordinate system and that the target positions and orientations are the same for them all.

The choice of the common frame is usually the workpiece itself, in order to make the programming coordinates more intuitive. Cell calibration is therefore reduced to a frame identification problem. The procedure must be repeated for each new application if the cell components are rearranged or if new workpieces are introduced. If multiple workpieces are present in the cell, then one frame calibration for each workpiece is needed.

Cell calibration allows for offline programming of all the robots without direct teaching. It is an essential procedure for flexible lines with small batches.

The solution is easy and does not involve any complicated math. Essentially, we need to find a common frame for the entire cell and then reference all the robots to that frame, so that the offline programmed positions are equal for everyone. The easiest way to measure the location of a frame is to manually jog the robot to touch some reference points of the frame axes (see Figure 10-9).

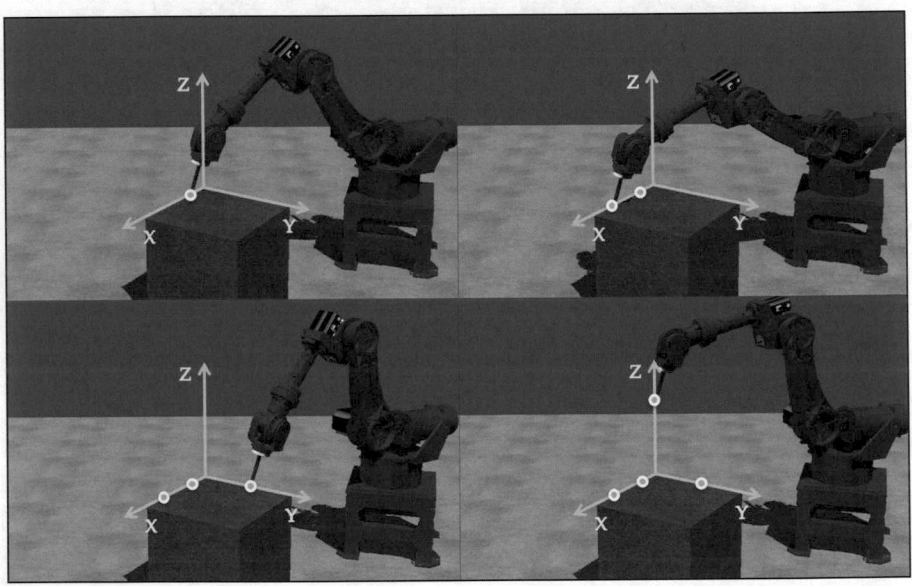

Figure 10-9. *Identifying the global frame through reference points: two points for the X axis (top), one point for the Y axis (bottom left), and one (optional) point for the Z axis (bottom right)*

Uniquely defining a coordinate system requires only three points: two for the X axis and then one for the Y axis. Two points define a unique line (e.g., the X axis), three points define a unique plane (e.g., the X-Y plane), and the orientation of the last coordinate (e.g., the Z axis) can be found automatically. If the user does not specify which of the two axes is X and which one is Y, then we need a fourth point to determine the direction of Z.

Let's see how to derive the position and orientation of a frame, given the position of three points along its X and Y axes. We call the three taught points P_0, P_1, P_2: they uniquely define a coordinate system, which we denote using the unit vectors V_X, V_Y, V_Z (see Figure 10-10). We also assume that the three points are noncoincident and noncollinear; otherwise, the frame cannot be identified.

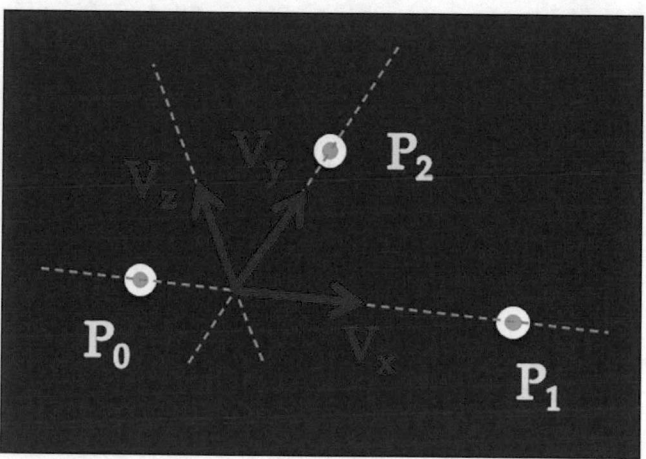

Figure 10-10. *Generating a frame from three points*

The goal is to find the distance and the rotation of the frame $[V_X, V_Y, V_Z]$ from the base frame of the robot.

V_X is quickly found from P_0 and P_1 and then normalized:

$$V_X = \frac{P_1 - P_0}{|P_0 P_1|} \tag{10-22}$$

341

Then we calculate the position of the frame's origin, i.e., the point where V_X and V_Y meet. The origin lies on the P_0P_1 line: to define its position, we use a parameter k, which can be either positive or negative, depending on the location of P_2.

Actually, k represents the projection of the segment P_0P_2 on the line defined by V_X, and a quick way to find its value is the dot product between the two vectors:

$$k = (P_0P_2) \cdot V_X \tag{10-23}$$

The position of the frame's origin is then as follows:

$$O = P_0 + kV_X \tag{10-24}$$

Since we have both O and P_2, we can immediately find (and normalize) V_Y:

$$V_Y = \frac{P_2 - O}{|OP_2|} \tag{10-25}$$

Finally, because of perpendicularity, V_Z is the cross product between V_X and V_Y:

$$V_Z = V_X \times V_Y \tag{10-26}$$

We have now identified the user frame with respect to the robot's base frame: its position is the frame origin O; its orientation is defined by the rotation matrix composed using the axes as vector columns (see Column Vectors Section in Chapter 2):

$$R = [V_X \ V_Y \ V_Z] \tag{10-27}$$

This matrix could then be decomposed in Euler angles if needed.

The same identification procedure must be repeated by each robot for all the workpieces in the cell. The frames poses can then be saved in the control program and automatically activated in the user program in each motion command. In our example robotics language, we used the command *Z* to activate a specific frame (see Interpreter Section in Chapter 9). For example, if the point P0 is defined as [X=0, Y=0, Z=0, A=0, B=0, C=0], then we can move the TCP of the robot to the origins of all defined workpiece frames (see Figure 10-11) with the following commands:

```
MJ P0 Z1 F100
MJ P0 Z2
MJ P0 Z3
```

Critically, the command *MJ P0* should not be executed without a specified valid frame, because it would result in an error, as the robot cannot physically reach its own base origin (default frame *Z0*).

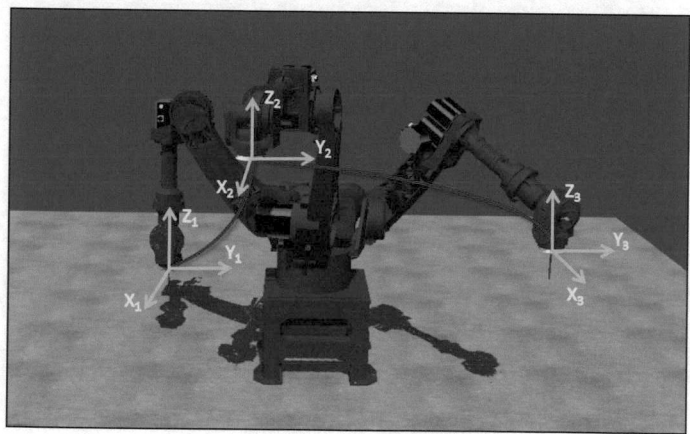

Figure 10-11. *Using the calibrated frames in a user program*

In terms of kinematic calculations, activating a zero frame requires adding an extra homogeneous matrix to the transformation chain. The details were described in Zero Frame Sections of Chapters 3 and 4.

Summary

A raw uncalibrated robot can perform all basic motion functions but has severe limitations in high-end applications where accuracy is critical. In this chapter, we briefly presented a few methods used to increase the accuracy of the robot's parameters, both those related to its mechanical body and those of the mounted tool.

We also described how to identify individual frames in a working cell, so that multiple robots can easily share the same workpieces and same user programs.

CHAPTER 11

Commissioning

Once the robot has been assembled and calibrated, it is ready to start working on its first application. The procedure followed by an application engineer to prepare the robot for a specific task is called *commissioning*. We focus here mainly on the software details and assume that all the hardware components have been wired correctly. It is always advisable to double-check all the connections before powering up the system.

The most critical step of a commissioning activity is making sure that all the safety aspects are correct and in place. The goal is to guarantee a safe operation of the robot regardless of possible errors committed by the operator while executing manual movements. The second step is the tuning of the controllers' parameters, needed to guarantee a smooth operation of the robot under all circumstances. Once safety and tuning are completed, the operator can jog and program the robot at will without further concerns.

Safety

Robots can be dangerous machines to operate. It is of paramount importance that you understand the possible sources of risk and put all measures in place to eliminate (or at least minimize) them.

Most industrial robots, with the notable exception of some cobots, have very rigid mechanical structures: collisions with the surrounding environment can easily cause damage to the robot itself and the hit objects. Therefore, robots are typically confined to work in safety cells,

© Fabrizio Frigeni 2023
F. Frigeni, *Industrial Robotics Control*, Maker Innovations Series,
https://doi.org/10.1007/978-1-4842-8989-1_11

in which human operators are not allowed during active operations (see Figure 11-1). The protective gates are connected to safety switches, which deactivate the robot as soon as the door is opened.

Figure 11-1. *Most industrial robots operate behind safety cages*

The protective cell should be well dimensioned to cover all the possible reaches of the TCP, including additional optional tools.

As an additional safety measure, the behavior of a new software version or a new user program should always be tested in a simulated environment before putting the real robot into action (see Chapter 12).

The robot's control hardware must have at least one emergency switch of the kind shown in Figure 11-2. There are different ways of wiring the **E-Stop** buttons, each with its own advantages. However, no matter what kind of stopping procedure is selected, the functionality must be tested at the very beginning of the commissioning phase.

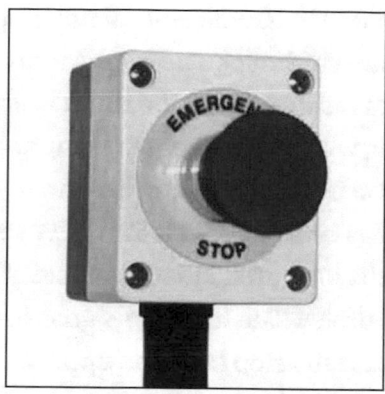

Figure 11-2. *Always test the E-Stop before running the robot*

The simplest way to connect an E-Stop to the control system is to make it shut the power supply off. While this approach is widely used mostly in cheap, low-quality, and hobby projects, it is unfortunately not a good way to stop the robot. One reason is that suddenly removing the power from the servo drives leaves the motors to spin out without active control in an unsafe way. The other reason is that the power loss causes the motors' brakes (if available) to engage while the motors are still running. The brakes are meant to be holding brakes, not halting brakes, so their life would be greatly reduced and their damage risks increased.

A slightly better, although far from optimal, way to wire the E-Stop is to connect it as a software input to the servo drives. When the drives' firmware detects a low-input value, it commands a movement's stop, which brings the motors to standstill within the configured deceleration limits. This approach has the advantage that the stopping is controlled and the brakes are engaged only after the motors have completely stopped their movement. On the other hand, the main disadvantage is that each axis stops independently, and the TCP's position likely violates the programmed path. That is not desirable when the tool works closely with the workpiece, as collisions might easily occur.

The most common way to handle emergency stops inputs is to wire them directly to the main controller. The application software triggers a stopping procedure in response, and the robot is brought to a halt along the planned path, with all the axes stopping at different speed rates following the active interpolated movement. In parallel to that, the same E-stop signal is also sent to a specialized HW safety module (see Figure 11-3), which waits for a predefined time delay before actively switching off the servo drives. The HW stop signal simply acts as a redundant backup in case the stop from the application software did not work properly, but it does not perform the stopping function in normal circumstances. The time delay is introduced to allow the soft stopping procedure to complete before the drives are powered off and must be configured according to the robot's dynamics specifications.

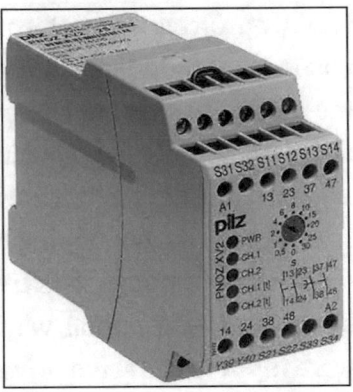

Figure 11-3. *Specialized HW safety module with configurable delay time (with permission from PILZ)*

Besides simple safety relays, a number of other safety hardware modules exist, from generic controllers to motor drives. Safety modules are easily recognizable because of their yellow color and are much more expensive than standard hardware components. The reason is that their hardware and software designs are all *redundant*: they use at least two identical controllers in parallel, running identical instances of the same program and cyclically checking on each other. For highest levels of safety requirements (e.g., in power plants), triple redundancy is often employed. Safety modules also require additional tests and special certifications, which all add to the final cost.

Safety controllers can be used to monitor individual axes limits (e.g., motor speed and torque), but also entire robotics functions, such as the TCP speed and its workspace. They run redundant software and can only be programmed and configured by safety-certified engineers. As the required safety level increases, so do the development and deployment costs.

The I/O channels and other communication lines (e.g., fieldbuses to the motor drives, encoder wirings) are also redundant: any discordance between parallel signals will trigger an error and stop all active movements.

It is important to understand that safety controllers are meant to work in parallel to the main controller and play a mere monitoring rule, without ever actively stepping into the control calculations (see Figure 11-4). Only if a violation of the specified safety rules occurs (e.g., axes limits violations, TCP workspace, etc.), then the safety network engages and immediately deactivates all motion by means of certified safety functions.

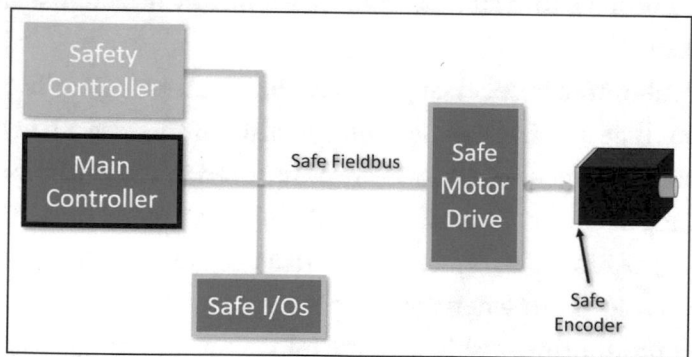

Figure 11-4. *The safety controller monitors the entire control system*

Since safety hardware is expensive, not all machines and robots rely on it, especially at hobbyist level. Nevertheless, motion safety functions should always be part of a robotics control software. All calculated outputs should be monitored by specifying sensible limits for all the robot's axes, both in position and speed, so to avoid possible violations and consequent damage to the mechanics. Robots, unlike planar CNC machines, do not typically mount hardware limit switches on their axes. Movement position limits must be monitored by the software, both in the robot's centralized library and in the individual motor drives.

Critically, axes limits can only work correctly if their reference zero value is set in the right position. Only then the joint limits are valid, and motion is safe. The referencing (or homing) of the robot's axes is normally done once during commissioning and does not need to be repeated anymore. That is because robot's motors typically mount absolute encoders with an internal memory, which can be used to permanently store the axis's reference position. If you work with incremental encoders or happen to replace one motor for maintenance reasons, then the axis needs to be referenced again.

Also, always make sure that the encoder unit scaling and direction are set correctly in the controller parameterization. Unit scaling should take into account gearbox ratios, which are typically provided by the manufacturer.

A particular note is necessary for axes that are mechanically coupled to other axes (see Mechanical Coupling Section in Chapter 3). In that case, the position limits must be necessarily monitored in the centralized robot's path planning software, instead of directly in their servo drives. That is because the position of a physical joint does not necessarily correspond to that of its driving motor when coupling is active.

Besides monitoring positions in the joint space, we must also closely monitor the path space to avoid collisions. The entire Chapter 6 of this book was dedicated to workspace monitoring functions, from safe and

forbidden zones to self-collision and multi-robot collision detection. All
these functions should be activated at all times to reduce risks caused by
errors in the user program.

Once all position limits are in place, we need to set the *dynamic limits*:
speed, acceleration, and jerk. The goal is to achieve maximum performance
from the robot without overloading the motors and the gearboxes.

The maximum speed at which a mechanical axis can move is normally
limited by the nominal speed of its driving motor. This value is specified
in the motor's datasheet and should not be overridden, because a quick
decrease in output torque will occur.

Setting acceleration limits is a bit trickier. Electrical motors can usually
generate very high accelerations safely, but the connected gearboxes (if
present) are not designed to handle very large torques, and care must
be taken not to exceed their limit value in order to avoid mechanical
damage. The relation between the torque τ applied to an axis and the
resulting acceleration a is given by the inertia of the mechanical link I_l. The
acceleration limit is then given by the following:

$$a_{max} = \frac{\tau}{I_l} \tag{11-1}$$

The link inertia can be estimated using the identification function for
the dynamic model parameters described in Chapter 8. Since robotics
joints see a variable load inertia according to the position of all the other
axes, the worst-case scenario should normally be accounted for. Finally,
remember to transform the inertia value from the link side of the gearbox
to the motor side.

In rare cases, the motor's maximum torque itself could be the limiting
factor to the movement's acceleration. That happens in cases where the
gearbox is not present (e.g., tripod robot with direct drive) or in cases
where the motors are undersized for the required application. Care must
be taken in all cases not to exceed the motor's maximal torque; otherwise,
demagnetization might occur, causing irreversible damage to the actuator.

It is also possible that torque limitations are actually imposed by the servo drives, if these are undersized for the application and cannot provide enough current to the motors during the peak acceleration phases. In that case, the alternative to replacing the servo drive is again to limit the movement's acceleration.

Lastly, we need to set jerk limits. These values cannot be derived from physical quantities and must be tuned manually, by monitoring the behavior of the axes during movements. High jerk values can cause the axis to oscillate, while low jerk values force the movement to slow down considerably: a good compromise must be found empirically by fine-tuning. A good starting point is about ten times the acceleration limits, and then increase or decrease from there according to the observed results.

Finally, using some good old common sense is always the safest option: all test movements should be executed at slow speed at first, to make sure that everything is configured correctly. Most control software packages provide a speed override parameter to run user programs at a reduced path speed. Once the system appears to work well at low speed, we can slowly increase step-by-step and reach full operating capacity.

The primary goal of the commissioning procedure is to provide a safe configuration of the robot for the operator to work along with. If all the points highlighted in this section have been verified correctly, it should be now safe to power on the robot and start real movements.

Tuning

A key step to obtain high-quality movements with our robot is to fine-tune its axes controllers, the internal components of the servo loop described in Chapter 9. There are several parameters that can be configured in a generic industrial servo drive, and that directly affect the motion's behavior: we will describe some the most important in this section.

Tuning the parameters of a PID controller is said to be more of an art than science: it takes experience and feeling more than plain calculations. Nevertheless, there are measurements and empirical observations that can be performed to improve the resulting control quality.

In short, low gains cause larger errors, while high gains correct quickly but increase the risk of overshooting and oscillating. Usually, a more conservative tuning is more robust, in the sense that it can safely adapt to various conditions of the environment and of the load. Aggressive tuning, on the other hand, might exhibit great performance in a particular setting but might quickly drag the system into an unstable region of operation when some external conditions suddenly change.

Modern servo drives feature autotuning functions that provide very good starting parameters. From there, fine-tuning is sometimes required, especially for high-end applications.

By far the most important step of the tuning process is the measurement of the servo loop *frequency response*. The resulting graph is called **Bode plot**, an example of which is shown in Figure 11-5 for a closed-loop speed controller. You can think of a motion trajectory as a time-varying signal, which can be decomposed in the sum of multiple individual sinusoidal signals with different frequencies, according to standard Fourier analysis. The Bode plot shows how each individual frequency component is transformed in amplitude and phase while going through the controller, from the planned trajectory to the actual motor. Since the graph in Figure 11-5 is recorded from a speed controller, different frequencies correspond to different acceleration values.

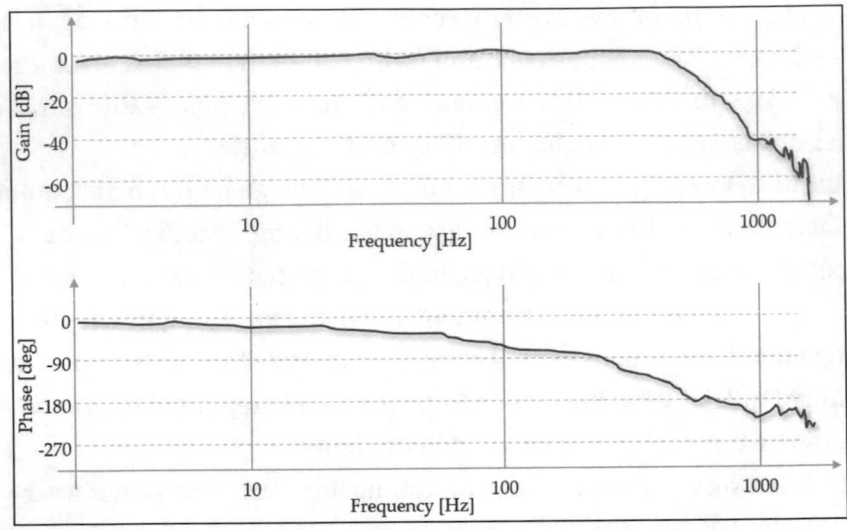

Figure 11-5. *Bode plot: the frequency response of a closed-loop speed controller*

The plot on the top shows the amplitude ratio between the output and input signals to the loop; the bottom plot shows their phase difference. In other words, the Bode plot describes how the controller is able to respond to signals of different frequencies. In a motion controller, frequency corresponds to variations of the signal in time: low-frequency signals are constant speed trajectories, while high-frequency signals correspond to quickly accelerating/decelerating signals. We expect a controller to be able to follow closely (with little tracking error) the slow varying movements and to have a harder time controlling the motor when swift excitations are applied.

As a matter of fact, the observed amplitude response in the low-frequency range, up to about 300~400Hz, is very flat (green area in Figure 11-6). This is the *bandwidth* of the controller, where the input signal is reflected almost 1:1 onto the output: e.g., if the commanded trajectory accelerates by 10 deg/sec^2, the motor rotation will exactly match the requested angular speed. In other words, if the planned trajectory requires only slow changing signals, i.e., movements with low acceleration and jerk, then the motor will be able to perform those movements accordingly, with no significant tracking error.

The response quality starts decreasing at higher frequencies (500Hz~2kHz), which means that the tracking error becomes quite large when the motor is asked to accelerate quickly (yellow area in Figure 11-6). That is because the servo drive is not able to accelerate the motor any more at such high rates. Most tracking errors occur when accelerating and decelerating, not when running at constant speed or at standstill.

Finally, at very high frequencies (>2kHz), the signal picks up a lot of noise and becomes unusable, to the point that the controller cannot work at all in that range (red area in Figure 11-6). The motor essentially does not even respond to any signal with such high-frequency content. There are two reasons for that behavior: the resolution and accuracy of the encoder might be too low to be able to deliver a clean feedback signal (especially when using resolvers); the controller sampling rate might be too low to be able to read high-frequency feedback signals.

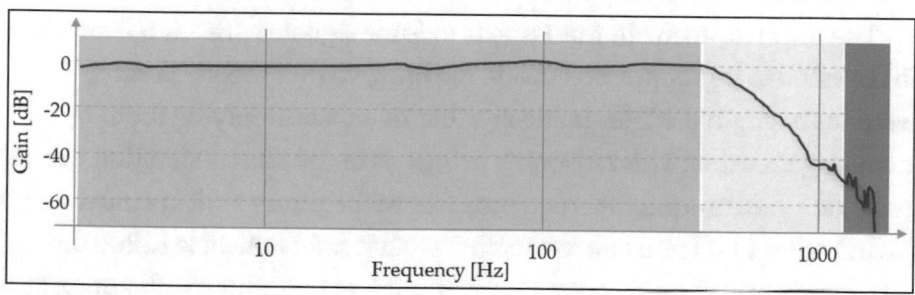

Figure 11-6. Limited bandwidth of the controller

When tuning a motion controller, our goal is to increase the gains as much as possible without running into instabilities, so that the actual motor tracking error is as small as possible but never starts to oscillate. What does it mean that the controller becomes unstable? If we tune the controller too aggressively (by setting very high gains), it will send too much current to the motor when trying to correct for an existing positive error. As a consequence, the error suddenly becomes negative. If we again correct too much in the other direction, the error once more flips sign, thereby causing the mechanical system to start oscillating. An unstable oscillating system is undesirable because it wears out the mechanics quickly and also because it emits noise at the frequency at which the stability region is exceeded.

Instabilities can be visualized in the Bode plot, especially when observing the phase response. The difference in phase between the set and actual values increases with the frequency of the signal. The reason was already explained earlier: the motor cannot keep up with the commanded trajectory at higher dynamics and starts lagging behind. When the phase of the output state lags by a complete flip of -180 degrees with respect to the input signal, the system reaches a critical point: the negative feedback to the PID controller now reinforces the input (i.e., they have the same sign), and the signal in the loop grows larger and larger. The motor and the attached load will start swinging loudly and wildly.

The only way to avoid this issue is to force signals with a large negative phase value to be suppressed inside the loop, so that their influence becomes negligible. Mathematically, that means multiplying them by a very low gain value. This reasoning brings us to the conclusion that the amplitude gain around the frequency where the phase shift is -180 degrees (f_{180} in Figure 11-7) must be well below 1. That safety offset is called the gain margin and should be kept large enough to guarantee robustness to the controller. Tuning the gain softer (green curve) assures a larger gain margin than a stiffer tuning (red curve).

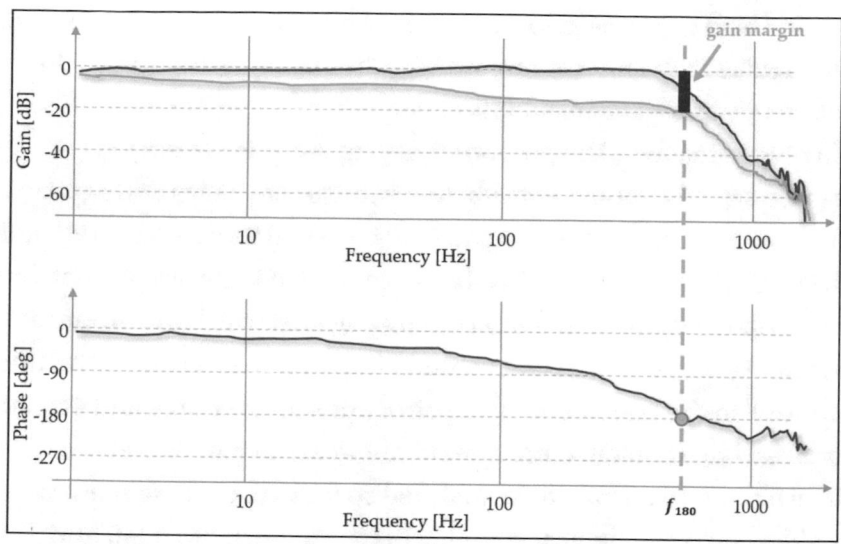

Figure 11-7. *A large gain margin ensures stability of the loop*

Because environmental conditions and actual payloads vary in the field, a robust control configuration is always preferable: a good margin from instability should always be ensured. However, the requirement for a large gain margin forces us to set a low controller gain at all frequencies, which in turn could result in a poor control behavior and large tracking errors even at low accelerations.

As usual, an optimal compromise between the two cases depends on the specific application: there are situations where we might want to leave some extra safety margin because the load conditions often change over time, and the absolute error during dynamic movements is not so critical (e.g., a palletizing robot). In other situations, however, we might need to push the controller to the limits of instability to guarantee highest accuracy in the working bandwidth (e.g., for laser cutting applications).

In advanced applications, the user might also decide to opt for an adaptive tuning approach. Instead of using a fixed gain value for all situations, the gain is actually adjusted according to the specific trajectory: softer settings to avoid oscillations at high speed and stiffer settings to

357

guarantee high accuracy at low speed. In the latter case, a low-pass filter is often added in the loop to remove high-frequency noise that could inadvertently introduce instabilities.

In other situations, the proportional gain alone is not enough to guarantee a good control behavior. Robotic axes are heavy and exert a torque on the motor because of gravity. The actual torque value depends on the configuration of the robot, but in general, it is always present and causes a stationary error in the servo loop, against which the proportional gain is not effective. The easiest solution is to activate the integral action of the controller: the integration of the error adds up over time and eventually generates a large enough correction to remove the error entirely.

The frequency response we analyzed so far is a best-case scenario for an ideal rigid mechanical coupling between the motor shaft and the arm link. However, that is not always the case in practice. One issue we often encounter is the presence of a *resonance frequency* in the working bandwidth of the servo loop (see Figure 11-8). A resonance effect exists for all coupled mechanical systems, and it results in increasingly large oscillations of the load when the system is excited at a specific frequency.

Typically, the effect happens at high frequencies, where the phase difference between input and output of the servo loop is already very large. The consequence is that the resulting amplitude peak is very likely to cause instabilities and to make the motor oscillate with a high-pitch loud noise. You might be tempted to think that the commanded trajectory signal does not have such high-frequency components and will never excite the resonance window of oscillations. However, in practice there is always a certain amount of noise at all frequencies in the loop, which will excite the resonance and will be strongly amplified.

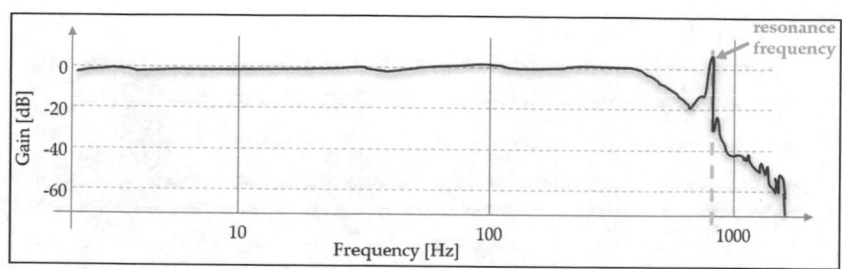

Figure 11-8. *The system can easily become unstable at its resonance frequency*

One common, though inefficient, option to mitigate the effects of a resonance is to lower the controller gain. A better solution is to filter out from the loop the unwanted window of frequencies while leaving the rest of the bandwidth untouched. Such a selective filter is called **notch filter** and is shown in Figure 11-9: the green band represents the range of frequencies affected by the filter. The resulting amplitude response is greatly reduced in that area (the effect is shown in the lower half of the plot). A minor drawback of the notch filter is that it introduces a further phase shift in the loop and therefore reduces the useful bandwidth of the controller at parity of gain margin.

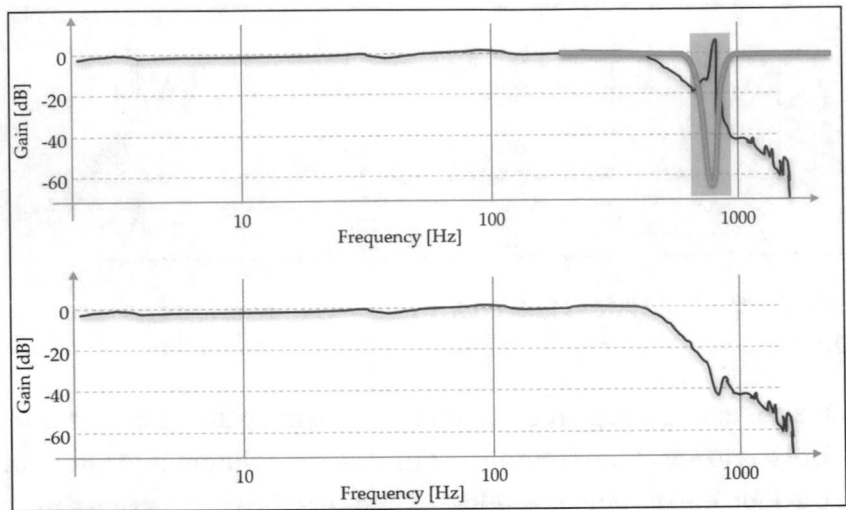

Figure 11-9. *A notch filter can be used to reduce the effects of resonance frequencies*

A second mechanical issue we might encounter in practice is that the coupling between motor and load could exhibit a certain degree of **compliance**, either because it was meant to be built that way (e.g., in some flexible cobots), or because it was worn over time for overuse, or even because of poor design choices, which resulted in a high-inertia mismatch between motor and load. The consequence is that the actual position of the load does not always correspond to that of the driving motor. The servo loop receives and controls the feedback position from the encoder mounted on the motor, but the position of the load itself could be significantly off. Their relative difference depends on the mechanical characteristics of the joint and on the frequency components of the driving trajectory.

A compliant system is much more difficult to control than a rigid one. In most cases, the useable bandwidth of the controller is significantly reduced: only slow and soft trajectories can be used before the actual tracking error builds up to be too large. Even high-speed controllers or high-resolution encoders offer no additional help, simply because a flexible mechanical system is not meant to work at high frequencies.

Physically, an elastic system behaves as a *two-mass oscillator with a springlike link* in between. Such a system can be quickly recognized by its peculiar frequency response, which features the presence of both a resonance and an antiresonance frequency (see Figure 11-10).

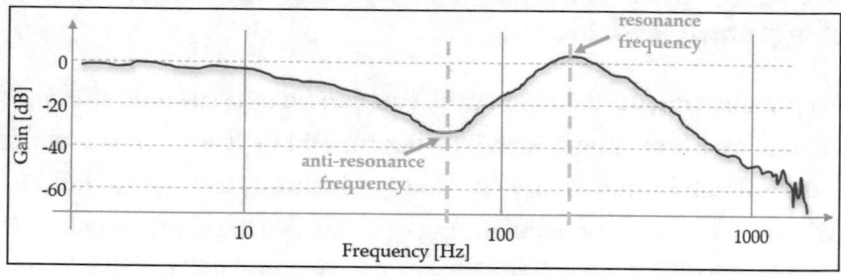

Figure 11-10. *Typical response of a compliant joint*

At the resonance frequency (f_{res}), the signal from the motor is strongly amplified into the load, which often starts to oscillate. On the other hand, at the antiresonance frequency (f_{ares}), the transfer function has a very low amplification, which means that whatever signal magnitude we send into the loop, the output will be strongly reduced. In practice, the load does not respond to that excitation and causes a large position error. The physical reason is simple: the load cannot follow the motor at that particular acceleration, either because the link is too flexible or because the motor has a much smaller inertia than the load.

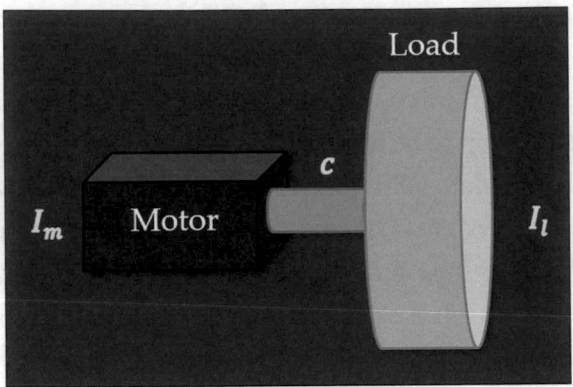

Figure 11-11. *Systems with large inertia mismatch and/or elastic couplings are difficult to drive*

To visualize an inertial mismatch, you can imagine a small slim motor trying to drive a large heavy load (see Figure 11-11). Even if the motor can produce enough torque to drive the load, it is simply too light to hold it stiff when accelerating or decelerating quickly. During sharp accelerations, the load inertia will resist the motor's movement and will cause a large tracking error between set and actual position. Similarly, when trying to stop abruptly, the large load inertia will drag the motor along and will oppose deceleration, also causing a significant position error.

The controller can only drive the system well in the range of frequencies lower than the antiresonance frequency. Any higher acceleration present in the planned trajectory could not be followed and would cause the load's actual position to lag significantly from the set position. That is why very flexible or mismatched joints, with a very low f_{ares}, are very difficult to drive accurately unless the commanded trajectory is very soft.

The position of the antiresonance frequency depends on both the load inertia I_l and the stiffness c of the link, as given in Equation (11-2). As expected, heavier loads (large I_l) and more flexible links (small c) bring the value of f_{ares} down and limit the useable controller's bandwidth.

$$f_{ares} = \frac{1}{2\pi} \sqrt{\frac{c}{I_l}} \tag{11-2}$$

Critically, the relative position of f_{res} and f_{ares} depends on the ratio between the inertia of the motor (I_m) and the inertia of the total system (motor plus load: $I_{tot} = I_m + I_l$).

$$\left(\frac{f_{ares}}{f_{res}} \right)^2 = \frac{I_m}{I_{tot}} \tag{11-3}$$

A significant consequence is that the two frequencies can be brought closer to each other by reducing the inertia mismatch between motor and load, either by employing a larger motor or introducing a larger gear ratio between the two (which would reduce the load inertia seen on the motor side). The result is that the two gain peaks in the frequency response tend to cancel each other out, thereby flattening the amplitude response curve. A system with a flatter response is much easier to control, as the load follows the motor's position more accurately even at higher frequencies (i.e., at higher acceleration).

However, in practice, replacing the actuators is not always possible, especially because of costs reasons, since motors with large inertia tend to be expensive. An alternative is to introduce software algorithms to compensate for the compliance and support the controller in its difficult task.

A common approach is to use a mathematical model to simulate the behavior of the system and predict the position of the load while the motor is moving. The employed model is a two-mass oscillator between motor and load. The output of the calculation is the predicted position of the load, as if an encoder was actually mounted on the load side. That is often called a *virtual encoder*, to distinguish it from the real encoder, which is mounted on the motor side (see Figure 11-12). Providing double feedback to the servo loop (a two-encoder drive system) improves the control quality significantly and reduces the tracking error of the load.

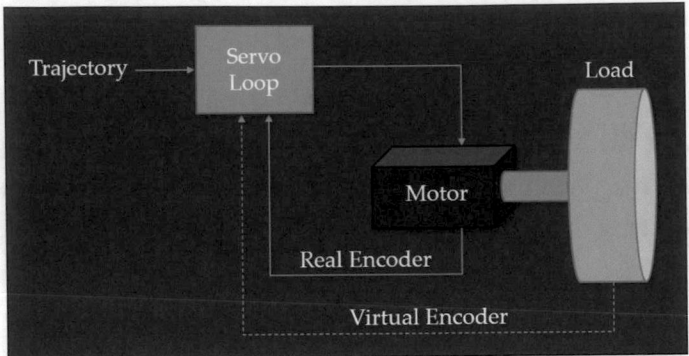

Figure 11-12. *Two-encoder controller loop*

However, as in all model-based solutions, the quality of the actual predictions depends on the quality of the model and its parameters. An accurate identification of the load inertia and the link's stiffness and damping through several empirical measurements is critical.

Additionally, while this control architecture is relatively easy to implement and is quite efficient for an individual axis, it becomes more difficult to apply to robotics systems. That is because robotic axes exhibit a variable inertia, which depends on the actual pose and dynamic state of the entire robot, so the parameterization of the oscillator model needs to be updated in real time.

Summary

The act of deploying a robot for a specific application and making sure that everything runs safely and smoothly is called commissioning. Indeed, the two major concerns for an engineer responsible for the commissioning phase are the safety and the fine-tuning of the robot's movements.

Safety issues can be taken care by careful parameterization of software constraints and also by adoption of redundant safety hardware modules. Make sure you start with slow velocities and keep at sensible distance from the robot.

Tuning the controllers is a more sophisticated procedure that requires a good understanding of the behavior of the mechanical structure under control, because signals at different frequencies generate different responses of the robot. The goal is to tune the motion control system to be accurate but also robust enough to avoid running into instabilities over time.

CHAPTER 12

Simulation

We already mentioned a few times in this book that the deployment of a digital twin is a necessary step to test, optimize, and program offline any robotics control system. This chapter briefly shows how to take advantage of the game engine Unity to create a virtual environment for your robots.

Unity 3D

The main reason I personally like and suggest Unity is because it is fast to learn and intuitive to use. You will be able to set up a simple scene with a working robot in matter of minutes. For our hobbyist purposes, it is also free of charge, unless you work in a large company and use it for commercial purposes.

Unity is a game engine, which means it provides very realistic 3D graphics, physics, and sounds: you can build and tour entire factories and production lines filled with robots and other machines working on different processes while hearing the screaming sounds of their motors. The results can be very impressive to show off in your portfolio. The virtual world you create is not necessarily only a playground simulation: you can actually use it as a complete HMI visualization system to monitor and control real robots.

Unity includes a powerful physics engine, which means that the simulation is not only limited to animating models of robots, but it also performs realistic interactions between them: manipulators picking or pushing objects, conveyors bands transporting material, collision detection between robotic arms, and much more. Figure 12-1 shows a few examples of automation processes simulated with Unity: from industrial robots, to humanoids, to wind power plants.

Figure 12-1. *Using Unity to simulate industrial applications*

While most of its interface is graphical, Unity is also fully expandable via script programming, offering C# and JavaScript support. The availability of a programmable interface opens up a world of opportunities to the creative user. You could program your robot control system directly in Unity, or better yet, you could set up a TCP/IP interface to the external hardware controller where the main control program runs. We will show how to implement this option in Section on Communication Functions.

Keep in mind that Unity is a simulation environment, not a CAD system. You could create and modify simple meshes directly in Unity, but it is a better idea to create your CAD models elsewhere and import them in the Unity scene in the form of 3D objects. Actually, most robots' manufacturers post their CAD files on their websites free to download. For instance, you could download the model of a Fanuc robot from their homepage and import it in Unity for your simulation purposes.

Another nice feature of Unity is that it generates cross-platform applications. Once you have built a virtual environment, you can release it for most existing platforms: several different desktop, mobile, and web operating systems are supported. You could run your own 3D robot visualization on your phone and wirelessly control a real robot to make it move. Or you could run an interactive VR scene for impressive marketing purposes.

Finally, because the use of Unity is so widespread, you gain access to a large community support and to a wide range of 3D assets to import in your simulated environment. The Unity forum and store are great places to find and share resources.

Building a Scene

Building a simulation in Unity consists of three main steps: create a static environment; bring it to life through programming; compile it for the platform of your choice.

Creating the environment is mainly a work of art. You need to import all the models of robots, machines, products, and everything else you need to make the scene as realistic as you need: do not forget to insert lights and sounds. Also, if your goal is building a full-fledged user interface (**UI**) to interact directly with the system, then this is the time to add buttons, input fields, sliders, and other widgets (see Figure 12-2 for a basic example and User Interface Section for some implementation details).

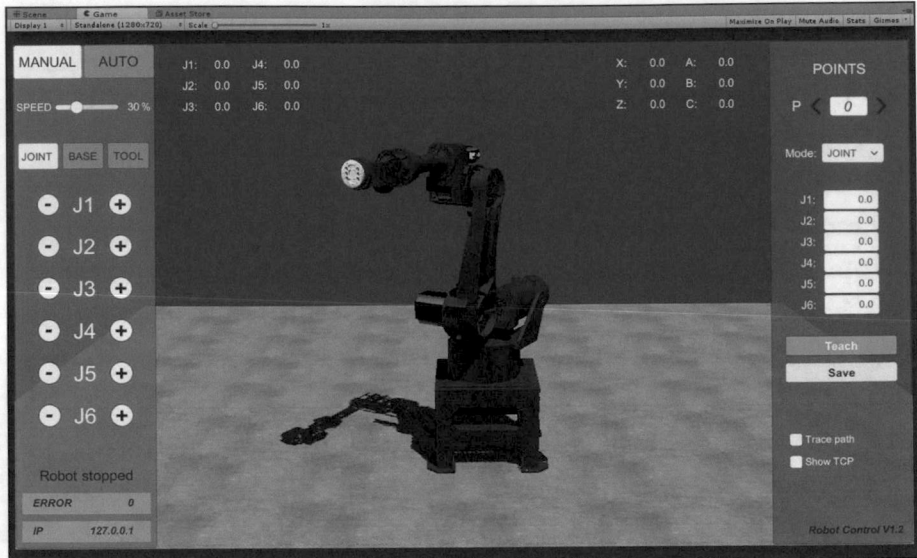

Figure 12-2. *Creating a simple UI in Unity with a few interactive controls*

The environment created so far is static: it does not move on its own. The next step is to bring it to life by defining its dynamic behavior. In Unity, that is accomplished by adding scripts in which you can program the motion control of each joint of the robot. For example, we could define a rule that takes as input the status of a button and generates a movement of the first joint of the robot when the button is pressed (as shown in Figure 12-3). Alternatively, you can program a communication interface to an external motion controller, in case you already implemented it somewhere else (running on a different hardware).

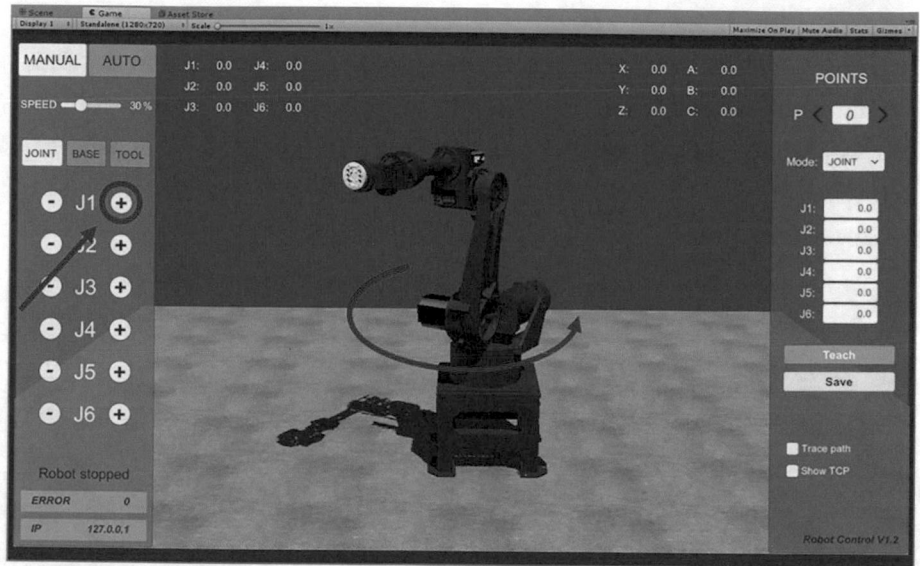

Figure 12-3. *Adding a script will make the robot move according to user commands*

Finally, when all the static objects are inserted and all the dynamic behaviors are programmed, you can compile the environment for a specific platform and generate an executable file that you can run and redistribute, independent from the Unity software.

Importing CAD Models

Unity supports a large variety of CAD formats (fbx, dxf, obj, dae, 3Ds, etc.). You can import essentially everything created with generic mesh editors like Fusion360, 3dsMax, SolidWorks, Blender, or even SketchUp.

You can download robots' models from constructors' websites or create them by yourself. In either case, before importing them into your Unity scene, you need to make sure that the model is subdivided in individual objects for each moving joint. Otherwise, if it comes as a single solid object, you cannot let Unity move each axis individually. When imported, the object will appear as a tree of sub-objects in the *Hierarchy* panel on the left side of the screen (see Figure 12-4). From top to bottom, you have nodes of children belonging to their parents. In our example case, we see the six axes of the robot: from *axis1* up to *axis6*. The properties belonging to each component are listed on the right side of the screen in the *Inspector* panel.

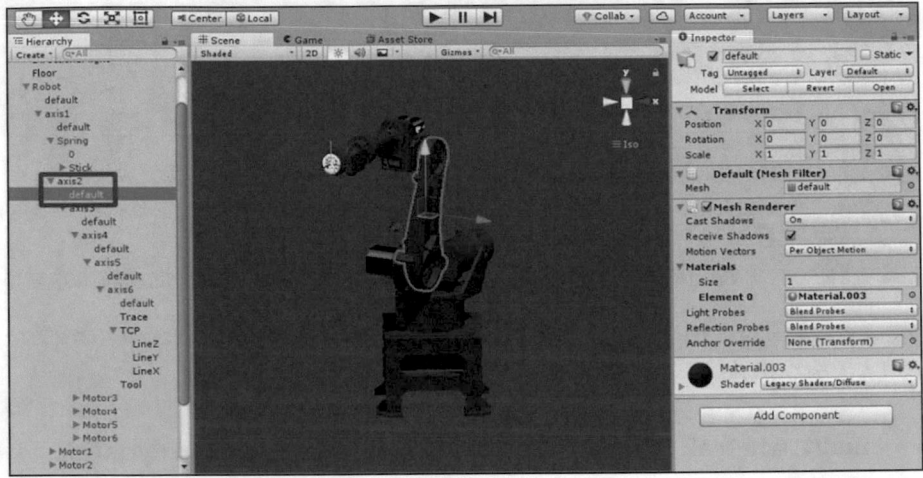

Figure 12-4. *Robot model imported in Unity as tree of individual axes*

In Unity, each individual object has a **Transform** property, with which the object's position, orientation, and scale can be modified. Keep in mind that when moving an object, all its children in the hierarchy are automatically dragged along. This is actually the trick we use to generate correct movements of kinematic chains. For example, if we select the component *axis1* and modify its orientation in the Transform window around its vertical axis, we observe the whole robot rotating, from the first axis all the way to the TCP (see Figure 12-5).

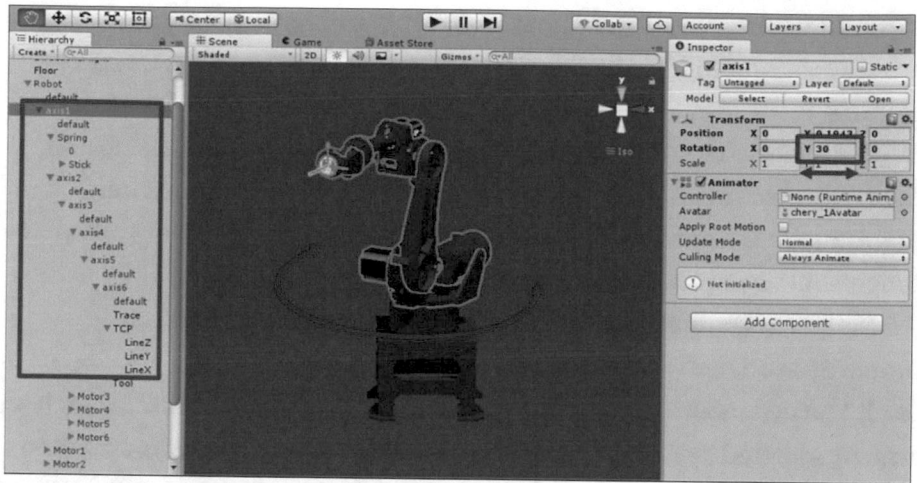

Figure 12-5. *Rotating the first axis causes the whole robot body to move along*

You can quickly jump to conclusion and realize that if we link the rotation property of an axis to a variable in a script, we can programmatically force the robot to move.

One small tip: When exporting meshed objects from CAD software, make sure that the pivot position is centered in the joint so that each link can be properly rotated. Otherwise, the rotation of the link will take place around a different point (away from the joint), and the robot will not behave correctly. Alternatively, you can add dummy joints between links in the robot's hierarchy (called *empty objects* in Unity) and use them to act as intermediate pivots.

Programming Scripts

The way Unity allows a program to control an object in the environment is to link public variables of a script to the object's properties. That allows you to control anything in the scene: from standard machine movements to dimming the intensity of lights, triggering sounds when some tools are at work, and modifying the color of an object while a robot is spray-painting it.

For a more basic starting point, we are interested in translating and rotating objects: either the robot's own axes to perform a movement or products on a conveyor sliding at a specific speed to simulate their progress along a production line.

Scripts in Unity are implemented as public classes, with their properties and methods (Figure 12-6 shows what a new empty script looks like). The two most important methods are *Start()* and *Update()*, which are already inserted by default when creating a new script. The *Start* method is where we put all the code that needs to be executed at startup in order to initialize the simulation correctly: for example, moving all robots to their home position. The *Update* method is where we put the code that runs cyclically during the whole simulation: for example, reading user inputs and calculating the target positions for all moving axes.

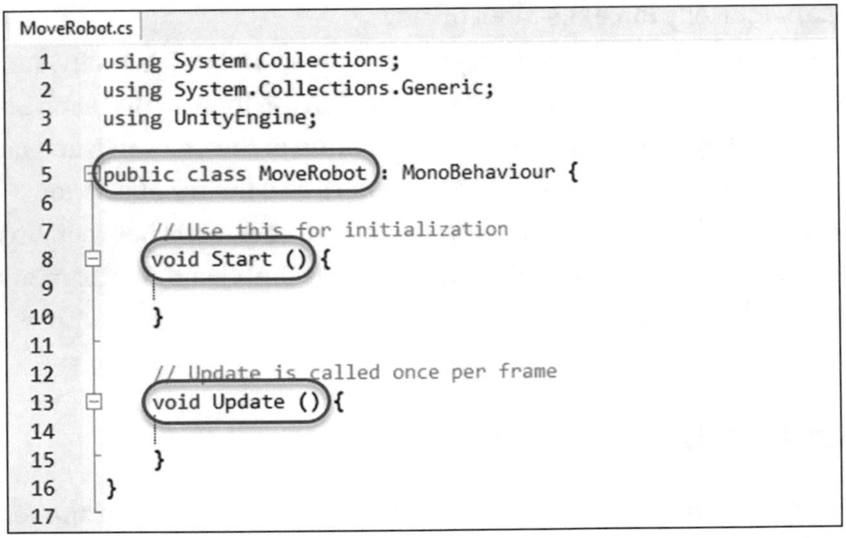

```
MoveRobot.cs
1    using System.Collections;
2    using System.Collections.Generic;
3    using UnityEngine;
4
5    public class MoveRobot : MonoBehaviour {
6
7        // Use this for initialization
8        void Start () {
9
10       }
11
12       // Update is called once per frame
13       void Update () {
14
15       }
16   }
17
```

Figure 12-6. *A new script comes with the Start and Update methods by default*

When assigning a new script to an object, the corresponding interface becomes visible in the editor under the Inspector view. That is where we find all the public variables defined in the script, so that they can be directly linked to physical objects present in the scene.

Let's look at an example in detail. Our scene contains a manipulator robot (shown in Figure 12-4), to which we assign the empty "MoveRobot. cs" script (shown in Figure 12-6). We can then define a public array of type *Transform* (see Figure 12-7), which represents the interface between the script and the scene. Each element of the array needs to be linked to one of the robot's joints, so that we can use the script to programmatically control all the properties of the six axes of our robot.

```
MoveRobot.cs
1    using System.Collections;
2    using System.Collections.Generic;
3    using UnityEngine;
4
5    public class MoveRobot : MonoBehaviour {
6
7        public Transform[] Robot = new Transform[6];
8        float J1 = 0;
9
10       // Use this for initialization
11       void Start () {
12
13       }
14
15       // Update is called once per frame
16       void Update () {
17           J1++;
18           Robot[0].localEulerAngles = new Vector3(0, J1, 0);
19       }
20   }
21
```

Figure 12-7. *Using public variables to programmatically control objects in a scene*

In order to generate a movement in the scene, we can cyclically increase the value of the first joint in the *Update* method. Specifically, we use the *localEulerAngles* property to modify the rotation of the object's Transform. In this particular case, we rotate the first joint of the robot around the Y axis, which is the vertical axis by default in Unity.

Finally, we need to link the public variable with the actual physical object. We do that by simply dragging the object *axis1* from the Hierarchy panel and dropping it on the corresponding variable in the Inspector panel, as shown in Figure 12-8.

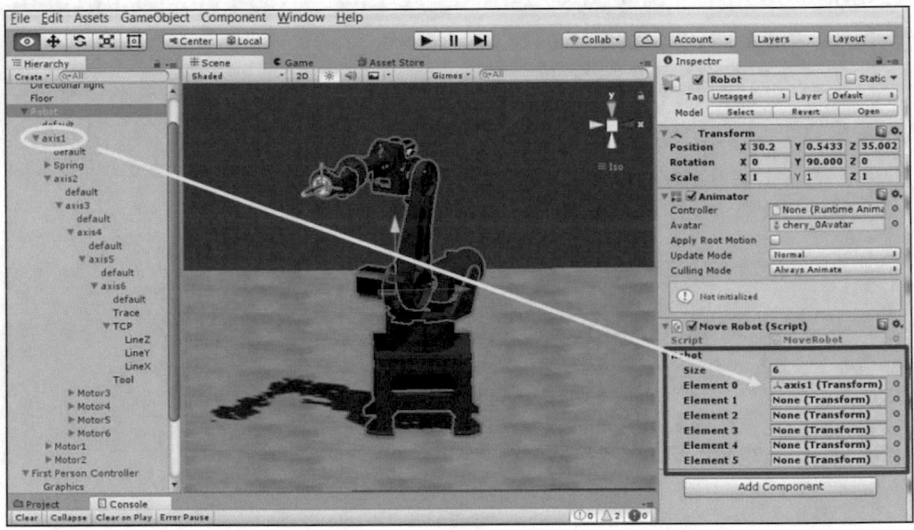

Figure 12-8. *Linking an object to a script variable to control its properties*

At this point, we are ready to run the Unity scene. Press play and observe the robot moving at constant speed around its first joint, just as we programmed in the script.

Communication Functions

Writing the entire robot control system in Unity is not always the most convenient option. In typical applications, the control software runs on the robot's main controller, while the Unity environment is only used for simulation, testing, or monitoring purposes. In that case, you need a way to let the two systems interact with each other: the robot's controller needs to send real-time positions of the joint axes to the Unity scene to visualize the actual movements; in the opposite direction, the UI needs to send parameters and commands to the controller to configure and trigger actions.

There are several different communication protocols you could use for that purpose. The limitation is normally not Unity, but the controller itself.

The industrial standard for data exchange is OPC UA (see Fieldbus Section in Chapter 18), which allows all different sorts of machines to communicate with each other independently of the device manufacturer. However, it is not always easy to find a simple and free implementation of server and client stacks, so it might not be the first option for hobbyists.

A quick valid alternative is the good old TCP/IP or even UDP if you need to go a bit faster. They are not deterministic protocols, but that is not a requirement for visualization purposes. As for the physical connection, you can use either Ethernet or Wi-Fi (see Figure 12-9).

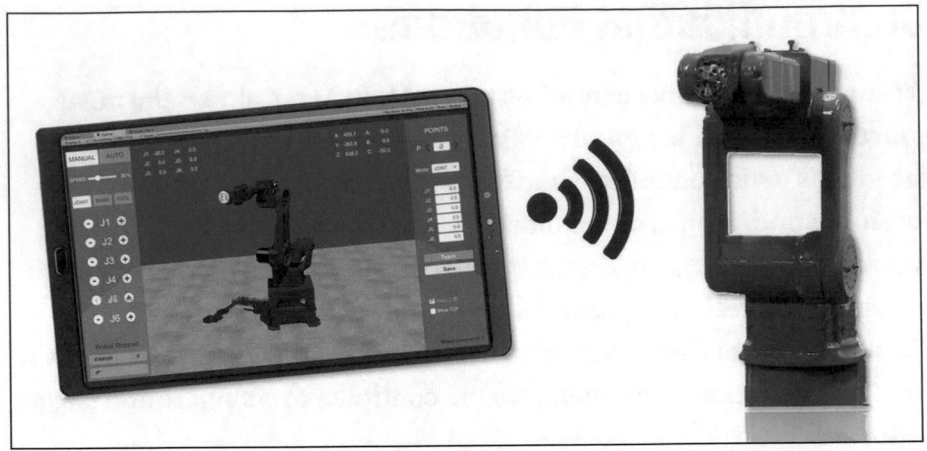

Figure 12-9. *Linking the Unity environment to the real robot control system*

Since Unity programming is based on *Mono*, which is an open source implementation of the Microsoft .NET framework, you have native access to a large number of standard libraries directly in your code. For example, the socket class can be used to set up a server/client TCP/IP communication between the Unity environment and the robot's controller.

Figure 12-10 shows the starting code to configure the Unity scene as a TCP client, which can connect to an external server. Using the *System.Net. Sockets* namespace, you can declare a private *TcpClient*, then connect to the server, and begin reading and writing data streams. You can find all the libraries documentation online.

```
//connect to TCP Server and wait for message
try
{
    client = new TcpClient(Server_IP, SERVER_PORT);
    client.GetStream().BeginRead(readBuffer, 0, READ_BUFFER_SIZE, new AsyncCallback(DoRead), null);
    Debug.Log("Connection OK!");
}
catch (Exception ex)
{
    Debug.Log("Connection ERROR!");
}

...

//new message received -> decode it and then write back
if (ReadOk)
{
    //decode message from Server
    DecodeMessage();

    //write to Server
    EncodeMessage();
    client.GetStream().BeginWrite(writeBuffer, 0, WRITE_BUFFER_SIZE, new AsyncCallback(DoWrite), null);
}
```

Figure 12-10. *Connecting to a TcpClient and then reading and writing data streams*

The memory buffers holding the data transmitted to and from the TCP stream need to be encoded and decoded correctly, because information is transmitted through the socket in byte format, while it is typically used locally in numeric format, either integer or floating point. The *BitConverter* class provides functions for you to do that.

User Interface

In case you are using the simulation not only for testing and monitoring purposes but also for direct control of the robot, you need to develop a user interface. The namespace *UnityEngine.UI* offers all sorts of widgets to do that: buttons, input fields, sliders, and many more.

You can use numeric input fields or sliders to configure the robot's parameters and jogging buttons to manually rotate the joints. You can also teach key points by saving various poses to an internal data structure and finally write and execute a user program to let the robot move through the taught points. A *trail renderer* object can be added to the TCP in the Unity editor in order to visualize the trace of the programmed path in space (see Figure 12-11).

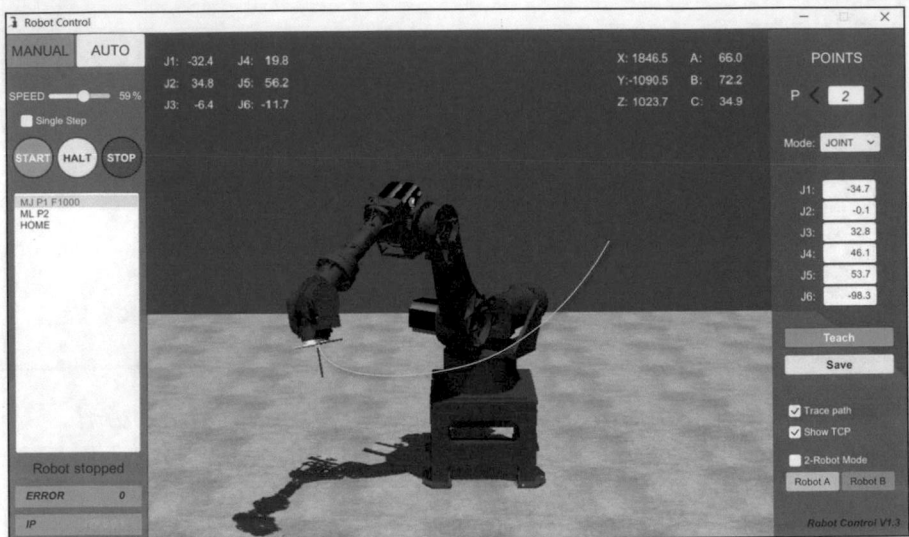

Figure 12-11. *Basic user interface to control the robot and show its path in space*

At this point, you have a nice tool to showcase the functions of your motion control library and a useful HMI to remote control a real robot.

Machine Learning

One additional interesting feature of Unity is the *machine learning toolkit*, which can be used to solve complex robotic problems. Machine learning is a branch of artificial intelligence that uses information from data to learn ways to solve tasks.

Typically, in robotics, we build models from deterministic formulas to predict the required position or torque of a motor. However, the more details we add, the more complicated the models get. Some physical behaviors are actually very difficult to write down into equations, while some others are simply impossible to model. That is exactly where machine learning comes into play: models are not described by analytical equations; they are learned from examples.

In other words, the main difference between traditional control and machine learning approach is that in the former case we understand the physical process and we can model it with exact equations. The correlation between input and output is clearly defined. However, when we are not able to build a model, either because it is too complicated or because we do not understand the physical system that we are trying to control, then we can learn the model from data, from observations, and from examples, just like children learn by watching adults. We do not write down the rules in a closed-form mathematical way; the relation between inputs and outputs is simply learned through experience.

Also, while traditional control is perfect for simpler problems with a low number of variables, it struggles when tackling large complex problems with a very high dimensionality. One of the strengths on machine learning, on the other hand, is the ability to learn correlations between a large number of variables, often discovering patterns that are not easily visible to humans.

This is not to say that machine learning is the best solution for all complex control problems. If you already have an approximate model of the process you want to control, then by all means use it as a starting point; you can then complement and optimize it further with some machine learning techniques.

Figure 12-12 shows an example of a task solved by machine learning techniques in Unity. Our six-axes manipulator is asked to balance a ball on a flat surface. The position of the ball is tracked by a camera hanging from the ceiling. The complexity of the task is twofold: first, we need to be able to extract the coordinates of the ball from the image captured by the camera (shown in the top left window); second, we need to adjust the robot's joint configuration in real time, by the correct angle and with the correct speed, in order for the ball to stay on the platform without rolling off.

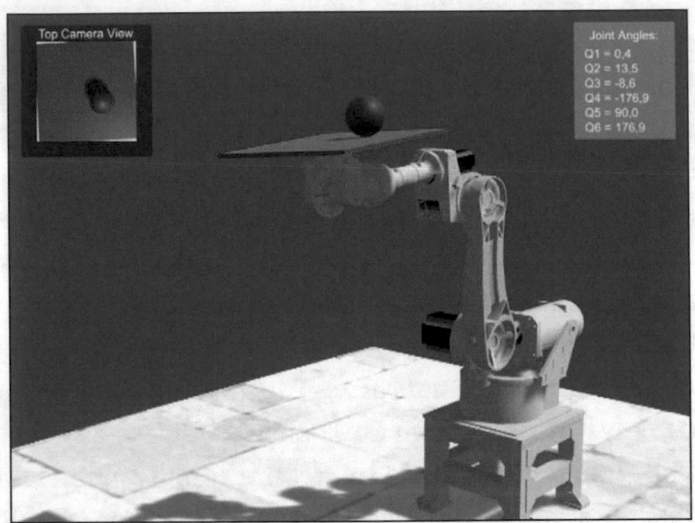

Figure 12-12.   *Example of a complex task solved using machine learning*

The actual task is clearly a playful game and has little practical application, but the underlying concept is indeed quite powerful: we use a simulation environment to make a robot control software learn to solve a motion task on its own, without providing any specific equation for it. Trying to do that in real life with an actual robot would be much more dangerous and would also require much more time. Unity offers the ability to recreate a very realistic environment with no safety concerns and *running it at an accelerated timescale* to speed up learning time considerably.

The first part of the problem is a typical machine vision application. Detecting and recognizing objects is a task that has a wide range of applications in industrial robotics. However, solving it with standard control and deterministic algorithms does not always work well. Detecting the position of a blue ball against a red background is a relatively trivial task. Sorting out products based on their labeling, shapes, or surface defects becomes increasingly difficult. We cannot easily derive mathematical models that classify objects from the content of the millions of individual pixels in the images taken by a camera. Vision is a typical problem where machine learning provides incredibly powerful solutions (see Chapter 13 for more details).

The second part of the problem is more related to motion control. A balancing task could be modeled with deterministic equations and directly coded into a controller. However, it would be very dependent on a series of parameters that are often difficult to identify in practice, i.e., friction coefficients. Letting the machine learn on its own, often by trial-and-error techniques, allows for optimal and robust solutions.

Unity provides a number of different machine learning techniques to learn from data. In particular, two distinct families of algorithms are offered:

- **Imitation learning**: This is a kind of supervised learning, which means learning from demonstrations provided by a human or a deterministic controller. In other words, supervised learning happens when a teacher, who already knows how to tackle a problem, shows the solution to the student through examples and the student learns by observing what the teacher does. The advantage is typically a fast-learning path. The disadvantage is that the student's ability will be inevitably influenced and limited by that of the teacher. An artificial brain using supervised learning can only learn to solve problems for which a solution is already known, albeit not analytically.

- **Reinforcement learning**: We all agree that students' knowledge and ability should not be limited to the level of their teachers. After all, progress comes from brilliant students who eventually exceed their masters. The idea of reinforcement learning is that the student goes on and learns on its own by exploring the environment, basically through trial-and-error strategies in the quest of tackling unsolved problems. In this case, the teacher cannot show solution examples, because the actual solution to the problem is unknown, but can offer feedback on the results: a positive reward for a good outcome and a penalty for a bad one. Positive rewards reinforce the confidence of the student showing that the learning process is going in the right direction: the actions leading to good results are prioritized in future trials; those leading to bad results are discarded. The main advantage of this technique is that the student is free to fully unleash its creativity and test new paths. Using reinforcement learning, artificial brains can learn to do things that no human can master and can also solve problems in new ways that humans have never thought about before. The disadvantage of independent learning is that it is time costly and possibly leads to dangerous actions (unless fully performed in simulation).

Technically speaking, the artificial controller is a **neural network** (NN). A neural network is essentially a function approximator: it takes some variables as inputs and generates some output values. The number of inputs and outputs can be arbitrary (see Figure 12-13).

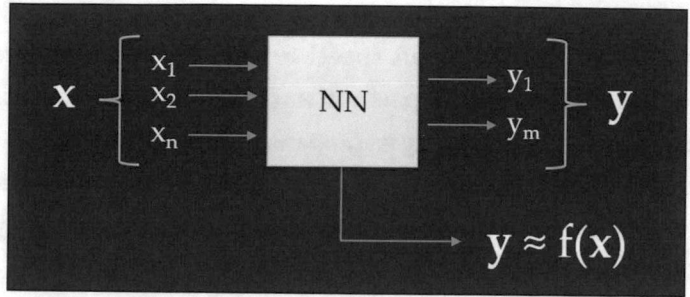

Figure 12-13. *A neural network is a function approximator*

If we name the input vector x, and the output vector y, then we can think of the neural network as an approximated function:

$$y \approx f(x) \tag{12-1}$$

Note the stress on *approximation*, because the network is not based on simple closed-form equations and does not provide an exact solution. Neural networks emulate complex functions that we are not able to express analytically. In supervised learning, given an input, we know what the output should be, but we do not have an easy way to write down that relationship in mathematical terms, so we let the artificial brain learn that pattern from examples. In reinforcement learning, given an input, we do not know what the output should be, but we can judge whether the outcome of the actions is good or bad, so we let the artificial brain learn the correct strategy by providing meaningful feedback.

In the image recognition task, we use supervised learning. Given an input image, the neural network needs to output the kind of object detected and its absolute position in the frame. We train the network by providing a large number of examples where the objects and their positions are known. Typically, such examples can be computer generated for free and in large number so to increase the training quality.

In the balancing task, we use reinforcement learning. Given the position of the ball (and maybe its speed vector), the neural network needs to output an action for the TCP position and orientation, in order to hold the ball on the surface for as long as possible. We assume that we do not know how to solve that task, so we cannot provide examples to the network to learn from. But we can judge whether the actions performed by the controller lead to a good or bad outcome. For example, we can provide a small positive reward for each extra second of time that the ball spends on the surface and a large negative reward (a penalty) when the ball is dropped. Given enough training time, the network learns to solve the problem and is able to balance the ball correctly. The technique learned to master the task is technically called a **policy** and is modeled by the neural network.

For safety reasons, we must always put some limits in place for the position and speed that the controller can test during its learning phase. Fortunately, since we are using a simulation environment, no real danger can occur. Learning through simulation also offers the opportunity to quickly modify certain parameters and let the model learn how to handle various scenarios. After all, reality represents just one of the many possible parametric sets of a model, so learning from many different sets makes the model more robust and more likely to work on the real robot.

Usually, the training process takes a long time, especially for reinforcement learning, but Unity offers the feature to speed up the physical clock of the environment and collect more training data in a shorter time. Once the artificial brain is trained to a satisfactory degree, we can test its results on the robot. The neural network runs in real time: it receives new input data from the camera and sends its output values to the robot's motion control software.

The reinforcement learning algorithms implemented in Unity can be used to let our robot controller find its own way to solve a variety of other tasks. For instance, interesting applications are the optimization of the path-planning and the trajectory generations algorithms. The controller could find new ways to increase the production output or lower the energy consumption of a given industrial process.

Similarly, we could run the same machine learning algorithms to let a robot find the best set of parameters for its dynamic model. We could start by training a neural network on the identified set of parameters, as described in Chapter 8, and then applying reinforced learning to improve the learned policy over time using batches of online data. The reward in this case would be based on the value of the tracking error of all joint axes.

Another task where machine learning often finds application in industrial automation is *condition monitoring*. A neural network learns to model and predict the status of a machine under standard circumstances and is then able to identify abnormal situations, where the actual behavior of monitored quantities (motor currents, oscillation frequencies, etc.) is not typical. The results are early warnings of possible excessive wear and damage of the machine.

In general, the process of learning a solution for a complex task should ideally be broken down into smaller progressive simpler goals. If the final target is too difficult to reach, the artificial brain might take an extremely long time to find a solution or might possibly never find one. A step-by-step approach is much more efficient and likely to deliver a positive outcome. In our balancing example, we could first let the robot master balancing objects with higher friction on a larger surface. Once that task is accomplished, we could then ask the robot to improve the learned policy and balance smoother objects on a smaller surface or even balance more objects at the same time. The technique of learning by steps of increasingly complexity is known as *hierarchical learning* and is featured in Unity under the name of curriculum learning.

Finally, one more powerful thing you can do with Unity is run a number of different robots in parallel, each collecting different experiences but all sharing the same artificial brain. This kind of communal learning brings in more exploration of the environment from different perspectives and usually improves the quality and speed of the whole learning process. A typical application where a large number of machines are parallelized is a production factory where many robotic manipulators all perform the same task to speed up the common learning curve. Another example is a wind power field, where all turbines share the same learning process to fine-tune the underlying control algorithm with the goal of always pointing in the optimal direction to maximize energy production. Disturbances between adjacent turbines can also be identified and modeled in order to increase the output power generation of the whole field.

Summary

Taking advantage of computer simulation to replicate a robot in a virtual environment offers great advantages in terms of convenience, cost-saving, and safety. In this chapter, we introduced the game engine Unity as one of the possible tools to develop a digital twin for our automated work cell.

Using Unity, you can quickly build a nice portfolio of robotic projects at zero cost, using virtual hardware to replace both the mechanical robots and their control systems. Unity also offers machine learning algorithms in order to train the robots in solving tasks that cannot be easily modeled analytically.

CHAPTER 13

Machine Vision

Adding an electronic eye to a robot opens up a new world of possibilities in terms of functionalities and practical applications. Machine vision is the branch that studies how images captured by a camera can be analyzed to extract useful information. It is a complex and fascinating topic at the same time, and entire books are dedicated to it. We can only touch the very basics here, showing how you can start setting up a simple system to recognize features in images and use them in different ways to increase the capabilities of your robot.

The general idea of driving a robot according to the input of an imagining device is called *visual servoing*. The schematic architecture is shown in Figure 13-1: a camera is used to capture images; different algorithms are implemented to extract certain features from the image and locate specific objects; the position of the observed target is sent to the motion control software, which moves the robot to reach the desired location detected by the camera.

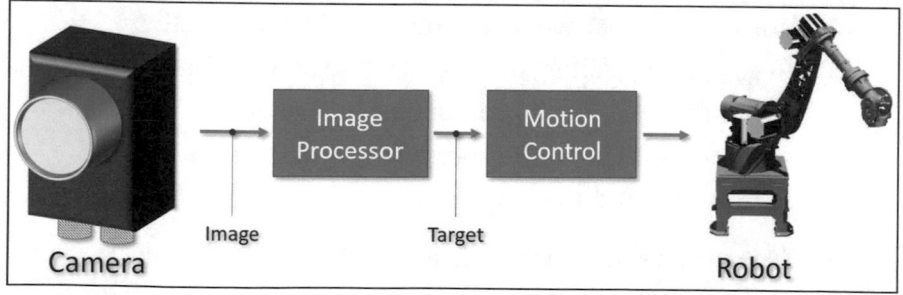

Figure 13-1. *Visual servoing architecture*

© Fabrizio Frigeni 2023
F. Frigeni, *Industrial Robotics Control*, Maker Innovations Series,
https://doi.org/10.1007/978-1-4842-8989-1_13

A large number of industrial applications take advantage of machine vision to solve tasks of increasing complexity. Some examples are as follows:

- **Identification**: This is probably one of the earliest use cases for machine vision. Instead of employing human operators to sort through products, a camera scans a barcode (either 1D or 2D) printed on a label and automatically classifies the object. The procedure is used for tracing and sorting purposes, in logistics and any production line.

- **Positioning**: We have discussed in Chapter 10 how to calibrate the coordinate system of a workpiece, so that a robot can perform actions on it. However, the manual procedure is lengthy, and in most cases, it can be sped up considerably using a camera: the vision software automatically detects the position and orientation of the workpiece and allows the robot to correctly align with it. This technique is used in several manufacturing processes: e.g., PCB assembly, welding, dispensing, etc.

- **Tracking**: Similar to positioning, but with a moving workpiece. The product is first detected by the camera and then dynamically tracked, either visually or with a position encoder, while the robot performs the action (e.g., painting, filling, packaging). The major advantage is that the production process can be executed synchronously to the line's motion, with the result of an increase in productivity.

- **Measurement**: Also known as gauging, this technique replaces manual measurements to assess the size of an object, either for sorting or for inspection purposes.

Optical measurement has the advantage of being a high-precision, high-speed, and no-contact procedure. Typical examples are monitoring the output of a manufacturing line to confirm the correct size of the final products, measuring the size of fruit or vegetables in agricultural outputs for sorting purposes, and checking the level of liquid content in bottle-filling applications.

- **Quality control**: Visual inspection of output products is a critical feature of modern production lines. Verification of integrity and defect detection are nowadays performed by machine vision systems in a fast and reliable way. Object visual analysis can be used to detect scratches, abnormal edges, chipped surfaces, and dirt particles.

In the next sections, we describe the main features of a typical industrial camera and its software used in common machine vision applications.

Smart Camera

Most machine vision algorithms involve the manipulation of images and tend to be computationally intensive. For that reason, industrial manufacturers decentralize the vision software to an external processor instead of running it directly on the robot's main controller. The result is a device often dubbed *smart camera*, which includes the vision sensor itself, plus the lens, some LEDs to provide illumination, and a powerful CPU (or FPGA) dedicated to the computation. The outputs of the vision functions are then sent back to the main controller using a fieldbus. The calibration and configuration of the camera and its algorithms can be performed remotely.

Figure 13-2 shows a typical smart camera for industrial automation. There are several options on the market, with different sizes and resolutions, but they all look quite similar to each other and basically all offer the same functionalities.

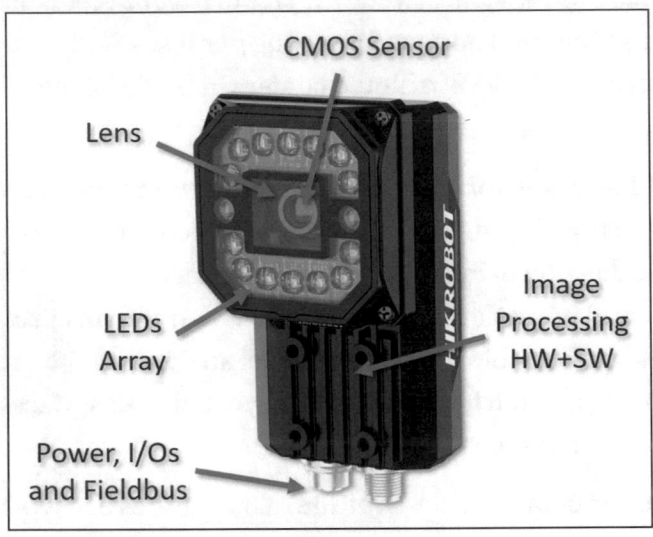

Figure 13-2. *A typical smart camera (with permission from HIKROBOT)*

The core component of the camera is a standard **CMOS sensor**, either monochromatic or in RGB format. The ability to resolve finer details depends on the number of pixels in the sensor. More importantly, the quality of the recorded image strongly depends on the size of each individual pixel: larger pixels offer much better signal-to-noise ratio and higher dynamic range (i.e., better contrast ratio between lightest and darkest regions).

On the other hand, larger sensors are expensive and not always required for all applications. Also, while a larger number of pixels offer higher output resolution, it quadratically increases the size of the output image. That is not desirable for applications that require high-speed data processing: we always need to find a good compromise between accuracy and speed. Unlike consumer cameras, whose goal is to make beautiful photography, industrial cameras are meant to optimally solve a technical task. Efficiency, not beauty, is the final goal. For instance, most vision problems can be solved in a gray color space and do not require color sensors. Besides saving costs, a monochromatic system is also much faster because it produces much smaller images (since each pixel only needs a single gray intensity channel instead of three RGB channels). The result is a further optimization in production speed.

A **lens** is mounted in front of the sensor to correctly focus the light to be captured. Different types of lenses exist for different applications. From a practical point of view, the most important parameters of a lens are the *focal length* and its *aperture*. The focal length determines the size of the observed field: either a wide bird's-eye view of the entire scene or a zoomed-in view focused on a narrower region. The aperture of a lens determines the amount of light that is allowed to reach the sensor. Depending on the intensity level of the scene illumination and on the required exposure time, different apertures can be selected.

In most situations, lenses for industrial cameras are fixed and have predefined configurations optimized for the specific application they are deployed to. In some cases, more flexibility is required, and dynamical configurations are called upon: zoom lenses with variable aperture exist for that purpose. All lenses regardless of their optical parameters offer an electronic focus function to achieve fast focusing of the observation target via software control.

An array of **LEDs** is typically placed around the lens to provide proper illumination to the scene. Algorithms for image analysis are fine-tuned and optimized for specific chromatic scales and brightness levels expected during normal operation. If the light conditions suddenly change, then the vision software might not work reliably anymore, and critical information is lost. Therefore, the camera must provide its own constant source of light instead of relying on ambience illumination, which can vary wildly over the course of a day both in intensity and in color content. In fact, not only absolute light intensity should be kept constant, but also the color of the scene must be controlled individually for each application. Both these features are provided by the LEDs, which can be easily configured and tuned via software.

Different light colors should be used to highlight specific features depending on the colors of the inspected part and of the background. LEDs are available at several different wavelengths to cover the entire visible spectrum. Additionally, filters can be added to the lens to enhance detection of specific color regions. Besides the camera's integrated LEDs, external lights can also be added for further illumination requirements: e.g., side lighting to avoid reflections or backlighting to better outline the silhouette of the inspected part.

Image processing is performed by a powerful CPU of FPGA, which runs demanding vision algorithms. The advantage of integrating the image processing hardware in the camera is that the transfer time latency from the sensor is minimized. The fieldbus connection from the camera to the main robot's controller only needs to transfer low-content information (e.g., position and orientation of the detected object). Besides the serial network, cameras traditionally also provide some I/O interfaces for basic control and monitoring purposes. However, all the software and hardware configurations are normally performed via fieldbus: exposure setting (from a few microseconds to a few seconds), color of the LEDs, selection of image's ROI (*region of interest*, to focus on a specific area of the image and speed up calculations), and selection and parameterization of the feature extraction algorithm.

Once the camera has captured and analyzed the image and transmitted the target location to the robot's controller, then the path-planner can start working out the best way to reach it. This step requires special attention because of the particular geometry of the work cell: the coordinate system adopted by the camera (CCS) is not the same as the global one used by the robot (GCS). The first difference we notice is a possible translation and rotation of the two coordinate systems with respect to each other (see Figure 13-3). We already know how to solve such frame operations by means of an additional homogeneous matrix in the kinematic chain, as described in Zero Frame Sections of Chapters 3 and 4. Keep in mind that the camera might be mounted on the robot's arm, in which case its reference frame needs to be dynamically updated.

Figure 13-3. *Camera frame vs. robot frame*

The second complication is introduced by the lens. First, the size of the recorded image in pixels must be translated in mm in order to correctly locate and size the detected objects. However, that translation depends on the focal length of the lens currently in use. More critically, *the scaling introduced by a lens is not always linear*. In other words, the same object placed in different parts of the image might appear with different sizes (see Figure 13-4). The distortion introduced by a nonuniform scaling is called stretch and must be avoided, either by careful alignment when mounting the camera or by software compensation. In general, the scaling is linear only if the camera sensor plane is exactly parallel to the product plane and if the focal length of the camera is long enough relative to the distance of the observed object.

Figure 13-4. *Uniform vs. nonuniform scaling of an image depends on the angle between the camera and the scene*

The process of measuring the relation between the CCS and the work-cell GCS is called **calibration** and is similar to the procedure described in the Cell Calibration Section in Chapter 10. In particular, the following parameters need to be evaluated:

- *Scaling* [*mm/px*], the linear size of each pixel. In case the scale factor is not uniform across the image, then a stretch matrix also needs to be calculated.

- *Reflection* [±X or ±Y], in case the camera flips one or both coordinate directions.

- *Translation* [ΔX, ΔY], the distance between the origins of the two coordinate systems.

- *Rotation* [$\alpha = atan2(\Delta Y, \Delta X)$], the relative orientation between the two coordinate systems.

Finally, we might face more complex applications that require the measurements of the distance of an object from the camera or the depth of the inspected part. An individual camera flattens a three-dimensional world into a two-dimensional image and is not able to directly extrapolate the third missing dimension. The solution is adding one or more extra cameras and use 3D geometry to recover spatial information. Two cameras observing the same object from different angles allow the software to determine the exact location of the target with respect to the camera and, consequently, with respect to the robot's TCP.

In general, setting up the vision system requires great care to guarantee long-term reliability and robustness. It is easy to get the process up and running to test a few sample images in a controlled environment, but it is much harder to generalize and make sure that the system still behaves with the same accuracy in an industrial setting, where light, temperature, vibrations, and other variables can strongly influence the results.

In order to capture good images and simplify the work of the underlying algorithms, we should provide constant and uniform lighting, beware of high-distortion lenses, and find an ideal trade-off between speed and accuracy by reducing the size of the captured image.

Vision Functions

A large number of different processing techniques exist to extract features from an image depending on the specific goals and requirements. We provide here a brief overview of the main functions offered by most camera manufacturers on the market:

- Edge and corner detection, to simplify the image content while preserving its main structural features

- Contour finding, by analyzing the edges and segmenting the image

- Blob detection, to isolate individual regions of similar color or brightness

- Optical character recognition (OCR), to detect and decode text and numbers

- Barcode detection and decoding, for product identification purposes

- Object localization, to find the position, orientation, and size of a specific object by feature matching

- Object detection, to distinguish between different classes of objects

Some of these functions are implemented using traditional analytical algorithms; others rely on machine learning techniques. We will describe the latter in the next section.

Many commercial software packages exist to offer the previously listed functions and many more. If you want to start implementing a real-time machine vision system by yourself with an industrial camera (or simply with your phone's camera), then a good place to start is **OpenCV**: an open source computer vision library, free also for commercial use, available for most operative systems, and with programming interfaces in C++, Python, and Java.

Let's look at a simple example to see how feature extraction from a digital image works in practice. We start with the base image shown on the left side of Figure 13-5, which contains an object to be detected and localized. Notice how we purposely chose a background color strongly contrasting the object's color to simplify the operation.

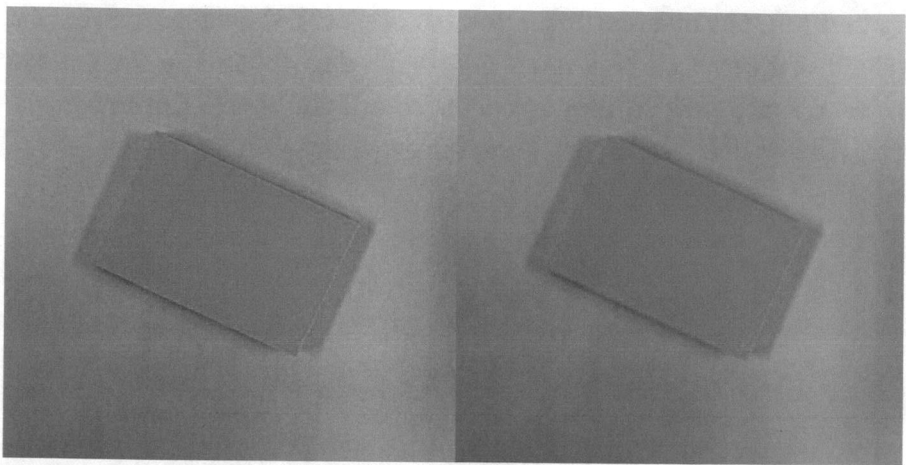

Figure 13-5. *Object to be localized in base image (left) and after preprocessing by blurring (right)*

The feature extraction function that we will use in this example is edge detection. However, applying it directly to the base image does not always return reliable results. A better approach is to slightly preprocess the image before analyzing it, in order to reduce noise and increase contrast. The most common preprocessing operation is **blurring**, which filters and smoothes out the noise introduced by the camera sensor. The result is optically very similar to the original image but numerically more stable for feature extraction algorithms to work on (see right side of Figure 13-5).

Different blurring algorithms are available in machine vision software. Some examples in OpenCV are the functions *GaussianBlur()* and *bilateralFilter()*. The first is a 2D spatial convolution of the image to increase the signal-to-noise contrast. We already described Gaussian filtering in Chapter 7 when smoothing speed trajectories in the dimension of time. Blurring an image is essentially the same concept, only extended to the two [X, Y] spatial dimensions. Essentially, the intensity of each pixel is replaced by a weighted average of the intensities of the surrounding pixels, where the weights depend solely on Euclidean distances (see

Figure 13-6). The bilateral filter is very similar, with the difference that the weights of the average between nearby pixels depend also on distances in the color space besides distances in the coordinate space. The advantage of this idea is that it better preserves the edges in an image while reducing noise.

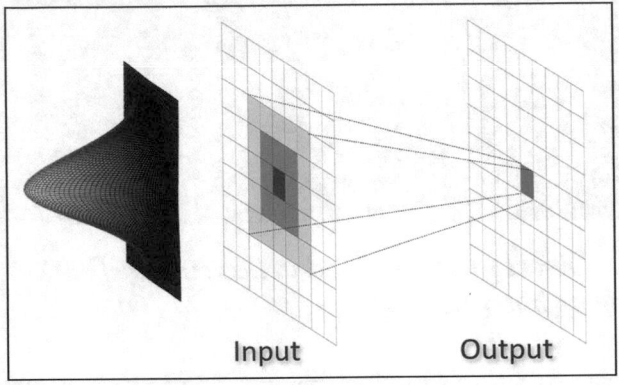

Figure 13-6. *2D convolutions are used to filter noise in an image*

Besides filtering, another preprocessing operation quite often used in practice is converting the image color space to grayscale and then to binary (black-and-white) via thresholding. The reason is that contrast between different intensity areas is much better highlighted in binary images instead of RGB color spaces.

When the image is finally ready for further analysis, we can apply an edge detection algorithm. Different alternatives exist, each with its own strengths and drawbacks: one of the most common choices is the **Canny** edge detector. Finding an edge in an array of pixels is equivalent to searching for sudden jumps in the intensity values. Mathematically, the function first calculates the gradient of the intensity profile ∇I along the two dimensions of the image and then searches for local maxima (see Figure 13-7).

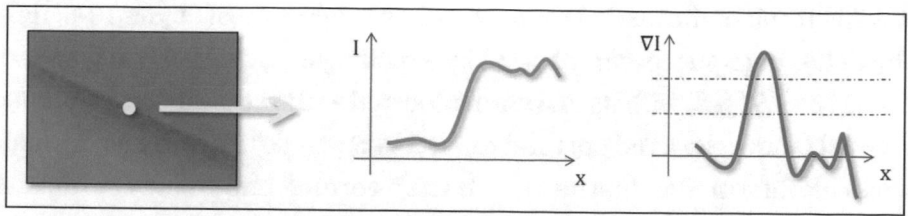

Figure 13-7. *Edge detection requires calculating the gradients of intensity and finding its peaks*

The result of the operation is shown on the left side of Figure 13-8: the edges of the object to be localized have been clearly identified. The last step is to perform a structural analysis of all the discovered edges in order to find specific contours. OpenCV offers a *findContour()* function, plus several additional options to define geometrical sets enclosing the possibly identified closed contours: rectangles, circles, convex hull, etc. In our simple example, we know that the shape of the object to be localized is a rectangle, so we use the *minAreaRect()* function to properly identify it.

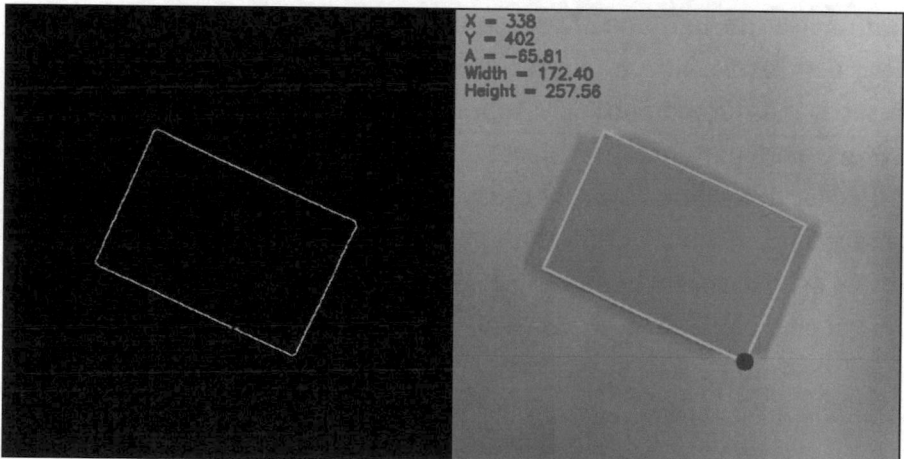

Figure 13-8. *The result of the Canny edge detection (left) and of contour detection (right)*

The result of the analysis is shown on the right side of Figure 13-8: the object has been successfully bound by a rectangle (highlighted in green) of size [172×257 px], with its first corner located at the coordinates [$X = 338$, $Y = 402$] (shown by a red dot) and rotated by $\theta = -65°$ from the horizontal axis. This information first needs to be transformed into global coordinates and can then be passed onto the robot's controller, which will plan the appropriate path to reach the detected object and perform the required operation.

Despite being a very simple example, this kind of procedure is very similar to many practical applications seen in real industrial settings. For instance, I once had to program the control system for a glue-dispensing machine: the dispenser tool was mounted on the robot, and the TCP needed to follow along the contour of the workpiece to deposit liquid glue (see Figure 13-9). The problem was that several different kinds of workpieces could be placed by an operator in random positions on the working table, and manual calibration of the coordinate system and size of each workpiece was a major bottleneck in the production line. The solution was the addition of a machine vision system, which automatically identified the workpiece, localized it in the robot's coordinate system, and correctly measured its size. The robot would then simply run a predefined user program parameterized according to the correct size and location of the identified part. Needless to say, the increase in productivity was massive. The vision algorithm was programmed with OpenCV, and a simple TCP/IP socket was used for the communication with the motion control part. Notice how the background color was purposely chosen green to strongly contrast with the white workpiece to simplify the detection process.

Figure 13-9. *A glue-dispensing machine with automatic workpiece detection*

Another application where I had to use standard contour finding techniques was the automatic visual inspection of metal pipes. The goal was to detect misalignments between the external profile and the internal hole in a quick and contactless way. The solution was simply to generate a high-contrast image from the top and fit the two edges with circles to check for position offsets between their center points (see Figure 13-10).

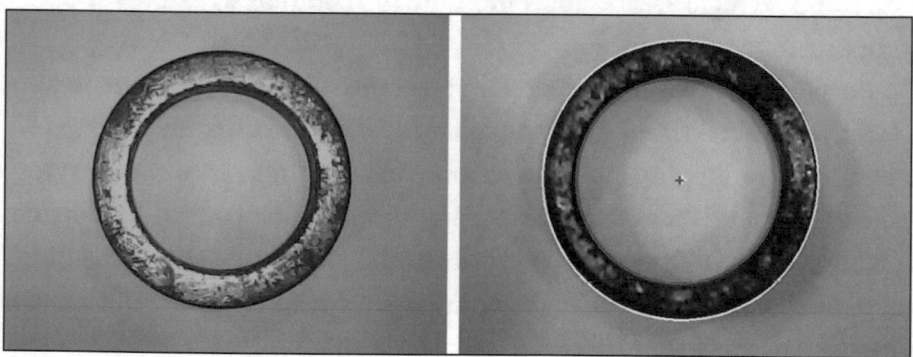

Figure 13-10. *Quality inspection on metal pipes: a good part on the left and a bad one on the right*

A more playful but more complex example I worked on was a robotic arm used to play ping-pong. Identifying the ping-pong ball moving at high speed was no easy task: a powerful processor for high frame rate operation was required. Also, the color of the ball was chosen orange to better extract it from the background image. The round shape and constant size of the ball also helped to isolate it from other possible false positives reported by the detection algorithm. Finally, a small ROI was selected out of the whole image to optimize calculations and reduce the margin of error. A sequence of images over time provided enough data to calculate the vectorial speed of the ball and predict a rough trajectory to help the robot premove to the correct direction and prepare to reach the ball. The results were less than ideal in terms of ping-pong playing skills, but the whole project was for sure great fun to experiment with.

Deep Learning

Some of the functions we presented in the previous section can be realized with standard logic algorithms; others are too complex to express with analytical models.

Recognizing a white rectangular object on a green background is a relatively easy task; but picking out an individual cookie from a group of mixed and overlapping ones can be more challenging. More complex yet is the detection and recognition of text on product labels or the identification of minor details in the shape and surface of objects for quality inspection purposes. For these sorts of tasks, building models on the base of analytical equations is impossible.

Machine learning offers a valid alternative by providing a way to generate models directly from data, as we already described at the end of Chapter 12. Instead of writing down equations to define the relation between an input image and the desired output (e.g., type and location of an object), we train a model to automatically identify patterns in images and extract the desired output information following given examples.

The idea is simple: humans can easily identify objects in an image but can hardly describe how to do that in mathematical terms. For instance, a VGA image has 640x480x3 input intensity values (where 3 is for the RGB channels), and there is no way to write down an equation that given those almost 1 million input values can output the type of object detected and its position in the image. However, a neural network can easily do that. Modeled directly after our brain, neural networks are a collection of nodes (also called neurons) connected to each other in successive layers (see Figure 13-11).

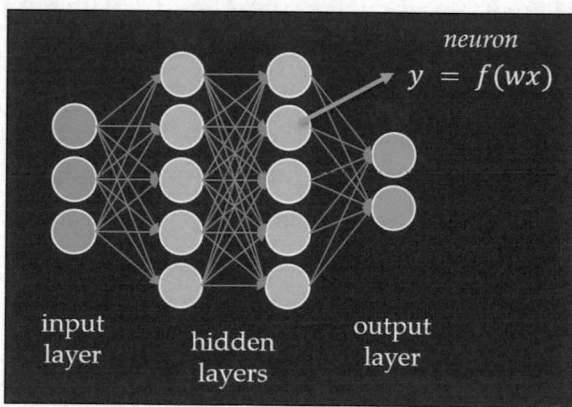

Figure 13-11. *A simple neural network: the information flows through layers of neurons*

The information flows from the *input layer* through all the *inner layers* all the way to the *output layer*. Every time a numerical value x goes through a node, it undergoes a nonlinear transformation $y = f(wx)$, where w is a weight unique for each node and f is the activation function, which can be anything nonlinear, usually a simple rectifier for simplicity and speed.

The resulting output of the network is a function of the input through a combination of all the internal weights of each neuron. You can imagine that a larger network with several layers and many nodes per layer can build up a very complicated function. Actually, given enough nodes, you can approximate any possible input-output relation. The reason why the individual activation function of each node must be nonlinear is that only so can a neural network as a whole approximate a nonlinear function.

The size of the network is called *capacity*: the larger the capacity, the more complex the functions the network can learn. On the other hand, larger networks require much more time and data to learn something useful. For smaller tasks, smaller nets suffice; for massive tasks, deeper nets are better. The term **deep learning** comes for the large number of layers used by some large networks to solve complex tasks, like object recognition. Those are deep nets, as opposed to shallow nets used for simpler tasks.

Neural networks can generate different kinds of outputs. For example, when trained to detect animals in an image, they can return some probability values for each different known class (e.g., cat, dog, horse, etc.). Selecting one output over the all the possible ones is called a *classification problem*. Depending on the dataset we use to train the network, we can have a different number of possible output classes: from just 2 to hundreds of them. It is said that humans can identify approximately 30,000 basic object categories.

We could also ask the network to find the exact position of the identified animal in the image. Generating real values as output is normally called a *regression problem*. Classification and regression can also be combined together by using a network with two separate output layers: a classification head and a regression head (see Figure 13-12).

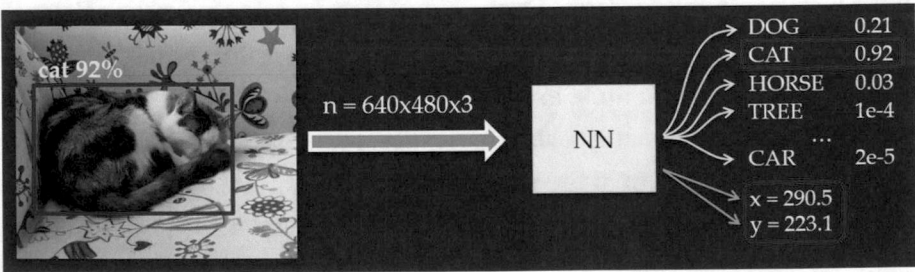

Figure 13-12. *Combining a classification head with a regression head*

Neural networks can also return the same number of outputs as their inputs. For example, autoencoders are used for dimensionality reduction or for semantic segmentation to divide images into sections. The output is an image of the same size as the input one (see Figure 13-13).

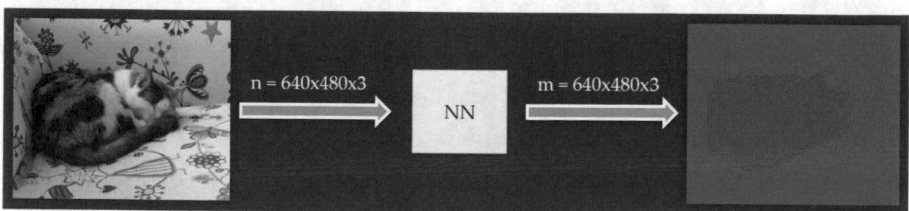

Figure 13-13. *Autoencoders have as many output as input nodes*

Now, the question is how can an artificial brain learn any function at all? The answer is conceptually simple, if we avoid all the mathematical nuances. Recall from the Machine Learning Section in Chapter 12 that there are two possible cases we consider: supervised and reinforcement learning. The first is used when we already know what output must be generated given a specific input. We collect data, from which we derive examples of the input-output relation, and we train the network on that dataset. A typical application of supervised learning is teaching a network to detect and localize a specific object in an image. Another application is classifying and identifying abnormal products for defect detection.

The second option is reinforcement learning, which is used when we do not know what the best output corresponding to an input should be. We can only observe the outcome of different actions and judge the result; we provide that feedback to train the network, and we let it choose the best output on its own. This technique is mostly used in control, where the robot performs certain actions or corrections based on the visual image of the environment. The feedback is based on physical quantities that must be optimized: e.g., total energy consumption for a trajectory, positioning error, and execution time. Typical applications are the optimization of pick-and-place strategies for a line of multiple robots or the improvement of the trajectory for mobile robots.

We focus here on supervised learning, which is the most common approach in most machine vision tasks. The goal is to teach the artificial brain a function $y = f(x)$, where x is the input vector of observation (e.g., an image) and y is the output vector (e.g., class and position of an object). The network initially knows nothing, which mathematically means that it is flooded with random weights, and for every given input, it returns a totally random output. Now we need to adjust all those parameters somehow, so that the output is what we want it to be.

The solution is simple: we show the network many different (x, y) tuples, sampled from a valid dataset, and let it slowly adjust its weights so that its calculations converge as close as possible to those data points. Mathematically, the process of adjusting the weights is essentially an optimization problem: we need to find the optimal weights (i.e., the best set of weights) that fit the given (x, y) dataset. The problem is commonly solved via *gradient descent* techniques.

You can think of it as a curve fitting problem: if x and y were both one-dimensional, we could plot them on a plane and search for the best curve to fit those points. Then, given a new input value x_0, we could use the learned function to predict an output value y_0. In reality, x and y are multidimensional, so the details are a bit more complicated, but the concept is absolutely the same.

Let's consider a simple network with only two hidden layers (just as the one shown in Figure 13-11). Given an input x, we can follow its path through the network: the value will become $h_1 = f(w_1 x + \dots)$ at the first hidden layer, then $h_2 = f(w_2 h_1 + \dots)$ at the second hidden layer, and finally $y = f(w_3 h_2 + \dots)$ at the output. The vector y is the prediction of the network using the current weights; if we modify the weights, then the output will change as well.

The goal of the training process is to adjust all the parameters of the network so that the predicted output y is as close as possible to the desired output $\hat{y}$ given by the training dataset, also known as *ground truth*. The error $\varepsilon = \hat{y} - y$ is a function of all weights of the neurons in the network. We can calculate the derivative of ε with respect to the weights and find the values of w_i for which the error is minimum.

In practice, the typical *error function* that we want to minimize is the Euclidean distance between the two vectors:

$$\varepsilon = \frac{1}{2}(\hat{y} - y)^2 \qquad (13\text{-}1)$$

We first observe how the error function reacts when changing the weights w_3 of the output layer. The derivative $\dfrac{\partial \varepsilon}{\partial w_3}$ can be easily found, given that $y = f(w_3 h_2)$ and considering that h_2 does not depend on w_3:

$$\frac{\partial \varepsilon}{\partial w_3} = \frac{1}{2}\left(\hat{y} - f\left(w_3 h_2\right)\right)^2 \qquad (13\text{-}2)$$

We can then go deeper and observe how the error function changes when modifying the weights w_3 of the second hidden layer:

$$\frac{\partial \varepsilon}{\partial w_2} = \frac{1}{2}\left(\hat{y} - f(f\left(w_2 h_1\right))\right)^2 \qquad (13\text{-}3)$$

The process continues through all the deeper layers all the way back to the input one. Mathematically, calculating the derivatives is a rather simple application of the standard chain rule.

Since we work in a multidimensional space, all those partial derivatives specify the *gradient of the error function* $\nabla \varepsilon$ with respect to the weights of each layer. In other words, they describe how the output error changes when applying small perturbations to the weights. Visually, the gradient shows in what direction each weight should be modified so that the total error function becomes smaller (see Figure 13-14).

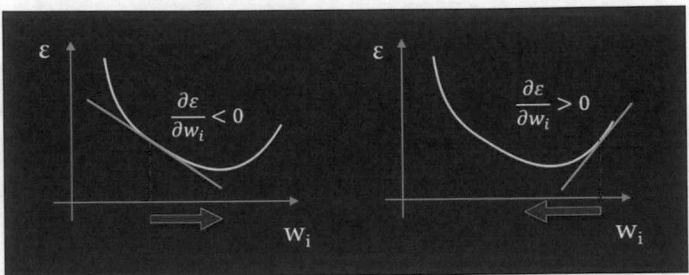

Figure 13-14. *Gradient descent is used to update the weights of a network during learning*

The required update for each weight is simply proportional to the value of the partial derivative:

$$\Delta w_i = -\alpha \frac{\partial \varepsilon}{\partial w_i} \qquad (13\text{-}4)$$

If the derivative is 0, then we have reached the (local) minimum possible error as a function of that specific weight. If the derivative is not 0, then we slightly modify the weight with the intention of reducing the total output error. Note the opposite sign between derivative and update: if the derivative is negative, we need to increase the value of weight to reduce the error; if the derivative is positive, we need to decrease the weight. The technique is known as **gradient descent**.

The proportionality factor α is called *learning rate*, and it regulates how fast the weights are updated. Low α values make learning converging slow; a large α makes the learning oscillate and possibly diverge.

Keep in mind that the error function depends on a huge number of variables and there are likely several local minima, not only one global minimum as we plotted in this example. In other words, the error function is not convex. However, modern optimization techniques are particularly good at finding best results and not get stuck in poor local minima.

The process of calculating the derivatives and updating the weights is called *back-propagation*, because it propagates the error from the output back through all the layers of the network and adjusts all the weights accordingly. The training process is made up of many small consecutive updates of the weights until the error between predicted and desired output is small enough and not further improvement is possible.

Training can take a long time, depending on the size of the network and the size of the training set. By plotting the error function during the training process, we typically observe a behavior similar to the one shown in Figure 13-15.

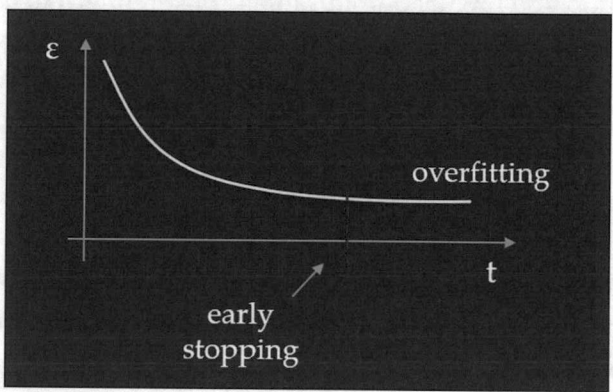

Figure 13-15. *Typical behavior of the error function during training*

Make sure you do not run the training for too long! As every good child educator can tell you, repeating the same exercise over and over for too long does not add any extra benefit, and it actually brings a negative effect in. The child learns the trick by heart but is not able to generalize to other problems. In machine learning, that effect is known as *overfitting*: the network learns the examples in the dataset by heart but then fails to generalize its function to unseen data. It is better to stop the training earlier.

There is a whole topic called *regularization* used to avoid overfitting issues. You can find all details in any standard machine learning course. We focus here on the application and assume that the optimizer to train our network has already been well designed in the standard libraries we will use.

Once all the weights of the network have been trained, we can assume that the target function has been learned and we can now use it to generate valid outputs given any input. This process is called *inference* and is much faster than the learning process. You only need to train the artificial brain once, but then you use it as many times as you want to execute the learned function and predict output values. Most modern networks can run inference for visual tasks in real time (>30fps) on simple standard hardware.

Convolutional Networks

In order to pass information from one layer to the next inside a neural network, the individual neurons of a layer connect to those of the next. In most cases, each neuron of a layer is connected to all neurons of the next: the two layers are said to be *fully connected*.

This architecture is not always optimal though. Imagine you have a VGA image as input: each node of the first hidden layer is reached by almost a million connections from the input nodes. Multiply that by the number of nodes in the layer and then by the number of layers in the network, and you end up with a number of weights that is impossible to train in reasonable times. In other words, this architecture does not scale well to high-dimensional dataset, like high-resolution images.

Another issue of fully connected layers is that they do not take into account the spatial property of the input. When looking for an object in an image, it does not matter where the object is physically located: the result of the classification should be the same. What is important is that certain input pixels are located near each other to build a feature in the image (e.g., the eyes, nose, and mouth in a face), while pixels located far away from each other have no mutual interaction and should be treated separately.

The solution to both problems comes by introducing **convolutional layers**. The connectivity between these kinds of layers is reduced and concentrated in a local region called the receptive field of the neuron (see Figure 13-16). The region also extends in-depth in case each pixel has more dimensions, as in the case of the three RGB channels. The output of the receptive field is not connected to one single neuron but rather to a set of them, known as the *kernel*. Each neuron in the kernel is called a *filter* and is responsible for detecting a specific feature in the input receptive field.

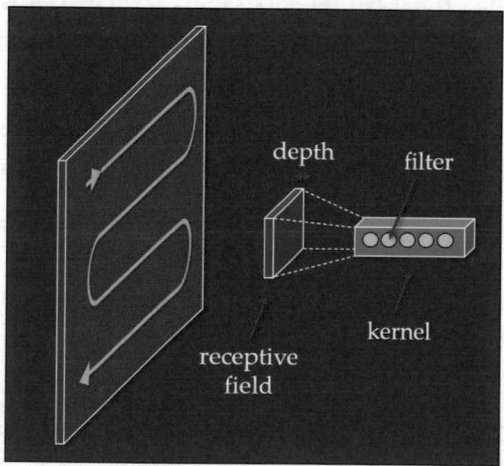

Figure 13-16. *A convolutional layer slides across an input layer*

Now, the key idea is that we only need to learn one kernel of filters for the entire layer and slide it around the image, as shown by the blue line in the draw. That is so, because the function to detect features should be the same for the whole image, regardless of the spatial location. Because we

413

use the same filters across the entire layer, this technique is called *weight sharing*, and it helps to considerably reduce the number of parameters in the network.

During the inference process, we simply slide the kernel over the layer and dot-multiply the input vector times the filters to generate the output volume. This operation is mathematically defined by a convolution, which is why these kinds of layers are called convolutional layers and the network is called convolutional neural network (**CNN**).

The convolutional approach significantly reduces the number of connections between neurons (so that training and inference are much faster) and also makes the operation translation-invariant. The results are impressive: deep CNNs regularly achieve better than human levels at object detection and classification tasks.

However, despite the advantages of using convolutional layers, the networks used in practical applications still have quite a large number of parameters to tune because they are very deep (see Figure 13-17). The training process takes advantage of graphics processing units (GPUs), which were specifically designed to parallelize a large number of simple operations much faster than CPUs (e.g., graphics rendering for each pixel when playing videogames or weight update when training a neural network). Even then, the whole process of training a very deep net on large datasets can take hours if not days.

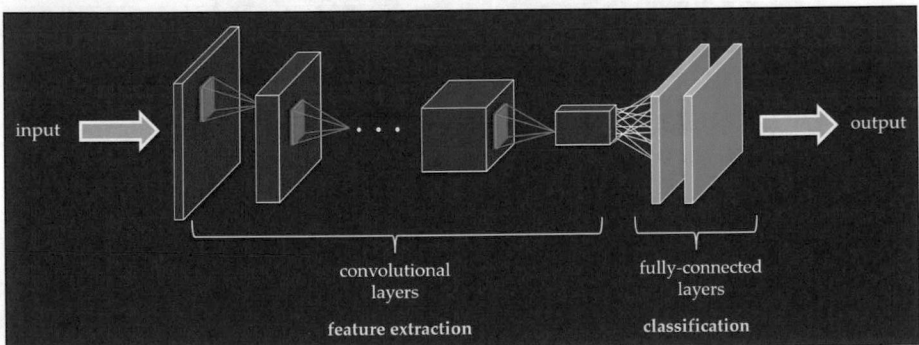

Figure 13-17. *Typical structure of a deep CNN*

414

In practice, instead of training a fresh new network for each application, it is much more convenient to reuse models that have been already trained by others (usually by the researchers and organizations who introduced and implemented the specific architectures) and customize them to our specific needs.

After all, when performing object recognition, all CNNs do essentially the same thing: they use all the deep convolutional layers to extract meaningful features from the image, and only at the very top they have a couple of fully connected layers to recognize objects from those features (classification head) or to localize and measure them (regression head).

What we do is the following: we take pretrained deep networks; we freeze the weights of the convolutional layers for feature extraction and then use our own customized dataset only to train the top classification head, which is quite shallow and converges fast. This technique is called **transfer learning**, because it transfers knowledge directly from one artificial brain to another. The advantage is a considerable speed up in the learning process, since the new network receives a massive knowledge kickstart and only needs to fine-tune a very small part of its parameters to its specific application.

Some traditional CNNs that offer pretrained weights and that can be used for transfer learning are VGG, ResNet, and Inception. On the other hand, if you want to experiment with ready-to-use networks, then some of the best examples these days are R-CNN, SSD, and YOLO. Each network differs in size, accuracy, and speed. Finally, the software libraries of choice to develop and deploy full-fledged machine learning applications are *TensorFlow* and *PyTorch*.

Keep in mind that when training a network to solve a specific task, providing a high-quality dataset is often more important than the network architecture itself. Sample data must be provided in abundance and rich in diversity. For example, if we want the network to reliably recognize between good and faulty products, we need to prepare a large set of images of both classes of objects, as seen under different conditions: viewing angles, illumination, texture, background, etc. The larger the diversity of examples in the training set, the better the network will be able to generalize to unseen cases from real life.

Often though, collecting and labeling sample data can be a strenuous task. In other cases, data is either not available in large quantities or maybe too expensive to collect. A common solution is to artificially generate new data by taking a few given example images and slightly modifying them in lots of random different ways: e.g., resizing; rotating; translating; stretching; cropping; retuning brightness, contrast, and color; injecting noise; and so on. Using a simple automated script (e.g., with the help of OpenCV), we can quickly generate a massive number of random samples to significantly increase the size of the training set and improve the performance of the detection algorithm. The procedure is known as **data augmentation**.

Another common way to automatically synthetize images is the use of generative adversarial networks (GANs). One network, called the generator, takes random noise as input and generates a realistic output image. In order to train that function, we use a second network, called the discriminator, which is a basic classifier and provides feedback to the generator by observing and judging the output images. After some training time, the generator will be able to output better and better images resembling the desired examples and therefore providing a large augmented dataset for free.

Augmenting the training dataset is also very useful when the available samples are not well distributed between the possible classes. This condition is called *class imbalance* and is typical of situations where one example is much easier to collect than another (e.g., lots of good products against only a few faulty ones). When the classes are imbalanced, the prediction accuracy of the network decreases considerably, as the classification process becomes more biased toward the more common category. Artificially generating more sample data for the missing class(es) restores balance and improves performance.

Let's now look at a few examples where convolutional networks are useful in practical industrial applications. OCR is one typical application and one of the easiest to solve, also with relative shallow networks. Reading digits and text from an image can be performed at real-time speed by a small microcontroller. In industrial settings, this feature is used by any machine needing identification and sorting functionalities based on product labels. On the fun side, I once built a robot able to read a handwritten algebraic expression, solve it, and speak out the result: faster than typing it in a calculator (see Figure 13-18). Similarly, text can be easily extracted from an image, translated to a different language, and pasted over in real time.

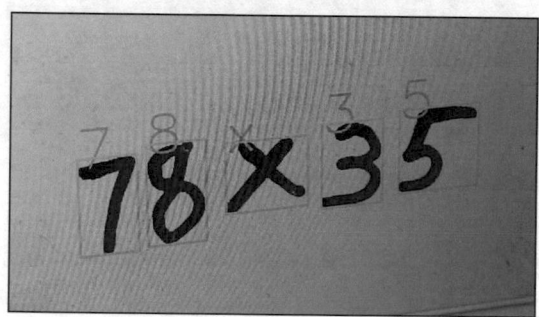

Figure 13-18. *A CNN can easily read digits and text*

A step further is the recognition of complex objects: products on a conveyor to be picked and sorted by a robot, screws to be driven in a panel, human operators to avoid collisions, and familiar faces to activate security gates. I once built an automatic pet feeder to identify approaching cats and activate the food and water dispensing motors accordingly.

More difficult yet is the requirement of distinguishing between different objects belonging to the same class: for example, evaluating the ripeness of tomatoes to sort them through different channels, or recognizing cracked cookies to discard them from a packaging line (see Figure 13-19), or identifying chipped edges on metal parts during standard assessments. Preparing a high-quality training dataset is critical for these kinds of applications.

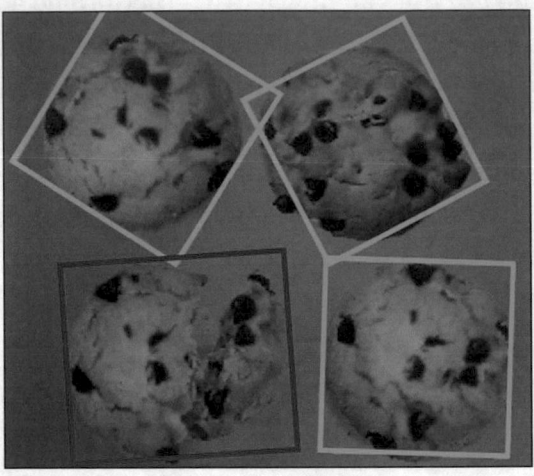

Figure 13-19. *A CNN can classify objects*

Finally, measuring the cross-size or the surface area of products is also an often-required task. Besides using traditional contour finding techniques, the task can be also solved by convolutional networks, which are able to automatically subdivide an image in different sections, with each pixel assigned to a specific class. The procedure is known as *semantic segmentation* and produces an output equivalent to the original image but modified in the form of a heat map (see Figure 13-20). The result allows to quickly measure the number, size, and orientation of specific objects. Because these networks only have convolutional layers by having ditched the fully connected layers at their heads, they take the name of *fully convolutional networks* (FCNs).

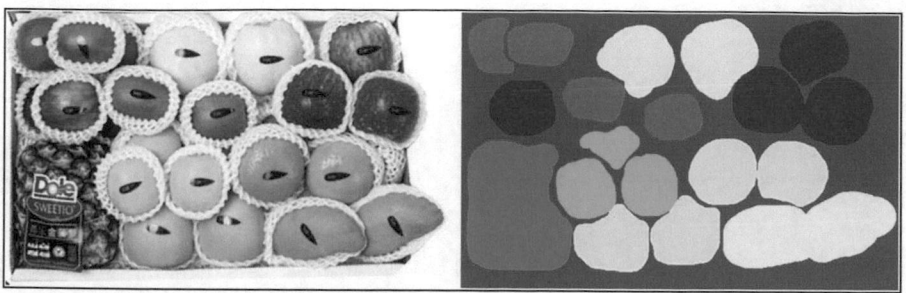

Figure 13-20. *Semantic segmentation can be used to identify and measure objects*

Apart from industrial applications, this technique is also used in many other fields: from autonomous driving, to agriculture, to medical image diagnosis, to automatic photo and video editing. For instance, any phone's camera software can pass the real-time image through a segmenting network to isolate the main subject and blur out the background, simulating shallow depth of field photography.

Summary

Machine vision is a hardware and software package that most robotic projects these days rely on for high-automated work processes. The availability of powerful and cheap hardware components, together with the support of feature-rich open source software libraries, allows even the humblest hobbyist projects to perform impressive visual control functions.

In this chapter, we learned how industrial smart cameras are built and what functionalities they offer to application engineers. Some functions rely on exact mathematical models; others take advantage of deep learning algorithms to extract patterns from large datasets and solve increasingly complex tasks with superhuman accuracy and speed.

PART IV

Robot Hardware

We now come to the last part of this book, where we learn how to design and build the electronic hardware required to control and move a robot. If you are a hobbyist and a maker, then you will likely try to build everything from scratch by yourself, which is always the best way to understand how things really work. If you are a robotics service technician, you will be more likely to purchase industrial-grade PCs and servo drives already available on the market. Nevertheless, this chapter can still be useful to gain a better understanding on the internal details of the hardware you use and hopefully get the best performance out of it.

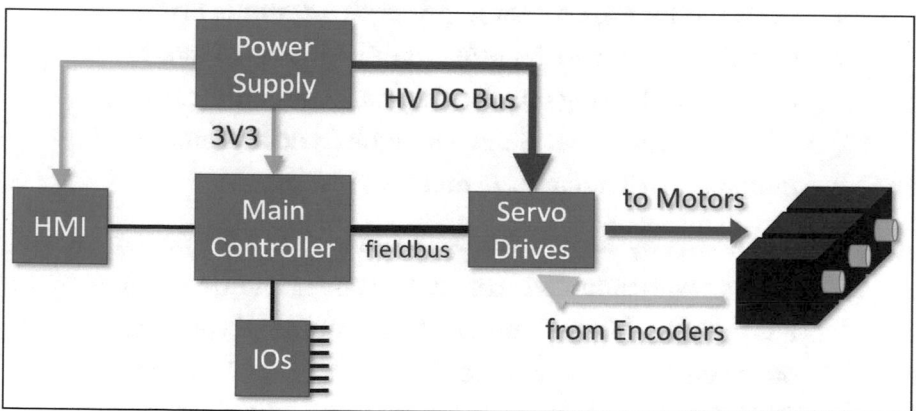

Figure IV-1. *Basic hardware configuration required to drive an industrial robot*

The essential hardware components needed to drive a robot are shown in Figure IV-1:

- **Motors**: While technically any kind of actuator can be used to bring the joints into motion (e.g., hydraulic actuators are often used for large hexapod platforms), the vast majority of industrial robots rely on electric motors to do the job. We will look at different kinds of electric motors and explain how to choose the right one for the application.

- **Encoders**: High-quality motor control requires the knowledge of the rotor's position (see servo loop structure and FOC commutation technique described in Chapter 9). The position feedback to the closed-loop controller is provided by an encoder attached to the motor shaft. We will describe some different kinds of encoders and their communication interfaces to the servo drives.

- **Drives**: Running a motor requires an electronic drive, and different kinds of motors require different kinds of drives. Considering that the vast majority of industrial robots work with permanent-magnet synchronous motors, we will focus here on servo inverters.

- **Power supply**: All electronic components require power to run. Microcontrollers run at 3.3V and consume much less than 1W. Motors can run at a few hundred volts and often consume power in the kW region. We will analyze the different power requirements of all the electronic parts in a robotics control system.

- **Controller**: The main controller is the brain of the robot, where most of the control software runs. It includes some IOs to connect to external equipment and a fieldbus to connect to the drives. Industrial robots' controllers typically rely on powerful multicore CPUs and large memory banks. However, today's processors are cheap and powerful, so even a small MCU can achieve great results by running complex motion calculations in short cycle times at a fraction of the cost. We will show how to design a simple MCU-based robot controller.

- **HMI**: As a developer, you normally program and deploy your code to the controller. However, you also need to provide a simple interface for the operator to configure the required parameters and send commands to the robot. That is usually done by a visualization panel, where hard and soft keys allow for configuration and operation of the robot. Older-style robots used an external teach-pendant hardwired to the main controller. Modern solutions rely on consumer tablets with touch screens and Wi-Fi connection to the robot.

At the end of the hardware section, we also add some details about practical fabrication: we describe how to design your own PCB and have it manufactured in a quick and cheap way for series production; we also suggest some ways to design and produce mechanical parts in case you want to build the robot's frame body on your own.

Traditionally, all robot's driving electronics (power supply, main controller, and motor drives) were confined in an electric cabinet, with long thick cables reaching out to the robot's body, where the motors and the encoders are placed. A more modern and convenient approach, especially for low-power robots and all cobots, is to integrate the electronics inside the robot's mechanical structure: the servo drives are directly attached to the motors and the main controller embedded in the base. The resulting solution considerably saves space and wiring costs.

CHAPTER 14

Motors

Electric motors are clean and efficient and can be driven with very high dynamic profiles. They outperform all other kinds of actuators, unless extremely high forces are required, in which case hydraulic actuators are preferred (e.g., presses). Since the advent of high-capacity and fast-charging batteries, electric motors have also replaced combustion motors in mobile vehicles.

Electric motors convert electrical power into mechanical power. The input electrical power depends on the applied voltage and the drawn current. The output mechanical power is a combination of speed and torque. The input voltage is directly related to the output speed. For example, running a DC motor with no attached load at different voltages modifies the rotational speed of the shaft. Current, on the other hand, is directly related to torque. A motor drawing more current produces more torque. Equivalently, a motor running at high speed does not necessarily draw high current, particularly if the load is light and little torque is required.

In fact, two motors with equal rated power can be built differently (with different internal windings structures) to achieve either higher speeds (allowing for higher voltages to be applied) or higher torques (allowing for higher currents to flow through).

All electric motors have a stationary section, called *stator*, and a moving section, called *rotor*. How exactly these two sections are controlled to move with respect to each other depends on the specific kind of motor. We will explore some of them in the rest of the chapter. In general, some

© Fabrizio Frigeni 2023
F. Frigeni, *Industrial Robotics Control*, Maker Innovations Series,
https://doi.org/10.1007/978-1-4842-8989-1_14

coil windings are used to generate an induced magnetic field, which interacts with the magnetic field of the permanent magnets placed in the motor and generates torque.

Electric motors can be modeled with a series of a resistor, an inductor, and a power source (see Figure 14-1). The resistance of the windings is typically very low and allows high currents to flow through the motor to generate torque. The inductance also comes from the windings and is a property that opposes the change of current flowing through the coils. The power source is caused by the rotation of the motor itself, which generates a voltage proportional to the rotating speed. This voltage is called *back electromotive force* (BEMF) and opposes the supply voltage applied externally to the motor.

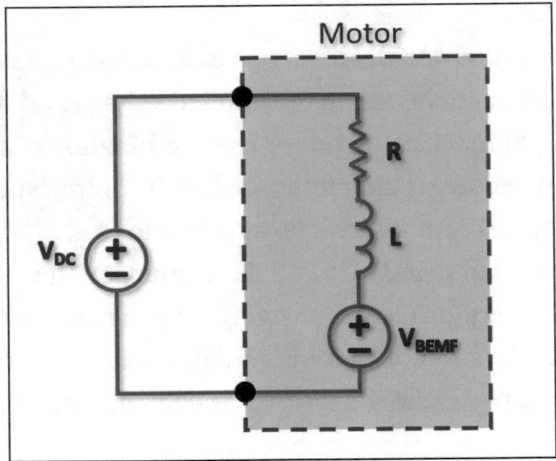

Figure 14-1. *Equivalent circuit of an electric motor*

The BEMF is an inherent property of all motors, and it practically limits the maximum rotational speed. At higher speeds, the V_{BEMF} increases, which means the remaining net voltage available to generate torque gets smaller and smaller. At a certain speed, the V_{BEMF} and the V_{DC} supply balance out with the generated torque needed to overcome friction and drive the load. That is the maximum speed of the motor at that given supply voltage. Higher speeds can be achieved with a higher supply voltage, as long as the motor can support it without burning out.

The existence of the BEMF also explains the dynamic braking effect that was briefly described in the Section on Electronic Commutation in Chapter 9: shorting the terminals of a motor together essentially forces the BEMF to 0. In other words, the rotor is not allowed to move anymore, and the movement rapidly decelerates to a stop. After that, the current still stored in the inductance will slowly decay while burning off as heat across the resistance.

Besides the electrical model, we also need to understand the mechanical structure of a motor. The most important components used in a typical electric motor are shown in Figure 14-2 (in a very simplified way). In this particular case, the rotor is on the inside and the stator on the outside. Other motors are built differently and might have the rotor on the outside (like the one showed earlier in Figure 9-23). Also, the magnets are the moving part here, while the windings are the stationary part. Other motors (notably most linear motors) are built with fixed magnets and moving coils. The principles of operations, however, are the same in all cases.

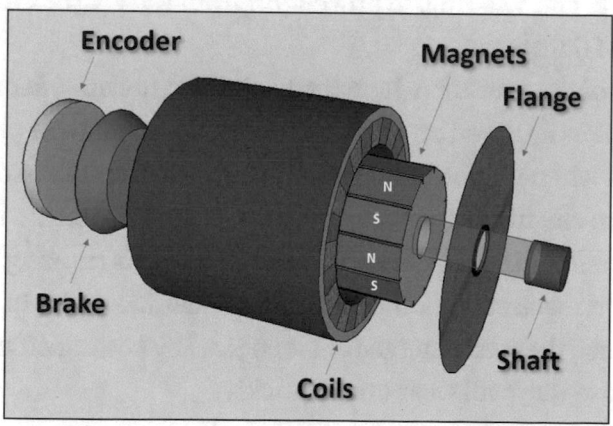

Figure 14-2. *Simplified assembly of a typical electric motor*

The *shaft* is shown very short in the drawing for clarity, but it typically goes across the whole motor for best mechanical support. If the rotor is on the inside, then the shaft also rotates and transmits torque to the load. If the rotor is on the outside, then the shaft is stationary and holds the motor in place.

The front face of the motor is called *flange* and is typically mounted on the mechanics of the robot (either directly on the body or on the gearbox, if present). Its main goal is to dissipate heat through conduction. That is a very important function, because motors can generate a large amount of heat and their operating temperature can easily exceed 100°C. They can overheat and damage if not cooled properly. Most motors used in robotic applications have an internal temperature sensor to allow for monitoring and warning from the servo drive.

In the specific case of cobots, which are meant to work in close contact with people, motors are designed to operate at lower temperature, typically not exceeding 80°C. Besides protecting people, lower operating temperature also reduces the risk of damage to the electronics of the motor drive, which in cobots is often mounted directly next to the motor itself.

When attaching a gearbox to the motor, care must be taken that the shaft mounts a safety O-ring, to prevent oil from the gearbox to leak inside the motor and damage it.

The back of the motor is where the brake and the encoder are normally mounted. Brakes are used on motors to make sure that their position does not shift when no power is applied. They use friction discs to provide torque and are electromagnetically (de)activated. These are called *holding brakes* and are not meant to stop the motor while it is running. They should only be released after the mechanical axis has come to a complete stop. Do not use them as emergency stops (i.e., by cutting off the main power) because they will wear down quickly.

Brakes are engaged when no power is fed to the motor, so that the axis does not move and the robot does not fall down under its own weight. However, a short time overlap should always be applied between electric torque and friction torque to guarantee that no position shift happens at startup and shutdown. In other words, the controller should activate the brake moments before electric power is completely removed from the motor; similarly, the brake should only be disengaged moments after the controller energizes the motor. The time overlap between brake and current controller is usually in the range of 0.1~0.3 seconds.

There are several types of electric motors on the market that you could choose from to actuate your robot. The overwhelming majority of industrial robots are fitted with *permanent-magnet synchronous motors* (often simply referred to as brushless motors) and mount high-resolution optical encoders for servo loop feedback. These systems are highly accurate and offer great dynamic performance but also come at a high price tag and require complex controllers. For simpler applications with lower requirements, you could start experimenting with DC brushed motors and then move on to stepper motors, before starting designing servo controllers.

DC Motors

The simplest and cheapest kind of electric actuator is the DC brushed motor (see Figure 14-3). Control of this motor is very easy: applying a DC voltage to its two terminals makes the motor spin with a speed proportional to the voltage (under no load condition). This simplicity and its very low cost make the brushed motor a favorite among hobbyists. Indeed, it still finds application also in low-end household appliances and basic industrial machines or automobiles parts.

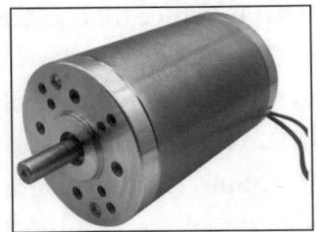

Figure 14-3. *A DC brushed motor*

The structure of a brushed motor is composed by an external stator, where the permanent magnets are fixed, and an internal rotor, where the coils are wound (see Figure 14-4). The coils are connected to the external wirings via brushes, which make and break contact with different windings as the motor rotates. As soon as the rotor comes into alignment with the next magnet, the brushes come in contact with the opposite wiring, which in turn energizes the next winding, allowing the rotation to continue indefinitely. This convenient mechanical commutation eliminates the need for electronic commutation and makes the control of brushed motors very convenient.

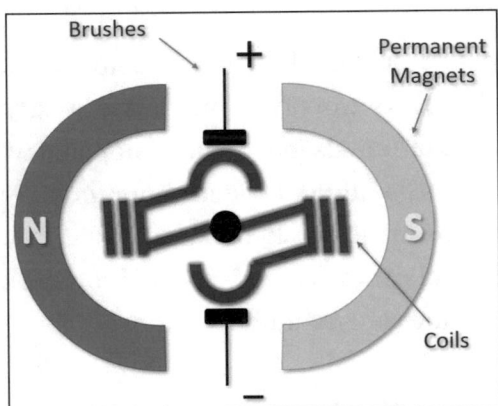

Figure 14-4. *Simplified internal structure of a DC brushed motor*

Since commutation is entirely mechanical, all we need to do to drive a brushed motor is apply a DC voltage to its terminals, proportional to the desired speed and with a sign equal to the desired direction of rotation. The standard way to do that is use an H-bridge with PWM modulation, as we already described in detail in Chapter 9.

Furthermore, DC motors can be driven with or without encoder, depending on the required accuracy. When working in open-loop, the controller generates a voltage proportional to the trajectory speed, without knowing the actual status of the motor. On the other hand, when a position (or speed) feedback is provided, the controller can work in closed-loop and increase the control quality by adjusting the voltage according to the relative error between set and actual speed.

Despite simplifying the motor's control, the brushes are also the weak spot of the DC motor, because their constant physical friction and electric switching wear them out with time: as a consequence, the lifetime of brushed motors is much lower than that of other types of motors.

Another issue with DC motors is their poor thermal management. The coils, where the current flows and heat is generated, are placed in the inside and have little chance to dissipate heat by contact with the environment to cool down.

Stepper Motors

The next step-up from DC motors, in terms of performance, driving complexity, and price, is stepper motors (see Figure 14-5).

Figure 14-5. *A stepper motor*

The internal structure of stepper motors is shown in Figure 14-6. The coils are mounted on the external stator, while the magnets are arranged on a pair of toothed wheels in the inside. Notice the absence of brushes, which means that commutation of the winding is done electronically. However, driving stepper motors is fairly simple: energizing the coils in sequence will make the motor turn one step at a time.

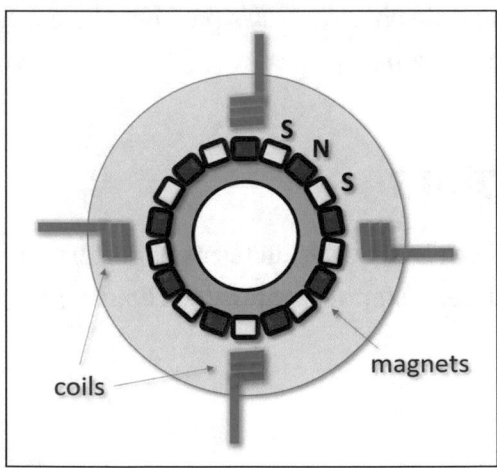

Figure 14-6. *Simplified internal structure of a stepper motor*

By generating a correct sequence of pulses, we can drive the motor at the desired speed in either direction. On the other hand, applying a constant voltage to the coils will hold the rotor still. The resolution of the movement depends on the step size, which is a structural characteristic of the motor. A procedure called micro-stepping is typically implemented to achieve higher resolution and decrease the noise and the torque ripples. The idea consists in applying modulated voltage values to the coils instead of full pulses.

Stepper motors are excellent for simple positioning applications, given their low cost when compared to servo systems. Also, they provide impressive high torque values at low speed. On the other hand, their torque decreases significantly at high-speed ranges, and in those cases servomotors performance becomes dominant.

Just like DC brushed motors, stepper motors can also be driven both in open-loop and closed-loop configurations. Open-loop is more common because encoders are expensive and because the very nature of steppers is to advance one step at a time, which provides a good positioning accuracy for most applications. Things can always go wrong though, and steps can be skipped at times, especially when the driven load requires more torque than the motor can produce. In that case, the rotor is overloaded and it misses out several steps, with the result that its position can no longer be determined. To prevent this situation from creating safety concerns, closed-loop operation is preferred.

Brushless Motors

We now come to the best performant motors on the market used in virtually all industrial robots. These motors are brushless (i.e., require electronic commutation of the coil currents); they are built with permanent magnets on the rotor; their stator and rotor magnetic fields rotate with synchronous speed; and they are controlled using a servo loop. Because of all these overlapping characteristics, this class of motors is often addressed with several names: brushless motors, or permanent-magnet synchronous motors (PMSM), or servomotors. We use all three names indistinguishably in this book. There might be minor internal differences between some of them (e.g., trapezoidal vs. sinusoidal windings, outer rotor vs. inner rotor), but the general working idea is the same for all of them.

Figure 14-7 shows a typical servomotor used in industrial machines. This particular kind is an *inrunner* motor, with the rotor and its permanent magnets located in the inside. Its internal structure is shown in a simplified way in Figure 14-8: The magnets are fixed on the rotor, and the coils are wound on the stator.

Figure 14-7. *A permanent-magnet synchronous motor (with permission from FinePower)*

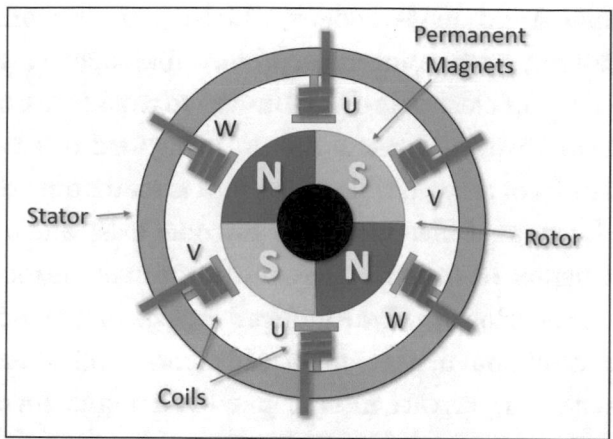

Figure 14-8. *Simplified internal structure of a brushless motor*

Each magnet is called a pole, and each couple N-S is called a pole pair. The number of pole pairs is an important feature of a servomotor and must be correctly parameterized in the control algorithm.

The places where the coils are wound on the stator are called slots, and the number of slots in a brushless motor is always a multiple of 3 because there are three phases to be wound (U, V, W). The number of stator slots may be higher or lower than the number of individual magnetic poles. All the windings of the same phase (e.g., all the U windings) are connected together, so that only three wires come out of the motor.

The most important feature of servomotors is that they exhibit a relatively flat torque-speed curve. In other words, they can produce the same torque performance at low speeds as well as at higher speeds, up to the nominal speed of the motor. After that point, the torque will decrease quickly, and operation in that region is not recommended. A constant output torque performance is very desirable for many applications: robotics in particular will benefit from that characteristic, as we know that joint axes often move with nonlinear speed profiles.

Unlike brushed and stepper motors, which can easily work in open-loop configurations, servomotors must provide their actual position to the controller in order for electronic commutation to work (see Chapter 9 for details on how to drive synchronous motors using field-oriented control). In the vast majority of cases, position feedback is read from an encoder mounted on the motor. Different kinds of encoder exist, and they will be described in Chapter 15. In some cases, the position of the rotor can be estimated by monitoring the BEMF induced voltage on the motor phases (sensor-less control), but that technique only finds application in simple machines that typically run at constant speed and require low accuracy (e.g., conveyors) and cannot be applied to robotic axes.

The fact that electronic commutation allows to maintain the magnetic field of the stator always perpendicular to that of the rotor optimizes efficiency by generating the desired torque with the least current and also provides a smoother output torque profile when compared to brushed or stepper motors.

Servomotors can be directly connected to the robot's axes (direct drive), although more often they are linked through gearboxes (either planetary or harmonic) in order to increase output torque and decrease the apparent load inertia.

Table 14-1 provides an overview of the motor features we analyzed in this chapter so far: servomotors are the clear choice for industrial robots, although they might be too expensive and too complex to drive for hobbyist projects. You should choose the right motor depending on the technical requirements and available resources.

Table 14-1. *Comparison between different kinds of motors*

	Brushed motor	**Stepper motor**	**Servomotor**
Cost	Very low	Low	High
Torque/speed	Low torque at low speed	High torque at low speed; low torque at high speed	Constant torque over entire speed range
Accuracy	Low	High	Very high
Driver	2-leg bridge	4-leg bridge	3-leg bridge
Commutation	Mechanical, no programming required	Electronic commutation with simple sequence of pulses	Electronic commutation with field-oriented control
Encoder	Usually not required	Usually not required	Typically required
Overall performance for robotics	Poor	Good	Excellent

Linear Motors

Although linear motors are seldom used in robotics, we briefly describe them here because they can be useful for miscellaneous applications where robots are also involved. One typical example are linear transport systems, which allow individual products in a production line to independently move on a rail and stop at different stations, where various tasks are often performed by robotic arms. Synchronization between the robots and the linear motor is simple to achieve if they are driven by the same central controller.

Linear motors (see Figure 14-9) are very fast and precise actuators. The load is directly mounted on the motor without gearboxes in between, so the actual acceleration is very high, and there is no backlash to account for. The movement accuracy is determined by the encoder (usually in the order of 1µm).

Figure 14-9. *A section of a linear motor, opened up to show the controllers for each individual cart and the permanent magnets fixed in the center of the rail*

The principle of operation of linear electric motors is exactly the same as the driving concept of standard servomotors: the movement of the cart with respect to the rail is achieved by the modulation of an induced magnetic field in the coils with respect to the constant flux generated by the permanent magnets.

There are two different configurations of linear motors available on the market: a first kind with the magnets fixed on the cart and the windings distributed along the rail (left in Figure 14-10) and a second kind with the magnets fixed on the rail and the coils wound on the cart (right in Figure 14-10).

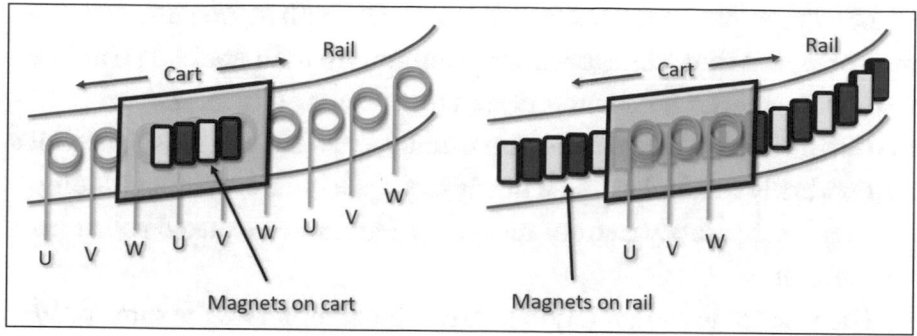

Figure 14-10. *Two different configurations of linear motors*

The kind on the left side is said to have a *passive cart*, in the sense that the cart is entirely electrically isolated from the environment and it is driven by the rail. Each cart simply acts as a passive transport platform without any ability to perform other functions. Furthermore, the wiring and control of the windings in the rail are very expensive and complicated, because every small section of rail requires its own servo drive.

On the other hand, the kind on the right side is said to have an *active cart*, because the cart itself is in control of its position while the rail is simply a passive stator. The advantage of this solution is that wiring and control are enormously simpler and cheaper, since only one servo drive per cart is needed. Also, each cart carries electric power and connection to the main controller, so it has the ability to mount its own sensors and actuators to further increase the potential range of applications. The only disadvantage is that the carts are moving and the wirings need to follow along. For large systems, the typical solution is to mount the control electronics directly on the cart and let it communicate with the other carts (or to an optional central controller) via wireless signals. Electric power is drawn from a common low-voltage DC bus on the side of the rails by means of brushes or wheels, much like a subway train is powered.

Linear motors for transport systems can be built in long rails with open or closed rings configurations analogous to train tracks. The ability to control each cart independently is a powerful feature for modern production lines, where batches are usually small and often each product is individually customized, so it needs to stop at different stations along the line. Clearly, safety features must be programmed to avoid collisions between carts.

The electronic commutation to drive linear motors is the same used for synchronous motors. The only difference is that the units are linear instead of angular: the distance between pole pairs is measured in µm instead of degrees and is often referred to as *magnet pitch*. Similarly, the encoder value is also measured in µm instead of degrees. As in the case of rotary motors, the relative shift between magnets and coils is typically referenced to 0 in correspondence of the magnetic axis of phase U (shown in Figure 14-11). During operation, the distance between the current position (measured by the encoder) and the zero position is transformed into an electric angle in the 0~360 degrees range and is fed to the standard FOC algorithm to generate the phase voltages.

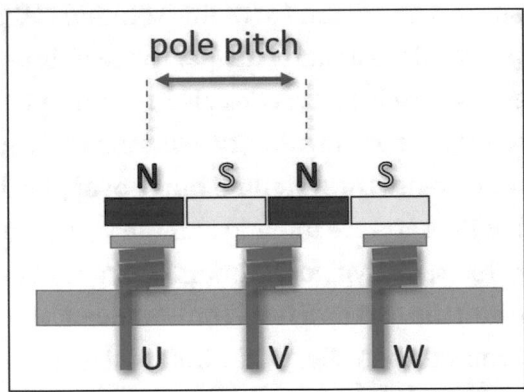

Figure 14-11. *Zero reference electric position of a linear motor*

Despite their great advantages, linear motors tend to be quite expensive and are used mainly in high-end applications. For cheaper and less accurate use cases, they are often replaced by simpler combinations of rotary motors and mechanical transmission systems (e.g., belts, ball screws, rack and pinion, etc.).

Motor Sizing

Once you have selected the best type of motor for your application, you need to pick the right size. Motor sizing is an important task for practical engineering: oversizing the motors for a robot means a wasteful increase in material cost; on the other hand, undersizing the motors might prevent the robot from reaching its required movement specifications.

Motors come in many different configurations according to their mechanical and electrical parameters. The three most important by far are speed, torque, and inertia. However, all parameters have their practical significance, and it is good to understand their meaning. We look at a typical datasheet of a permanent-magnet synchronous servomotor (see Figure 14-12) and describe all its listed parameters. We distinguish between *peak* values (i.e., the maximum critical values allowed before some sort of damage occurs) and *rated* or *nominal* values (i.e., the maximum operating values allowed for any length of time).

技术数据 Specification	单位 units	5FSNA83- [nnn]XabXe0.00-0		5FSNA84- [nnn]XabXe0.00-0		5FSNA85- [nnn]XabXe0.00-0		5FSNA86- [nnn]XabXe0.00-0	
[nnn]		220	300	220	300	150	200	150	200
极数 Number of poles		6							
额定转速 n_N Rated speed	min^{-1}	2200	3000	2200	3000	1500	2000	1500	2000
额定转矩 M_N Rated torque	Nm	31.0	27.0	51.5	48.5	77.0	72.0	97.0	85.0
额定功率 P_N Rated power	KW	7.1	8.5	11.9	15.2	12.1	15.1	15.2	17.8
额定电流 I_N Rated current	A	14.0	17.0	23.5	30.0	22.5	29.5	31.0	35.0
静态转矩 M_0 Stall torque	Nm	40		69		94		115	
静态电流 I_0 Stall current	A	18.2	25.0	31.4	43.1	27.6	39.2	37.1	47.9
转矩常数 K_T Torque constant	Nm/A	2.2	1.6	2.2	1.6	3.4	2.4	3.1	2.4
电压常数 K_E Voltage constant	V/1000min^{-1}	135	99	135	99	198	150	198	150
定子电阻（线）R_{2ph} Resistance (line)	Ω	0.39	0.21	0.15	0.10	0.30	0.15	0.21	0.11
定子电感（线）L_{2ph} Inductance(line)	mH	8.5	4.9	4.8	2.8	8.6	4.4	5.6	3.5
无制动器转动惯量 J Inertia without brake	Kgcm^2	58		107		142		182	
带制动器转动惯量 J_{Br} Inertia with brake	Kgcm^2	96		149		190		230	
无制动器重量 m Weight without brake	Kg	39		48.5		65.9		83.5	
带制动器重量 m_{Br} Weight with brake	Kg	46		55.5		72.5		90.5	
制动力矩 M_{Br} Holding torque of the brake	Nm	130							
电源电缆推荐横截面 Recommended Cable Cross Section	mm^2	4	4	10	10	4	10	10	10

Figure 14-12. *Typical motor datasheet (with permission from FinePower)*

- **Pole number**: This is the number of magnets attached to the rotor and must be properly configured in the servo drive firmware in order to calculate the correct electric angle during operation (see electronic commutation in Chapter 9).

- **Rated speed**: The maximum speed at which the motor can move before a quick decrease in torque will occur. This value should not be exceeded in practical applications.

- **Rated torque**: The torque value that the motor outputs at its rated speed when the rated current is being drawn. The motor can guarantee this torque output for an infinite amount of time without suffering damage, as long as the cooling requirements specified in the datasheet are met. In normal operations, we are allowed to use more torque than the rated value (e.g., during accelerations), but not for too long, because the motor draws much more current and starts overheating quickly.

- **Rated power**: The output power of the motor at rated speed. This parameter is commonly used in technical conversations to quickly define the generic size of a motor.

- **Rated current**: The amount of current drawn by the motor when operating at the rated speed and rated torque.

- **Stall torque**: The torque value that can be exerted by the motor at zero speed, i.e., to hold an axis still against gravity. This value is typically slightly higher than the rated torque.

- **Stall current**: This is the amount of current drawn by the motor to generate the stall torque at zero speed. This current can be drawn for any length of time, as long as the cooling conditions are met, and is an important parameter to size the electronics driving the motor, which also must be able to provide this much current indefinitely without burning out.

- **Torque constant**: This value shows how much torque (τ) the motor is able to generate for each Amp of current (i) flowing through its windings: $\tau = k_T i$. It is essentially an index of the motor's efficiency. Despite being called torque constant, its value is slightly dependent on the environmental temperature and shows that the efficiency of the motor decreases in hotter conditions. In other words, the same motor in a warmer environment generates less torque while drawing the same amount of current.

- **Voltage constant**: The amount of voltage generated by the back electromotive force at a given speed (1000rpm in this particular case). This parameter specifies the minimum amount of voltage required to drive the motor in order to reach a certain speed. For instance, if $k_T = 100V/1000min^{-1}$ and we run the motor at 300rpm, the motor generates a BEMF of 30V, which opposes the driving voltage. If we use a 36V power supply, we have essentially reached the maximum allowed speed (since we need an extra margin to overcome friction and load). On the other hand, if we use a 200V supply (as long as it is allowed by the windings), we still have plenty of speed to squeeze out of the motor.

- **Resistance/inductance**: These are basic electrical properties of the windings, measured across two terminals (phase to phase).

- **Inertia**: The moment of inertia of the motor's rotor with and without the holding brake. This is an often-overlooked parameter but can make all the difference in the quality of movement control, as previously explained in Tuning Section in Chapter 11. Note that mounting a brake adds significant inertia to the shaft and helps in stabilizing the controller.

- **Weight**: Mechanical weight of the motor with and without the holding brake. This parameter is important for the dynamic model of the serial kinematics robots.

- **Brake holding torque**: This is the maximum torque that the brake can hold before slipping. The holding brake is meant to hold an axis still against gravity, as in the case of robotic arms. It is not meant to slow down a moving axis to a stop.

- **Cable cross section**: This value represents the minimum recommended cable size to be used when wiring the motor to its driver and depends on the average and maximum current that the motor can draw. It is easy to overlook this parameter, but it is also easy to start a fire by burning a thin wire that heats up when too much current flows through it.

Besides the parameters table, the next important motor characteristic always shown in the datasheet is the speed-torque curve (see Figure 14-13). The graphic shows the three main operational ranges of the motors: safe (green in the drawing), temporary (yellow), and critical (red).

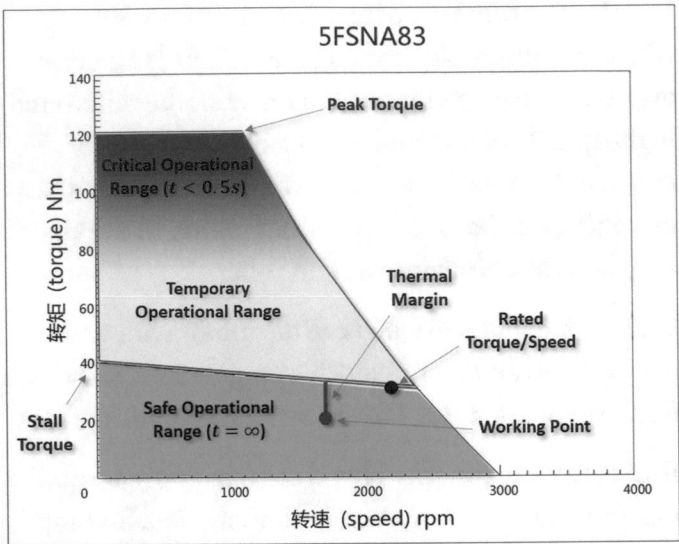

Figure 14-13. *Speed-torque curve of an electric motor (adapted with permission from FinePower)*

A motor can be operated in the green area for any length of time without damage, provided that the environmental conditions and cooling requirements specified in the datasheet are met. A higher torque is temporarily allowed for acceleration profiles up to the peak torque, above which the motor's magnets may be demagnetized and irreversible damage occurs. The amount of time the motor can spend in the yellow/red regions depends on the actual temperature, which should always be monitored by the servo drive. The motor overheats when generating high torques because large currents flow in its windings and therefore needs time to cool off to prevent damage. The cooling can either be forced (with external fans or water pipes) or can be natural via the flange to the mechanical parts of the machine. Operation in the green area can run for any length of time, because the currents there are low enough and the generated heat is lower than that naturally radiated by the motor. On the other hand, operations in the red area are only allowed for very short times (typically <0.5s).

In order to estimate the required motor parameters for a specific application, we need to simulate the programmed movement of the axis over a machine cycle and extract the RMS values for torque and speed. We call this the *working point* (shown in red in Figure 14-13). Both RMS values of torque and speed should not exceed the motor's rated torque and rated speed. Actually, it is wise to leave some torque margin because of unpredictable temperature variations and provide a safer configuration. This is often called the *thermal margin*.

Simulating the movement of a robot requires its kinematic model to calculate the positions of the joints (and from there derive their speed) and the dynamic model to calculate the corresponding torques of the motors. The idea of taking advantage of the dynamic model to size the motor torques was already introduced in Motor Sizing Section in Chapter 8.

When calculating the nominal speed required for an application, always remember to include the gearbox ratio, which reduces the actual speed of the actuator but increases its effective output torque. Also, make sure that the gearbox can handle the expected torque values. The gearbox is typically much more fragile than the mechanical arm of the robot and can be easily damaged if its allowed maximum torque is exceeded.

It is also important to size the stall torque correctly, i.e., the torque value that needs to be exerted at zero speed to hold the axis still. The required standstill torque of an axis depends on its mechanical configuration. For example, hanging loads require torque from the motors even at zero speed. In the specific case of a six-axes manipulator, the first axis is horizontal, which means it does not require any standstill torque. However, all other axes do, especially the second and the third, which bear most of the weight. Often, a large mechanical spring is added in parallel to the second axis to relieve the motor from some of its load. Using large gear ratios also helps.

Logically, an axis standstill torque must never exceed the motor stall torque; otherwise, the motor will overheat just by sitting there and holding the axis still without even starting any movement.

The mechanical standstill torque must also never exceed the brake's holding torque; otherwise, the axis will fall down when the motor is powered off. During the sizing procedure, make sure to test the stall values in the worst situation: the heaviest load mounted on the TCP and the arm fully extended.

Finally, we need to determine the required moment of inertia. This parameter is often overlooked but actually very important for good control quality. As mentioned earlier in the tuning section, an inertia mismatch between motor and load is the most common cause of poor control accuracy. Notice that inertia and torque are strictly related to each other: once the required motor torque τ has been selected, we need to choose the inertia I according to the application's maximum required acceleration:

$$I = \frac{\tau}{a} \tag{14-1}$$

At parity of output power, a high-inertia motor is physically thick and short and can only provide low accelerations (see Figure 14-14). A low-inertia motor is longer and thinner and generates a higher acceleration. In case of robot control, only the first kind works well. The second kind causes a large inertia mismatch which results in poor axis control quality.

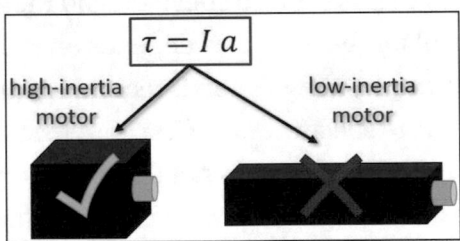

Figure 14-14. *Two motors with equivalent power can be built with different inertia values. Robotic applications usually opt for higher inertia to guarantee high control quality*

For example, in palletizing applications where large loads are carried at the TCP, we absolutely need high-inertia motors; otherwise, oscillations will easily occur when accelerating and decelerating quickly. Similarly, high-speed laser cutting applications are critical because oscillations at the load are amplified by the laser and result in poor cutting quality. In these cases, an inertia ratio of 1:1 between motor and load is typically required, which means that the selected motors are usually quite large.

Under-dimensioned motors do not allow the robot to move fast enough, either because not enough torque is generated or because the load cannot be well controlled if the motor's inertia is too small. In both cases, the robot is required to slow down, and the efficiency of the working cell is inevitably reduced.

Summary

Of all the electric hardware components required to build a robot, motors are the most important to guarantee best performance. In this chapter, we went through different kinds of options that you find on the market, from simple DC brushed motors to more efficient brushless variants, where an advance in performance is typically accompanied by an increase in price and complexity of the driving electronics and control algorithms.

Besides choosing the kind of motor you want to use, you also need to select its correct size. Typical datasheets include a large number of parameters that need to be understood in order to size each actuator according to the desired application requirements.

CHAPTER 15

Encoders

Identifying the rotor position is an important aspect of motor control: the servo loop needs to know by how much the actual and set positions (and speeds) differ from each other in order to calculate the right amount of current correction; the electronic commutation algorithm needs to know where the permanent magnets are positioned with respect to the windings in order to generate the correct voltage values for each phase. In general, the higher the resolution of the encoder, the higher the control quality that can be achieved.

There are several kinds of encoders on the market and also different kinds of communication protocols used to transmit the data back to the drive: some are open source, while others are proprietary.

One of the simplest feedback devices is the *resolver* (shown in Figure 15-1): an electromagnetic transducer that outputs a pair of analog electric signals (the sine and cosine of the actual rotation angle). The resolver is a rugged device, it has a very simple and robust design, it does not carry any on-board electronics, and it is therefore suitable for all tough and critical environmental conditions, as well as for military applications. They can tolerate mechanical shocks, vibrations, and high temperatures.

© Fabrizio Frigeni 2023
F. Frigeni, *Industrial Robotics Control*, Maker Innovations Series,
https://doi.org/10.1007/978-1-4842-8989-1_15

Figure 15-1. *Resolvers are simple and tough devices*

In order to read the incoming analog signals, the servo drive needs a pair of high-precision and high-speed analog to digital converters and can then calculate the rotation angle by using the arctangent function.

Encoders, on the other hand, carry the entire electronics inside their housing and transmit digital signals to the drives. This characteristic makes them less robust to environmental conditions but also makes them more flexible and typically more accurate. In some models, the on-board electronics also include a small flash memory, where all the motor parameters can be stored, so that the servo drive can configure itself automatically at startup.

There exist two main families of encoders, depending on the technology used: *magnetic* and *optical* (see Figure 15-2). The first kind uses a magnetic disc (or a strip) and senses the alternating N-S poles to detect motion. The second kind uses optical sensors (e.g., photodiodes) to detect a light beam produced by an LED as it passes through a marked disk.

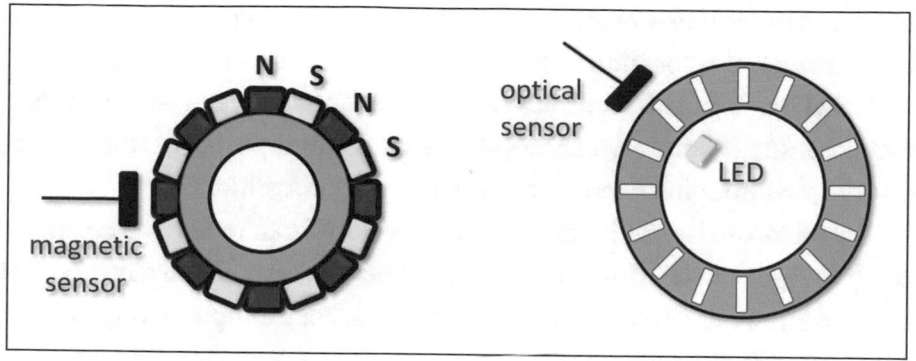

Figure 15-2. *Magnetic encoder (left) vs. optical encoder (right)*

Optical encoders are fragile devices: never hit the back of a motor; otherwise, you risk damaging the optical disk. Also, contamination by dirt or moisture can degrade their performance. However, they typically reach much higher resolutions than magnetic encoders and consequently allow for higher positioning accuracy, which is very important in robotic applications. The vast majority of industrial robotics motors mount optical encoders.

Magnetic encoders tend to be cheaper and offer lower resolutions but are often the best choice for hobbyist and low-end applications in general. When accurate position control is not required, magnetic encoders can be reduced to only the sensing component and rely on the motor's own permanent magnets for position detection. That is the case of Hall sensors used in brushless motors driven with low-resolution speed control (see Section on Hall Sensors).

Regardless of the technology used to sense position, the output signal can be of two kinds: *incremental* or *absolute*. Incremental encoders are the simpler option and only provide information about the relative change of position. They typically output a train of pulses in number proportional to the travelled distance (see, e.g., Section on Quadrature). Absolute encoders are more sophisticated (and also more expensive): they always output their exact position within a revolution (single-turn) or within a large number of revolutions (multi-turn). The main practical difference is

453

that a mechanical axis using incremental encoders needs to be referenced to a starting home location each time it powers up, before being able to start tracking its position correctly. Robotic axes usually mount absolute encoders, so the controller knows all the actual joint positions immediately at startup without the need for any referencing procedure.

Encoders can be used to detect positions along both rotary and linear axes (see Figure 15-3). The resolution of a rotary encoder is typically expressed in information bits transmitted for each reading. For example, a single-turn 10-bit encoder can read 1024 different positions per revolution. The actual angular resolution of the physical axis depends on the gear ratio between motor and load. In the case of linear encoders, where motor and load are directly connected to each other, the resolution is often specified in μm.

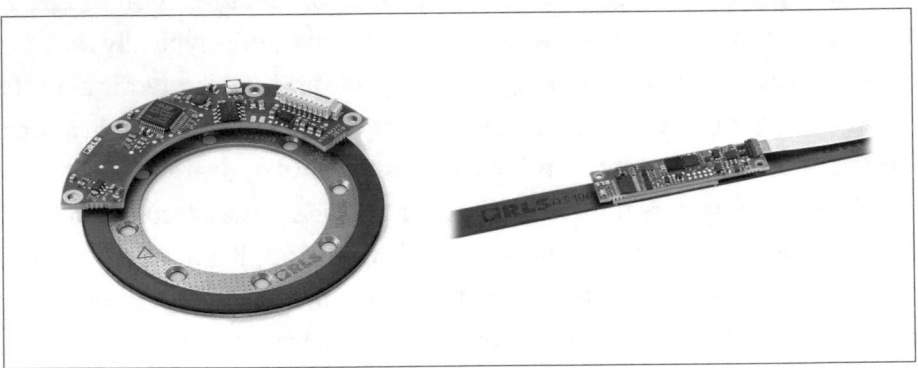

Figure 15-3. *Rotary encoder (left) vs. linear encoder (right) (with permission from RLS)*

When using encoders (or resolvers) with very low resolution, the servo loop sampling time might exceed their scaling. In other words, when moving at low speed, the signal will appear stationary because of oversampling, and the controller might think that the axis is not moving. As a consequence, it will largely increase the output current until the next encoder step arrives, at which time the controller needs to correct back in the opposite direction. This behavior causes instability and is best avoided with a smoothing average filter on the feedback signal.

Of all the several encoder interfaces options that exist on the market, we describe here some of the most commonly used in robotic applications: Hall sensors for basic speed control or auxiliary axes, quadrature encoders for incremental signals, and SSI and Tamagawa interfaces for absolute encoders. You are likely to encounter many other kinds, but the basic ideas are similar.

Hall Sensors

The most common method of identifying the rotor position in a brushless motor is by using Hall effect sensors. That is by far the cheapest solution and is a valid alternative to encoders when the axis is running in speed control mode (e.g., conveyors or spindle axes). For generic robotic axes with accurate positioning requirements, encoders are definitely preferred.

Hall effect sensors detect changes in the magnetic field produced by the motor's permanent magnets (see Figure 15-4). Their output signals are logic levels corresponding to the detected magnetic pole: e.g., 5V for an N pole and 0V for an S pole.

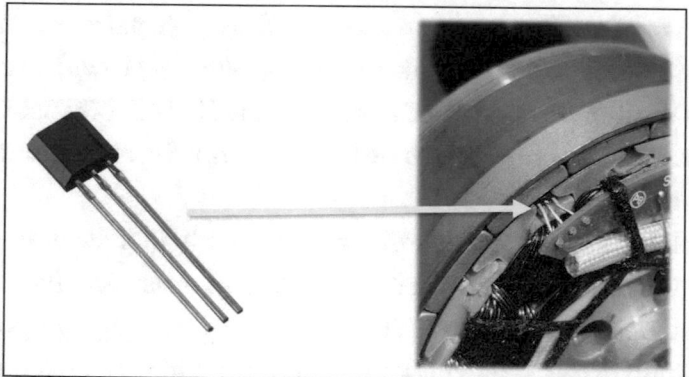

Figure 15-4. *Hall effect sensors are mounted next to the rotor's magnets*

By spacing three Hall sensors around the motor at an offset of 120 degrees of electrical angle, it is possible to detect six different positions (zones) within an electrical cycle (see Figure 15-5) and therefore allow the controller to energize the correct phases to move the rotor.

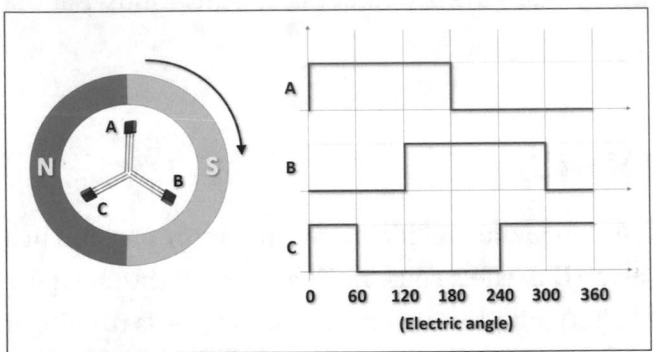

Figure 15-5. *The six possible Hall states based on the electrical angle*

The electrical angle is usually not equivalent to the mechanical angle, because the number of electrical cycles (0~360 *deg*) in one motor revolution is based on the total number of pole pairs in the rotor. The example shown in Figure 15-5 has only one pole pair, so the two angles are equal to each other. However, in a motor with 10 pole pairs, the electrical cycle is equivalent to 36 mechanical degrees (360 *deg* /10 *pp*), and the maximum position resolution achievable using Hall effect sensors is 6 mechanical degrees (36 *deg* /6 *zones*), which is much coarser than any encoder would offer.

Not being able to detect the rotor position with finer accuracy means that the magnetic flux generated by the windings is necessarily held constant in the entire window of 60 electrical degrees, regardless of the actual position of the rotor. This kind of commutation is called trapezoidal and is typically used for low-cost speed-controlled systems. The relative orientation between the two magnetic fields is constantly changing within an electric zone, instead of being held perpendicular as is the case of field-oriented control. The result is that the motor's generated torque is not constant during motion. This effect is known as *torque ripple* and is particularly accentuated in motors with a low pole-pair count running at low speed.

By measuring the time between each state change of any given Hall sensor, the angular velocity of the motor can be determined. In theory, that allows for more complex commutation schemes by interpolating the position between two Hall pulses, using the actual motor speed as a predictor. In practice, however, Hall sensors are mainly used with simple trapezoidal commutation for basic speed control applications.

One warning when using Hall sensors is that it is often necessary to add software filters to the incoming signals to reduce false triggering due to noise, which would cause false commutation and wrong speed readings, both of which significantly affect the quality of motor control.

Quadrature

An incremental encoder represents a significant step up from Hall effect sensors, in terms of both output resolution and cost. The increased accuracy in rotor position detection allows for precise position control and also for FOC commutation.

While the technological implementation of incremental encoders can be both magnetic and optical, the way they transmit the signal back to the servo drive is usually the same. The most common protocol used is an AB quadrature signal, where two trains of logical pulses are sent over two channels (A and B) separated by a fixed phase offset (see Figure 15-6).

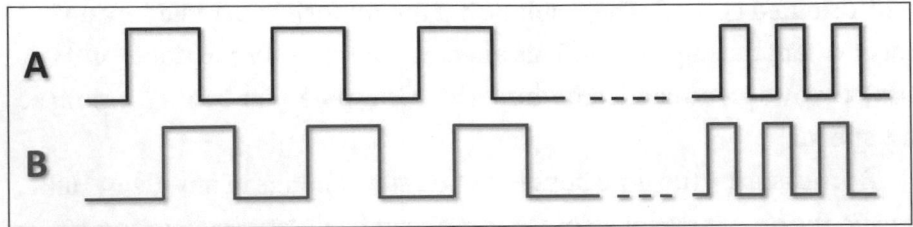

Figure 15-6. Quadrature AB pulses from an axis moving at positive slow speed (left) and negative fast speed (right)

Each encoder is characterized by a specific number of pulses per revolution. By counting the incoming pulses, the servo drive can detect the distance travelled by the rotor. Additionally, by detecting the relative relationship between the two channels, the direction of movement can also be inferred: for example, positive direction if A leads B; negative direction if A trails B. The meaning of positive and negative depends on the specific geometry: for rotary encoders, it corresponds to clockwise or anticlockwise rotations; for linear encoder, it represents the linear directions along the motor axis.

Furthermore, by measuring the temporal distance between two pulses, the controller can also infer the speed of rotation. For example, the signal shown in Figure 15-6 on the left side corresponds to a motor moving with slow positive speed; the signal on the right side is generated by a motor running with fast negative speed.

The events monitored by the controller are the pulse edge transitions, not the pulses themselves. A total of four distinct events occur for each encoder pulse: positive and negative edge of channel A and B. As a consequence, the position resolution achievable is actually four times larger than the specified number of pulses per revolution.

As in the case of Hall effect sensors, the electric signal is affected by noise and typically needs some software filters on the receiver side to avoid misreading the incoming pulses. However, much better noise isolation is achieved in practice using a redundant signal transmission.

Two parallel conductors (usually twisted wires) are used to send the same signal in complementary forms (*differential signaling*). The balanced line between driver and receiver (shown in Figure 15-7) allows for great noise and interference rejection, which in turns guarantees a much better motor control quality. Despite the drawback of needing additional wiring and hardware components both on the transmitter and receiver side, differential lines are the de facto standard in all industrial encoder transmission.

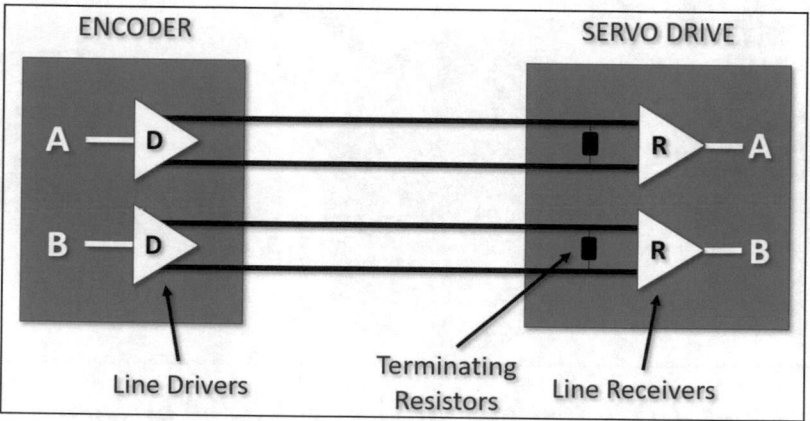

Figure 15-7. *Differential signaling over a balanced line improves signal noise immunity*

Impedance matching between the two sides is critical: if the impedance of the peripheral device does not match the impedance on the receiving side, signal reflections could result, thereby degrading the quality of the signal. A terminating resistance (R_{term}) must be added for that purpose, typically 120Ω, although the exact value is normally specified in the encoder datasheet.

Besides the two A/B channels, most rotary encoders also provide a reference pulse (often named R or Z) for each mechanical revolution. The reference pulse is sometimes also used for accurate homing purposes.

While simple incremental encoders can be found relatively inexpensive for the hobbyist market (though certainly more expensive than individual Hall effect sensors), you can also easily build one yourself using a multipole magnet and a sensing chip (see Figure 15-8). For rotary encoders, you need a magnetic ring, and for linear encoders, a magnetic strip, both of which are readily available and quite cheap.

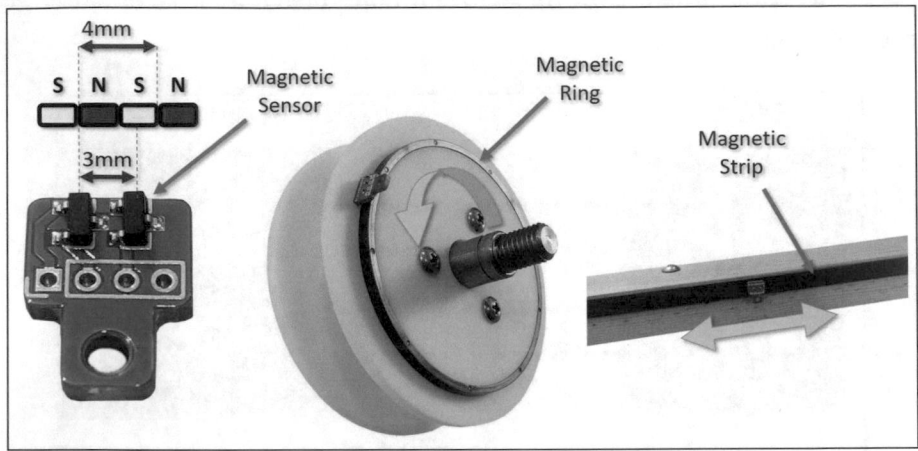

Figure 15-8. *DIY incremental encoder (left), for off-axis rotary (center) and linear applications (right)*

One sensor is enough to read a variation of the magnetic field, but two sensors are needed to detect the direction of movement. By spacing the two sensors apart by a fixed phase offset (e.g., 90 electrical degrees), we are able to generate two complementary sine/cosine waves with which the controller can accurately reconstruct the incremental position change from the previous sensing cycle.

Simple Hall effect ICs exist with an analog output interface (e.g., MT910X by MagnTek) or with a digital I2C interface (e.g., AS5510 by ams). More sophisticated chips that directly integrate two sensing elements also exists (e.g., AS5311 by ams, with both AB quadrature and SSI absolute output), but they are considerably more expensive.

As for the receiving end, most microcontrollers provide some sort of pulse counter module that can count signal edges and decode a quadrature signal into speed and direction output values. A glitch filter is also typically provided in order to remove unwanted short spikes in the signal, usually caused by noise, that would cause the counter module to misfire. Two common filtering techniques to consider a pulse transition valid are as follows: three high-speed (e.g., 100ns) readings in a row must be consistent; the pulse width must exceed a minimum threshold value (e.g., 1000ns).

In case the incoming signal is differential, we also need an intermediate receiver to translate it in a single line signal for the microcontroller. Any receiver designed for RS422/RS485 protocols will work fine, as long as the data rate is high enough for the expected axis speed and the output voltage matches that of the microcontroller input pin.

SSI

Incremental encoders are simple and inexpensive, but they suffer a major drawback: they are only able to measure relative distances. When the robot control system starts up, it has no idea where the joint axes are located and cannot perform kinematic calculations. One possible solution would be to mount home sensors and let the axes slowly move to search for the reference signals. A much better approach is to use absolute encoders, which always know the actual position of the axes, so that the controller can start moving the robot right away.

The actual axis position must be cyclically transmitted from the encoder to the servo drive. Several different protocols are used in the industry, one of the most common being SSI (Synchronous Serial Interface). The concept is very simple: the controller acts as the master and cyclically requests the position information from its slave encoder. It is a simple point-to-point communication based on two lines: one clock signal (CLK) and one data signal (DATA). The presence of a clock signal makes this protocol synchronous, as opposed to other protocols where no clock line is used.

SSI clock frequencies can exceed 1MHz, although the maximum supported data rate depends on the length of the cable connecting encoder and servo drive. The time needed for transmission also depends on the encoder resolution, because higher resolutions require more data to be sent.

The physical SSI cable uses six internal wires (CLK and DATA are differential), plus an external shield to prevent noise coupling and signal degradation (see Figure 15-9).

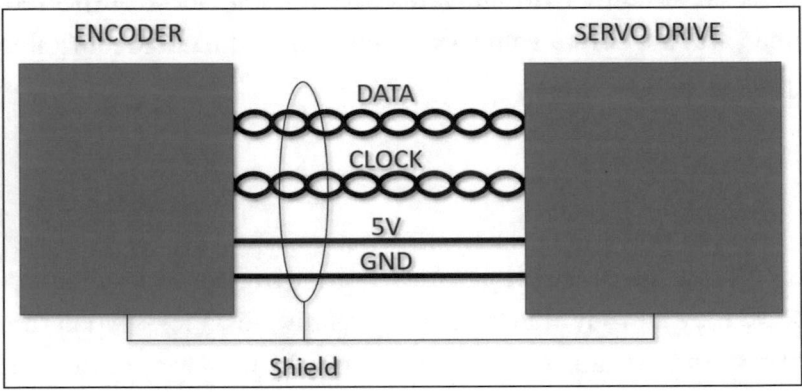

Figure 15-9. *SSI wiring details*

The actual communication protocol for SSI encoders works along the following steps:

- The master starts communicating by pulling the CLK line low (see Figure 15-10).

- The encoder freezes the actual position and starts sending it bit by bit on the DATA line. Specifically, one new data bit is set on each rising edge of the CLK line.

- The master reads every bit on the falling edge of the CLK line.

- After all the bits are read, the master needs to wait at least 30µs before polling the encoder again for a new position value; otherwise, the encoder starts resending the old value. It is actually possible to take advantage of this feature and read each position value twice, as a double-check safety procedure.

- All data bits are sent in order: first the multi-turn value and then the single-turn value.

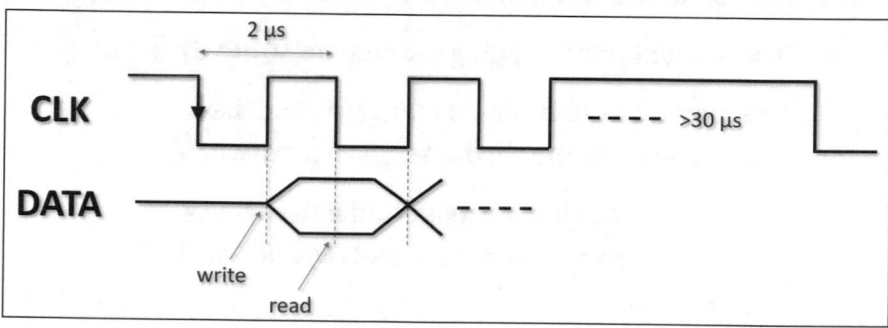

Figure 15-10. *SSI data transmission*

The hardware needed to receive the SSI data on the servo drive side is a pair of high-speed receiver and transmitter. Any IC designed for RS422/RS485 protocols will work, provided the data rate is high enough and the signal voltage matches that of the microcontroller. There are also full-duplex integrated transceivers that provide transmit (CLK line) and receive (DATA line) on the same chip.

Tamagawa

The last encoder interface we analyze is the Tamagawa protocol, which is a very popular brand in Asia. Specifically, we focus here on multi-turn absolute encoder, which provides high-precision absolute positioning for robotic axes.

Unlike the SSI interface, which uses two transmission channels (CLK and DATA), the Tamagawa interface only requires one communication channel, which is used both to send and receive data (half-duplex). Data transmission is not linked to a clock signal and is therefore called asynchronous. The transmission speed is fixed at 2.5 Mbps.

The actual communication protocol works along the following steps:

- The master starts by setting the channel to drive mode.

- The master transmits one request byte and then immediately sets the channel back to receive mode.

- After 3µs of receiving the request, the slave starts sending a response, whose content depends on the request byte.

Since there is no clock line to take advantage of, the timing of switching the transmission mode of the channel from driver to receiver is critical: if set too early, the request byte is cut and not properly sent; if set too late, some of the received bits might be lost.

In order to write and read serial data to and from the channel, a microcontroller typically uses a dedicated UART (Universal Asynchronous Receiver/Transmitter) port. In the specific case of the Tamagawa interface, the UART needs to be configured with: a baud rate of 2.5 Mbps, 8 bit of data per frame, disabled parity, and 1 stop bit. A word of warning for practical implementations: bits are sent from LSB to MSB.

When reading the actual encoder position, the master needs to send a specific request byte (see Figure 15-11), called control field (CF), which is delimited by a start and stop bit at the two ends. A total of 10 bits are sent, 8 of which contain the CF data: sink code (fixed, 3bits), data ID code (4 bits), and ID parity (1 bit). The ID code we use to request position data is 3.

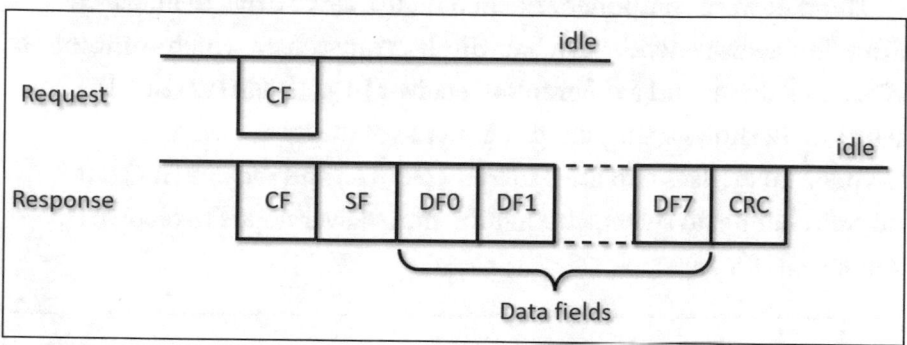

Figure 15-11. *Frame format for reading the actual position of a Tamagawa encoder*

The response from the encoder is a stream of 11 bytes (each of them individually delimited by start and stop bit):

- **Control field (CF)**: Sent back to confirm the request

- **Status field (SF)**: Includes encoder error (e.g., overheat, low battery) and communication alarms

- **Data fields (DF0...7)**: Includes absolute position in one revolution, multi-turn count, encoder ID and error

- **CRC field**: CRC-8 checksum for data verification (including all sent bytes, excluding delimiting bits and CRC byte)

The received encoder position can be extracted from the data field, and the specific resolution depends on the encoder: typically, it consists of 17bits of absolute single-turn position, plus 16bits of multi-turn counter (0~65535 revolutions). We can then convert the position into electric angle for the commutation algorithm.

The hardware component needed on the servo drive to interface with a Tamagawa encoder is a half-duplex transceiver, which combines a differential driver and a differential receiver (e.g., SN65HVD by TI). Figure 15-12 shows a simplified schematic of the interface. A microcontroller uses two lines to transmit (TX) and receive (RX) data and one channel to select whether the transceiver needs to receive or transmit (RTS).

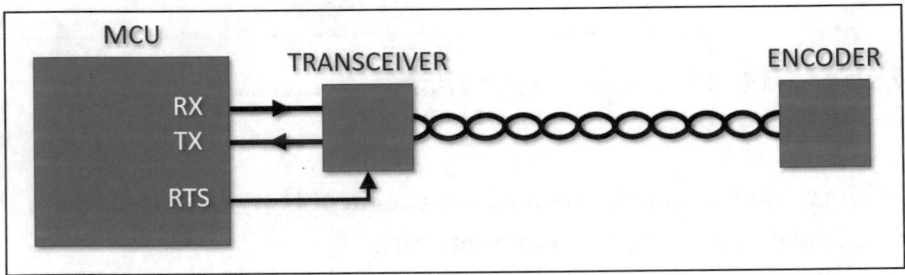

Figure 15-12. *Interfacing a Tamagawa encoder*

Receiving multi-turn counter data from the encoder is convenient in practical applications because the actual position of the axis is always up-to-date and no initial referencing is needed at startup. Note that a battery must be provided for the multi-turn counter to be operational at all times, even when the main power supply is switched off. Some manufacturers were able to eliminate the battery altogether and take advantage of the Wiegand effect to harvest enough power from the rotating encoder in order to update the counter value saved in memory.

Tamagawa encoders also have an internal memory allocation for user data, which the servo drive can read (and write) at any time. Most manufacturer use this feature to save all the motor parameters in the encoder memory, so that the drive can automatically configure its algorithms to the specific connected motor (e.g., zero reference electric angle for commutation; thermal model for temperature warnings; rated/ peak speed, torque, and current values for monitoring purposes; etc.).

Summary

No matter what kind of actuator you use to power your robot, the accuracy of its motion control is greatly enhanced when adding a sensing device to provide feedback on the actual position of the axes. A number of options are available to that purpose, some specialized for rugged environments (resolvers) and some for high-end applications where resolving power is more important than cost (optical encoders).

Most hobbyist projects compromise performance and cost by using magnetic encoders, which can provide either an incremental signal while rotating or an absolute value of the actual orientation angle.

Besides the physical implementation of the feedback sensor, you need to consider the actual data interface between encoder and microcontroller. We analyzed a few options in detail, from incremental quadrature signals to SSI and Tamagawa absolute digital values.

CHAPTER 16

Servo Drives

Motor controllers are a fundamental part of any electric machine. Being
able to understand their working principles in detail and, better yet, being
able to build one by yourself will provide a great advantage during the
operation and optimization phases.

Each type of motor we analyzed in Chapter 14 has different
characteristics and requires slightly different kind of driver. However, the
underlying concept is the same for all of them, and all designs are somewhat
similar to each other. Ideally, a motor driver should be designed to be
flexible enough to be able to drive different kinds of motors and interface
to several types of encoders, simply through software parameterization.
Customer would greatly benefit from such a convenient product.

Before diving into the hardware details, make sure you understand
well the software structure and algorithms of servo drives, which were
described in Chapter 9.

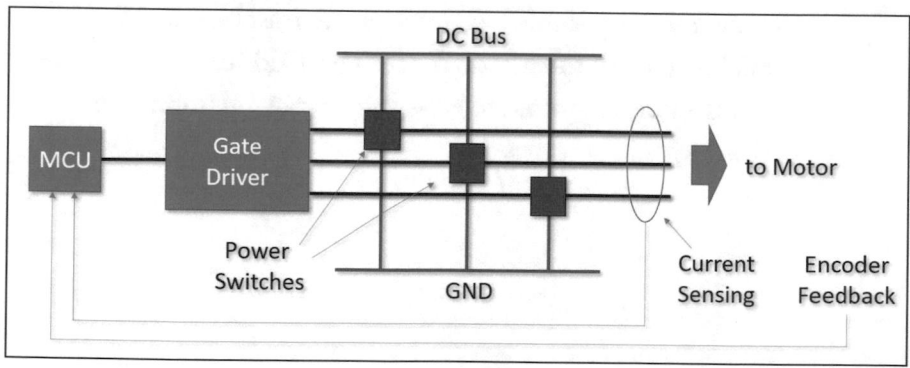

Figure 16-1. *Core hardware components of a servo drive*

© Fabrizio Frigeni 2023
F. Frigeni, *Industrial Robotics Control*, Maker Innovations Series,
https://doi.org/10.1007/978-1-4842-8989-1_16

The core functionalities of a typical motor drive are shown in Figure 16-1 and include the following:

- **Microcontroller**: Where the servo loop control and the electronic commutation algorithms are executed. The inputs required to run those two functions are set values from the main controller, actual position from the encoder, and actual current from the motor phases.

- **Gate driver**: The microcontroller generates the voltage levels for each motor phase, but it does not have enough power to directly drive the switches. That function is taken over by an intermediate device called gate driver.

- **Switches**: Delivering the desired voltage from the DC bus to the motor is accomplished by high-speed power switching components. Depending on the voltage level, either IGBTs or MOSFETs can be used.

- **Current sensing**: The actual current flowing through the motor must be sensed in order to control its quadrature component and minimize its direct component.

- **Power supply**: A stable DC bus voltage must be readily available to drive the motor. A number of critical protections are required to guarantee a safe operation of the device.

Note that in some cases, the gate driver and the switches are combined into a single integrated chip, useful for low-power systems (e.g., VNH7040 by ST). For high-power applications, the switches are left external.

In addition to the listed core functions, a typical servo drive also provides some I/O channels to interface with sensors or actuators that are directly related to the individual axis itself (e.g., referencing pulse, hardware safety limits). The drive's own microcontroller can run a number of additional functions and trigger auxiliary external actuators via digital output signals.

Furthermore, the servo drive needs a physical interface to communicate with the main controller in real time: the drive receives the set values (i.e., desired position and torque of the motor) and sends back its status for monitoring and diagnostic reasons (i.e., actual motor status, temperature, etc.). While in theory any kind of communication interface can be used, the common choice in industrial settings is a deterministic fieldbus. The critical issue is not so much the speed of communication but rather the synchronization of all drives with each other. We need to guarantee a constant and stable communication cycle time, during which all the drives receive the set positions from the main controller. If the drives receive the set position values all at different times, then the robot's movements will not be correct.

Servo drives are generally rated by the power they can supply to the motor, although we distinguish between nominal (rated) power and peak power. Motors require a certain power when moving a load with a specific speed profile. The drive is able to provide a constant nominal power for an indefinite length of time. However, when accelerating, the motor requires a higher peak of power, especially if a large load is attached and the required acceleration is high. That is the peak power, and the drive must be able to provide enough of it without saturating. Peak power values can only be delivered for very short amounts of time; otherwise, dangerous heating will quickly occur because of the large current flowing through the switches. All servo drives should provide safety monitoring functions that will automatically shut down the power stage before overheating and damage occurs.

A typical implementation of a servo drive consists of two separate stages: a logic board and a power board. The logic stage houses the microcontroller and its interfaces (I/O channels, encoder input, communication port). The power stage includes the power supply, the gate driver, the switches, and the current sensing.

Most industrial servo drives are powered by the AC main line and internally rectify that power to a DC voltage bus. Smaller robots work directly with a DC supply (e.g., 48V or 60V) as do all mobile systems because they are battery powered. Using low-voltage DC supplies increases safety but also decreases the maximum operating speed of the motors. Also, at parity of electric power, lower voltage translates into higher currents, which require thicker and more expensive cables.

Figure 16-2 shows a compact 300W servo drive built by the author. The device can drive a three-phase brushless motor with a maximum supply voltage of 60V and a rated current of 5A (given proper cooling). The top green board is the logic stage, while the bottom white board is the power stage. The power board is built on aluminum substrate to allow for efficient conduction cooling through the mechanical body of the machine. One peculiarity of this drive is that it uses a wireless communication to exchange data with external devices, thereby simplifying wirings considerably.

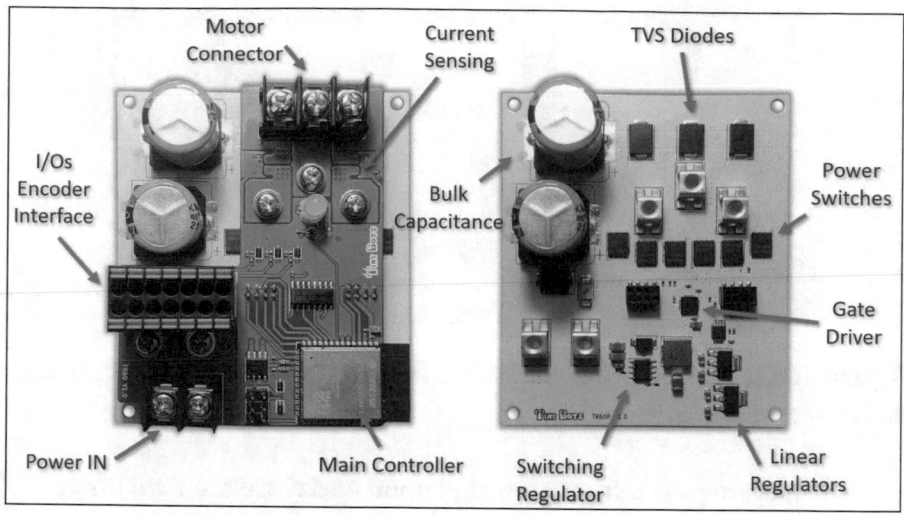

Figure 16-2. *A compact 300W servo drive*

In the next sections, we analyze each core component of the drive in detail.

Power Switches

If the microcontroller is the brain of the servo drive, the power switches are its heart: they direct the flow of current from the DC bus supply to the motor phases. The switches are typically arranged in three independent half-bridges (legs of the bridge), as shown in Figure 16-3. See Section on Electronic Commutation in Chapter 9 for more details on why a bridge configuration is used.

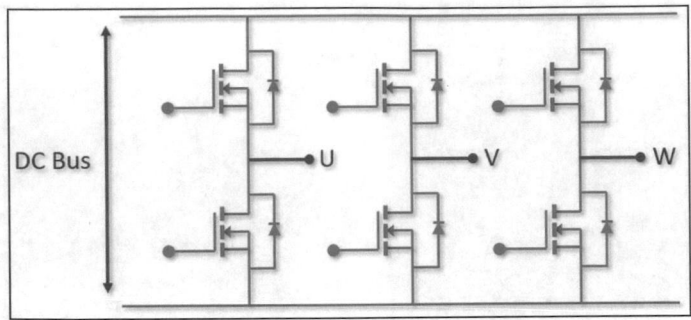

Figure 16-3. *Six N-channel MOSFETs arranged in a three-leg bridge can drive a brushless motor*

Switches are power transistors that open and close their channels according to the voltage applied to their gates. For high-voltage applications (DC bus >1kV), the devices of choice are usually IGBTs. Conversely, for DC bus voltages of lower-voltage levels, MOSFETs are typically employed, because they can switch faster and are more efficient (although MOSFETs that can handle well above 1000V also exist).

In particular, N-channel MOSFETs are almost always chosen because of their lower on-resistance. The bridge shown in Figure 16-3 employs six N-channel power MOSFETs arranged in three legs (three half-bridges). The three transistors on the top are called *high-side* switches, while the three on the bottom are called *low-side*.

An N-channel power MOSFET is shown in Figure 16-4 (left side). When a positive gate-to-source voltage (V_{GS}) is applied, the channel between the drain and the source opens up, and a current is allowed to flow from drain to source (I_{DS}). The electric resistance of the channel depends on V_{GS} (see Figure 16-4, right side): for very low values of V_{GS}, the channel is off and no current can flow (*cutoff region*); for increasingly higher values of V_{GS}, a proportionally larger current is allowed to flow (*linear/ohmic region*), until V_{GS} is large enough, well above the threshold value, that the device is fully turned on and a very large current can flow with minimal voltage drop, because the channel resistance is very low. If the voltage drop across the channel is forced to increase, then the current will not be able to keep up linearly and will eventually plateau (*saturation region*).

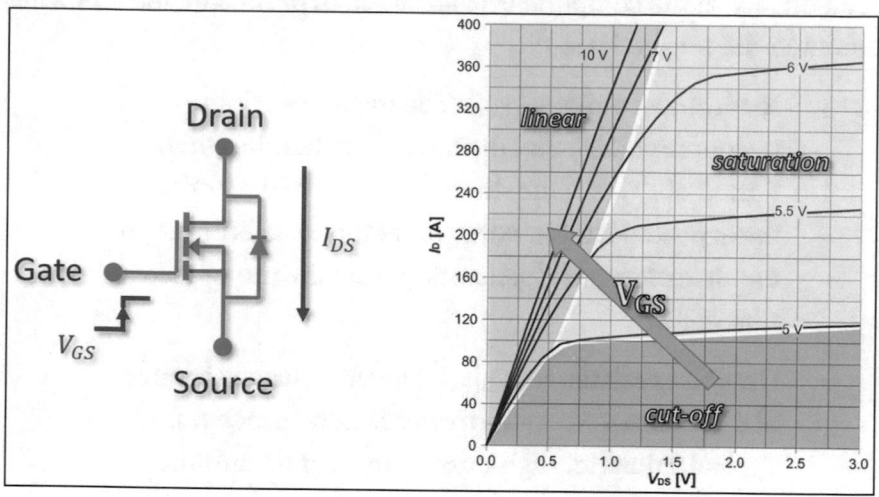

Figure 16-4. *N-channel power MOSFET and its characteristic curve*

While in some applications MOSFETs are used as signal amplifiers, here we use them exclusively as switches, turning them either completely off ($V_{GS} = 0$) or completely on (e.g., $V_{GS} = 12V$) transitioning from the cutoff region to the linear region. In practice, however, the gate of the device acts as a capacitor, which requires some small, but finite, time to charge and discharge. Therefore, the time it takes for the gate voltage to fully transition between the two opposite states depends on how much current the gate driver can supply to charge and discharge the MOSFET gate capacitance. A fast-switching ability, and therefore the availability of high gate currents, is critically important in order to provide a smooth voltage value to the motor phase, but especially to avoid operating the MOSFET in its high-resistive region, within which a large amount of power is wasted into heat.

Switches are by far the most expensive components in the material list of a servo drive and also the most likely to get damaged if not sized and protected correctly. Selecting the right switch for the application requires a detailed look at the component's datasheet. In particular, the following parameters are most relevant:

- **Breakdown voltage** (V_{DS}): The maximum drain-to-source voltage that the device can handle when turned off. This value should be chosen higher than the expected DC bus voltage, keeping a safety margin because of power regeneration from the motor (see Chapter 17).

- **On-state resistance** ($R_{DS\,on}$): The electrical resistance of the channel when the device is fully turned on. Typical values for high-current devices are around 1mΩ or less. Lower resistance values guarantee lower conduction losses due to power dissipation when current is flowing. However, very low resistance devices tend to be physically larger and more expensive. Note that electric resistance increases with the temperature of the device, which in turns leads to more dissipation and yet higher temperature (a positive feedback loop). The temperature should always be monitored and the current adjusted accordingly in order to avoid instability and thermal overrun.

- **Drain current** (I_D): The current that can flow for any length of time through the transistor without damaging it. This value strongly depends on the MOSFET operating temperature: the ability to keep the device cool plays a critical role in allowing more current to flow for longer time. Datasheets typically list a peak value (at 25°C) and then show a curve plotting how the rated current decreases with the increase of the case temperature (see Figure 16-5). MOSFET cooling can be achieved either through ambient air convection (e.g., using a fan) or through body conduction (e.g., using large copper areas on the PCB and a thick metal substrate). For yet larger current requirements, FETs can be easily parallelized to share the load.

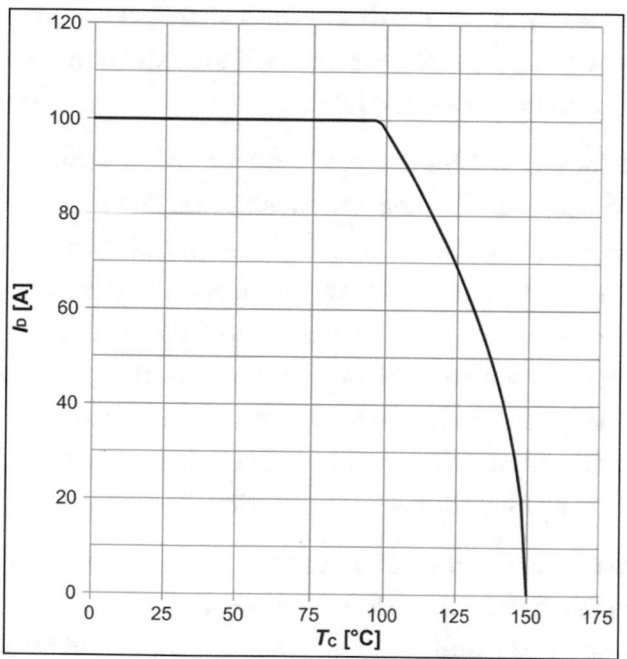

Figure 16-5. *MOSFETs rated drain current plotted against case temperature*

- **Total gate charge** (Q_G): The amount of electric charge required to charge/discharge the gate. This value can be used to calculate the maximum required current from the gate driver, given the desired switching time: $I_G = Q_G/t_{on}$. A lower gate charge is more efficient because it requires lower charging currents or, alternatively, allows for more rapid switching.

- **Diode current** (I_S): As a consequence of their internal structure, MOSFETs directly integrate a body diode between source and drain, which practically allows the device to conduct current in reverse even when the gate is turned off. The amount of current they can conduct is typically as large as that of the switch itself, but the diode path impedance is much higher, and it dissipates much more power into heat.

- **Packaging**: MOSFETs are manufactured in various packages with different characteristics. The two most important are the actual size of the footprint on the PCB and the ability to dissipate heat through its body, either to an external heat sink or to the PCB metal substrate. Notice that in some cases, the packaging is actually the limiting factor for the maximum allowed drain current, i.e., the frame legs could melt before the actual FET junction would experience any issue.

Figure 16-6 shows a group of power MOSFETs arranged in a simple H-bridge built by the author to drive a brushed DC motor. The board has an aluminum substrate and dissipates the heat generated by the transistors through body conduction. The motor draw peaks of over 150A under heavy load and high acceleration conditions.

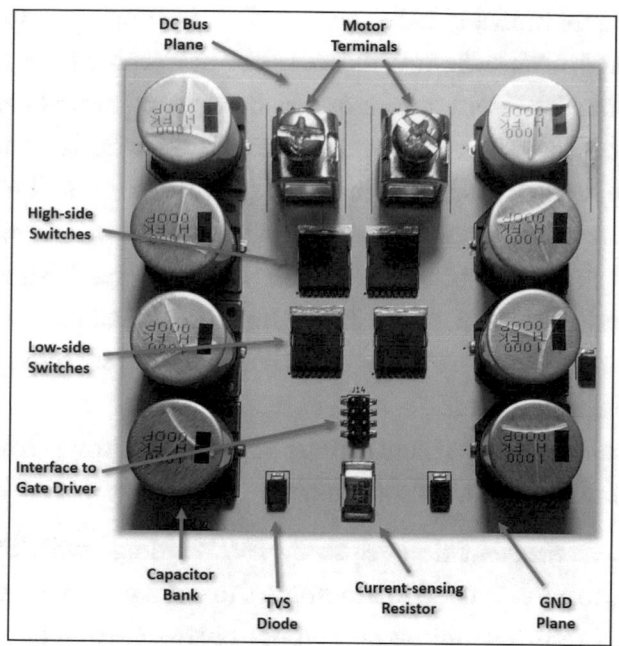

Figure 16-6. *MOSFETs arranged in H-bridge to drive a DC brushed motor*

By far the most important concern in the design phase of a motor driver is the amount of drain current that the switches can safely carry. Unfortunately, the nominal value I_D specified in the datasheet is not a realistic target for most practical applications, and efficient power dissipation is the limiting factor for current flow in practice.

Heat generation is mainly caused by two distinct processes: conduction losses and switching losses. The first are caused by current flowing through the channel when the device is fully turned on. The dissipated power can be quantified by $P = I^2R$, where the resistance is the $R_{DS\,on}$ value (derated at the actual temperature of operation).

Ironically, while larger MOSFETs with lower on-resistance provide lower conduction losses, they also feature a much higher gate capacitance, which in turn results in higher switching losses. In other words, there is a conflicting relationship between power losses during continuous operation and during switching times. In fact, switching losses can be the most serious issue with motor drives, dissipating more heat than conduction losses. They are caused by the transistor transitioning into its high-resistive region between being fully turned on or off. The transitions are usually quick (i.e., tens of nanoseconds) but still burn significant power into heat and should be minimized when possible. The dissipated power mainly depends on the switching frequency (how often the MOSFET needs to transition into its high-resistive region) and on the switching time (how long each transition takes).

Here are a few practical suggestions to minimize power losses and allow more drain current to flow safely:

- *Adjust the switching frequency.* PWM voltage switching at the gates is used to modulate the voltage applied to the motor's windings and ultimately the current flowing through each phase of the motor. Switching frequency determines how smooth the current is modulated: low frequencies cause current ripples, which in turn appear as torque ripples on the motor; high frequencies decrease the ripples and generate a smoother profile, at the cost of more heat losses. A good compromise is usually in the 10~100kHz range. Decreasing the PWM frequency below 10kHz is not a good idea, because an audible buzz is generated by the frequency entering the human hearing range.

- *Adjusting the switching time.* The time required by a MOSFET to transition from an off state into an on state (and vice versa) depends on how fast the gate driver is able to charge (and discharge) the gate capacitance. Using large gate currents (>1A) is a good idea but requires powerful and expensive gate drivers. Also, an issue with using too high current values is that they induce electromagnetic interference (EMI) into the circuit, often requiring large EMI filters to prevent violations of regulations. One approach that is often used in high-end applications is to actively control the charging and discharging gate dV/dt profile, reducing high-frequency components and therefore reducing the generated EMI.

- *Reduce the switching transitions.* One additional way to reduce the total number of required switching transitions is to cleverly rearrange the phase timings in the FOC implementation. Equations (9-11) and (9-12) presented in Chapter 9 can be slightly modified in a way called *alternating reversing sequencing,* so to reduce switching transitions while the actual phase timings remain the same.

- *Use GANFETs.* When efficient power management is the highest priority of the design, GANFETs can be used instead of MOSFETs. These transistors are based on gallium nitride instead of silicon and are able to switch at very high frequencies (well above 1MHz) virtually without losses: a motor driver using GANFETs can easily achieve 99% efficiency. Their drawbacks are that they are more expensive, do not inherently have a body diode, and also require a more powerful gate driver, because fast switching means driving a large gate current to charge the gate capacitors.

- *Reduce body diode losses.* We described the body diode when listing the switch parameter I_S (diode current): besides conduction and switching losses, a power MOSFET also dissipates heat when conducting in reverse. The body diode conductivity is lower than that of the switch's channel, so its use should be minimized. When programming commutation algorithms, make sure to activate the high-side switch when its low-side correspondent is turned off. The technique is called synchronous (or active) decay, because its timing is exactly determined by the switching activity. It allows any existing inductive current from the motor to flow through the switch instead of the body diode to reach the DC bus. For the same reason, the shoot-through protection time should be kept to a minimum, because it uses the body diodes to conduct the decaying current flow (known as asynchronous, or passive decay, because its timing is not related to the switching activity).

Gate Driver

Driving the switches in each leg of the H-bridge cannot be done directly by a microcontroller; a gate driver must be deployed instead. There are two main reasons for that: the current needed to quickly charge and discharge the gates of the switches is very large, much higher than a typical microcontroller output pin can provide; the voltage required to fully turn on both the low-side and, especially, the high-side switches is much above the levels typically handled by a microcontroller.

The solution is to use the microcontroller to generate the logic-level signals and then take advantage of the gate driver to transform those signals into high-power pulses for the gates of the switches. This technique is valid regardless of the kind of switches you use to drive the motor (MOSFETs, IGBTs, GANFETs, etc.), although current levels and switching frequency may vary.

The first requirement for gate drivers is to generate a high enough voltage to fully turn on the switches. The low-side is easy: 12V are typically enough and well above the threshold level to ensure full conductivity of the channel. The high-side is trickier, because when the switch conducts, its source voltage is essentially equal to its drain voltage, which corresponds to the DC bus voltage. Consequently, the gate driver needs to be able to generate a gate voltage for the high-side switches about 12V higher than the main bus voltage. This task is accomplished using a charge pump: a device that efficiently converts a low-voltage input into a high-voltage output using a *bootstrap* capacitor to temporarily store energy (see Figure 16-7).

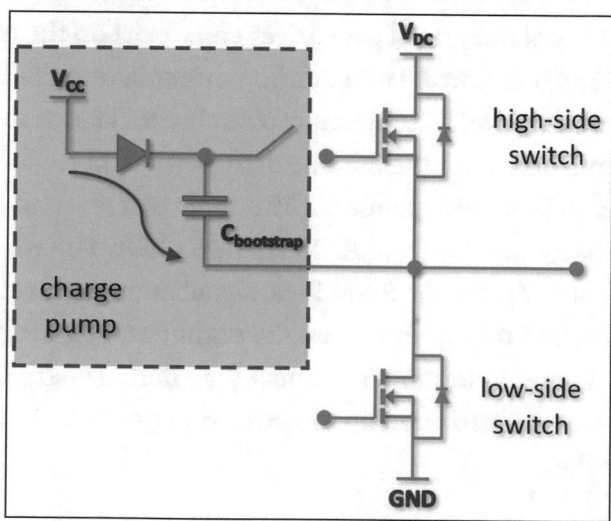

Figure 16-7. *A charge pump is used to generate the high voltage required to switch the high-side*

The capacitor is charged to a specified voltage (e.g., 12V) when its lower connector is grounded (e.g., when the low-side switch conducts). Then, when its lower connector is raised to the DC bus level, its higher connector will be at a voltage of $V_{DC} + 12V$ with respect to ground and will be able to fully activate the high-side switch when needed. A diode is used to prevent the higher voltage of the bootstrap capacitor to flow back into its power supply. Critically, the bootstrap capacitance should be high enough to provide the charge required by the gate of the high-side MOSFETs: larger switches require larger charge storages.

Another requirement of gate drivers is that they need to source and sink enough current to quickly open and close the switches and minimize transition time in their high-resistive region. The amount of current depends on the desired switching time and on the total gate charge of the transistors. Typically, gate drivers are capable to sink higher-current peaks than they can source (e.g., 1A source and 2A sink), which means that switching-on times are usually longer than switching-off times.

A large variety of integrated gate driver chips exist on the market, all sharing with similar features. Two common examples are the DRV83xx family by TI and the EiceDRIVER family by Infineon. Figure 16-8 shows the connection diagram for the EG2133 driver by EGmicro. This particular driver is powered by an independent 15V supply and can control three-phase brushless motors fed by a DC bus of up to 300V. The wiring is very simple: the H_{IN} and L_{IN} are the PWM logic signal lines coming from the microcontroller, and the H_O and L_O are the output gate controls for the switches. The V_B and V_S terminals connect the internal charge pump to the bootstrap capacitor to build up the extra voltage needed to turn on the high-side switches.

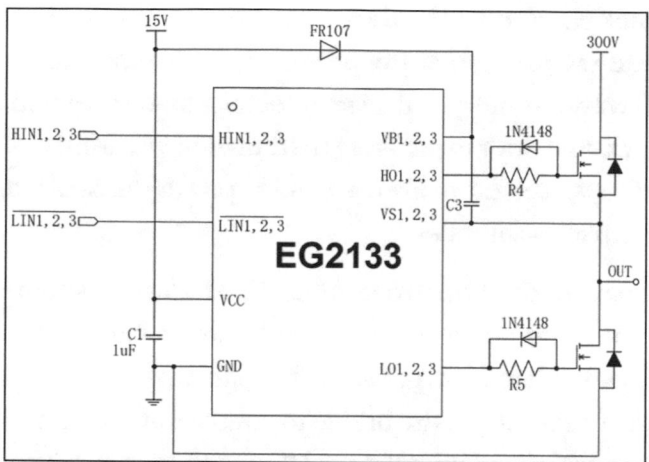

Figure 16-8. *Connection diagram of the EG2133 gate driver (with permission from EGmicro)*

In most cases, the input lines H_{IN} and L_{IN} for each individual half-bridge can be shorted together, since the low input side is inverted. That approach forces the high-side switches to be always in opposite state with respect to the low-side switches, which is a feature of the FOC commutation and at the same time saves wiring from the MCU.

Gate resistors are used to limit the peak gate current to the desired application levels, since this particular driver does not allow for internal tuning of the source and sink currents. Diodes can be added in parallel to the gate resistors to allow for faster discharge times.

In general, gate drivers dedicated to motor control come packed with useful safety features and configuration options. Here is a list of some of the most significant ones:

- **MCU interface**: The logic control signals coming from the microcontroller can be at 3V3 or 5V depending on the device. Gate drivers are usually flexible enough to allow for both voltage levels. Some gate drivers have a digital SPI interface to allow for detailed fault reporting and fine configuration of their internal parameters.

- **Buck converter**: The device shown in Figure 16-8 requires an external 15V power supply to operate. However, some gate drivers directly integrate a small buck converter to generate their own power from the DC bus, which can also be used to power the MCU for compact assemblies.

- **Cross conduction prevention**: Also known as shoot-through protection, this is a fundamental feature of all gate drivers that prevents the high-side and low-side of each leg of the bridge to conduct at the same time, which would short the DC bus to ground with catastrophic consequences. The protection is achieved by adding a small deadtime of a few nanoseconds (adjustable in some devices, otherwise fixed) to each side of the programmed transitions. However, that also introduces a drawback: the bridge is essentially turned off during the deadtime, but the motor inductance still forces the current to keep flowing. The only available path for the current is through the body diodes of the switches (asynchronous decay). Body diodes have higher impedance and dissipation than the switches themselves, so the shoot-through protection window should be kept as short as possible.

- **Overcurrent protection**: Some gate drivers directly monitor the V_{DS} voltage across the drain and source of the switches and can quickly react if that value exceeds a predefined threshold when the switch is fully conducting. The V_{DS} voltage builds up from the drain current I_D flowing through the low-impedance channel. The overcurrent protection triggers when that voltage exceeds the $V_{DS\,max}$ limit, determined by the $R_{DS\,on}$

resistance and the maximum desired drain current. The protection can either send a simple feedback warning to the microcontroller or also actively decide to switch off the MOSFET entirely.

- **Overvoltage and undervoltage lockout**: When the DC bus voltage moves outside predefined limits on either end of the safety range, the gate driver usually shuts off the entire bridge putting the switches in high impedance state. The charge pump is usually also turned off to avoid possible damage.

- **Overtemperature shutdown**: When the temperature of the chip exceeds a certain threshold, the gate driver shuts itself off, and operation usually cannot be resumed until the temperature has lowered below a safety level.

- **Slew rate**: Gate drivers can source and sink a maximum amount of current above which damage can occur. If the switches are particularly large and their gate charge exceeds the driver peak current capability, a gate resistor must be inserted in series to limit the maximum current flow. Gate resistors can also be used to filter ringing and noise. Some drivers can internally limit the source and sink current levels to predefined values. The result is a configurable slew rate, i.e., the rise- and falltime duration of the gate voltage transition.

- **Current-sense amplifiers**: Some gate drivers integrate amplifiers in order to simplify current measurement feedback circuitry and reduce external components count. The current can be measured using shunt resistors on each leg of the bridge (see Section on

Current Sensing for more details), and the signal is then locally amplified before being sent to the main controller. The integrated current-sense amplifiers often include advanced features such as programmable gain, offset calibration, unidirectional and bidirectional support, and a voltage reference setting.

Besides all these useful and fancy features, the fundamental requirements you need to keep in mind when selecting a gate driver for the application are as follows:

- Number of half-bridges to be driven (motor phases)

- Voltage rating (according to maximum DC bus voltage)

- Driver current rating (according to gate charges of selected switches)

- Maximum switching frequency (according to the switches and PCB cooling capabilities)

- Maximum operational temperature (depending on PCB cooling capabilities)

- Galvanic isolation (required when working with high-voltage buses)

Current Sensing

Measuring the current flowing through the motor windings is necessary both for monitoring and control reasons. Phase current monitoring helps preventing overheating in the windings and possible internal damage to the motor. The maximum allowed current is specified in the motor's datasheet, both the peak level and the average (nominal) value.

Accurate motion control also requires the value of the phase currents. The FOC commutation algorithm described in Chapter 9 generates PWM duty cycles to modulate phase voltages with the objective of inducing a stator magnetic flux perpendicular to that of the rotor. The magnitude of the commanded voltage values is adjusted by the current controller in order to minimize the direct component of the stator current and regulate the quadrature component according to the set torque (output from the speed controller). Since a motor is an inductive load and currents do not change instantaneously, the current control is best done in a closed-loop architecture taking into account the feedback of the actual stator current components and generating corrections accordingly.

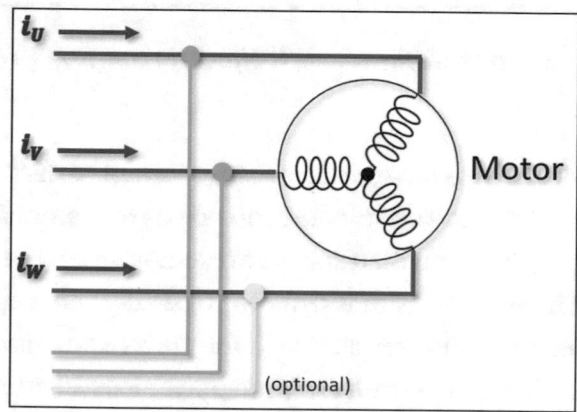

Figure 16-9. *Current sensing on motor phases*

Figure 16-9 shows the typical arrangement for sensing the current drawn by a motor. Measuring two of the motor's windings is technically enough, because the current in third phase can be simply found by Kirchhoff's current law, if needed. However, it can also be a good idea to directly monitor the third phase for safety reasons: if the three currents do not add up to 0 (bar a small tolerance), then there must be some current leakage to ground that could be a warning for a damaged motor.

In practice, sensing the current flowing through a PCB trace with high linearity and accuracy over a wide range is a difficult task. The two common methods consist in using either a shunt resistor to measure the current intensity directly or a Hall effect sensor to measure the strength of the magnetic field generated by the current flow (see Figure 16-10).

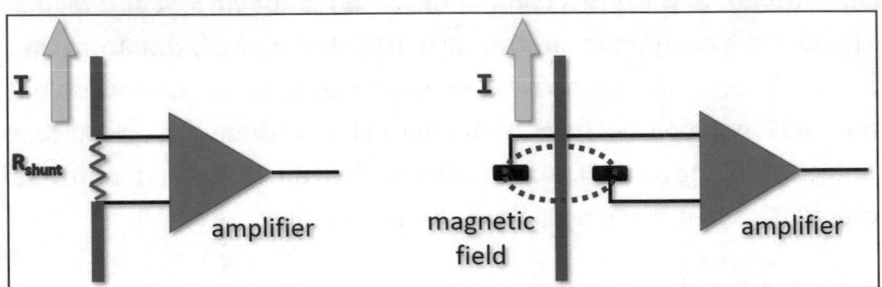

Figure 16-10. *Shunt resistor vs. Hall effect sensor for current measuring*

The first method uses an external sense (or shunt) resistor in the path to sense the load current. The voltage drop across R_s is amplified by a current-sense amplifier and then sent as an analog signal to an external ADC receiver. The sense resistor is usually physically very large because it needs to oppose very little resistance to the current and also needs to dissipate a lot of heat caused by the ohmic power loss. Figure 16-11 shows a 0.5mΩ 15W shunt resistor to measure the current in a brushed DC motor (on the left side) and a pair of Hall effect sensors to measure the current on two phases of a brushless motor (on the right side).

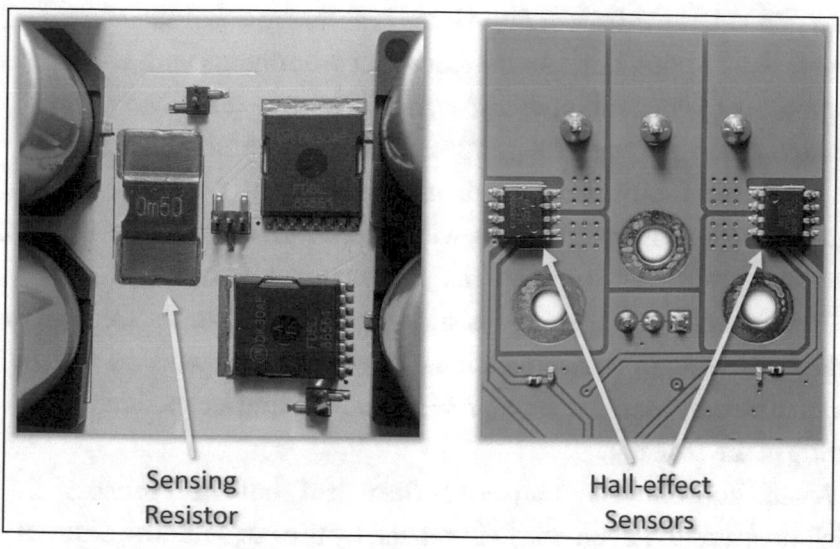

Sensing
Resistor

Hall-effect
Sensors

Figure 16-11. *Two different ways of measuring a current: shunt resistor (left) and Hall effect sensor (right)*

Calculating the size of the sense resistor requires making some compromises. We first consider the maximum expected phase current I_{max} and the maximum input voltage V_{max} allowed by the amplifier before it reaches saturation.

$$R_s = \frac{V_{max}}{I_{max}}$$

(16-1)

We could use an amplifier with a larger gain so to allow for a smaller sense resistance. The advantage would be a smaller power loss, but the drawback would be a less accurate and noisier signal. Keeping the PCB traces between the sense resistor and the amplifier as short as possible helps with noise reduction. Conversely, using higher resistor values allows for better sensing accuracy but generates more power losses.

In general, regardless of the selected value, the absolute tolerance of the resistor is very important. Choosing components with very low temperature coefficients guarantees more accurate readings within a given temperature range. Alternatively, the temperature drift can be compensated in software via modeling based on the datasheet curves. Similarly, high-quality amplifiers, with stable gains over a large range of input signal values and ambient temperature, should be selected.

The use of a current-sense resistor is a cheap and stable solution that works well in low-power applications. However, when working with high currents, power losses become problematic causing a significant self-heating of the resistor.

A valid alternative for that case is the use of Hall effect sensors, which measure the magnetic field caused by the current flow, without introducing additional resistance to the path and thereby avoiding power losses. The sensor and its electronics are entirely isolated from the main current path, adding safety to the solution, especially for high-voltage systems. The drawbacks are lower sensing accuracy at low current values, larger temperature drift, and possible interferences from external magnetic fields.

Figure 16-12 shows the typical wiring of a Hall effect current sensor: the IP+ and IP- on the left side are the traces where the primary current to be sensed flows. The output is a voltage linearly proportional to the flowing current. The signal is fed to an ADC, through an optional RC filter to smooth out noise, especially spikes generated by the high-frequency PWM transitions of the phase voltage.

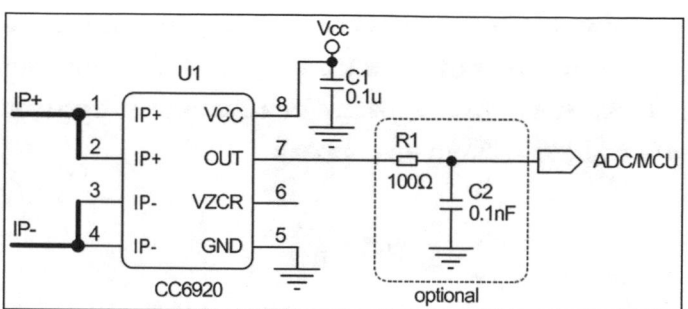

Figure 16-12. *Interfacing a Hall effect sensor for isolated current sensing (with permission from CrossChip)*

The CC6920 series by CrossChip can measure bidirectional currents up to ±50A. Other chips by different manufacturer can reach even higher levels, usually at the cost of decreasing resolution. For cases where an extended measurement range is needed, the current can be split in two paths, each carrying a percentage of the total current inversely proportional to the resistance of the branch (see Figure 16-13).

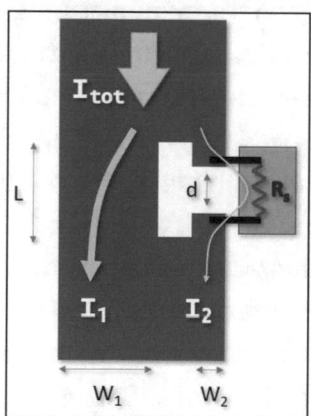

Figure 16-13. *Splitting the current in two branches allows for higher sensing ranges*

The total current to be measured is $I_{tot} = I_1 + I_2$. The resistance of the two paths is determined by their widths W_1 and W_2, their lengths L and $L - d$, their thicknesses T (e.g., 0.07mm for a 2oz copper trace), and their resistivity ρ (e.g., 1.68e-8 Ωmm for copper):

$$R_1 = \rho \frac{L}{W_1 T} \tag{16-2}$$

$$R_2 = \rho \frac{L - d}{W_2 T} + R_s \tag{16-3}$$

The internal resistance R_s of the current sensor (or current shunt) also needs to be accounted for. By designing the traces with appropriate widths W_1 and W_2, we can control the percentage of current flowing through the sensing branch and allow devices with smaller ranges to measure large current flows: e.g., using a 1:3 ratio between I_2 and I_{tot}, we can use a 100A sensor to measure a 300A max phase current.

Though very practical, the disadvantage of this method is that its accuracy depends strongly on the manufacturing tolerances and on the environmental temperature. Using calibration and software compensation increases the final measurement accuracy significantly.

Another issue that we need to keep in mind when measuring the motor phase currents is the sampling frequency. In other words, how often should we sample a current value? There are a few factors influencing this decision: the control loop cycle time, the ADC speed, and the PWM frequency. The two PIDs of the current controller run at a fixed frequency, typically around 100µs: that dictates the lowest sampling speed in order to have an output signal that allows for sufficiently accurate control. On the other hand, we cannot sample too quickly. The two limiting factors are the sampling speed of the ADC (in some slow devices, it can be well over 10µs) and the PWM speed (which in most cases is set to 20kHz, i.e., 50µs).

Since the voltage is applied to the motor phases via high-frequency PWM, the actual current levels are also rising and dropping accordingly, although filtered by the high inductance of the motor windings (see Figure 16-14). Each motor phase acts as an RL load and smoothes out the waveform profile, but there would still be a significant variation in measured values if the sampling would be done at a rate higher than the PWM frequency (yellow dots in the drawing). The common solution in practice is to add a low-pass RC anti-alias filter at the output of the current sensor to average out the measured value over at least one PWM period (green line in the drawing). The RC values must be selected so that the cutoff frequency $f_c = (2\pi RC)^{-1}$ is lower than half of the PWM frequency.

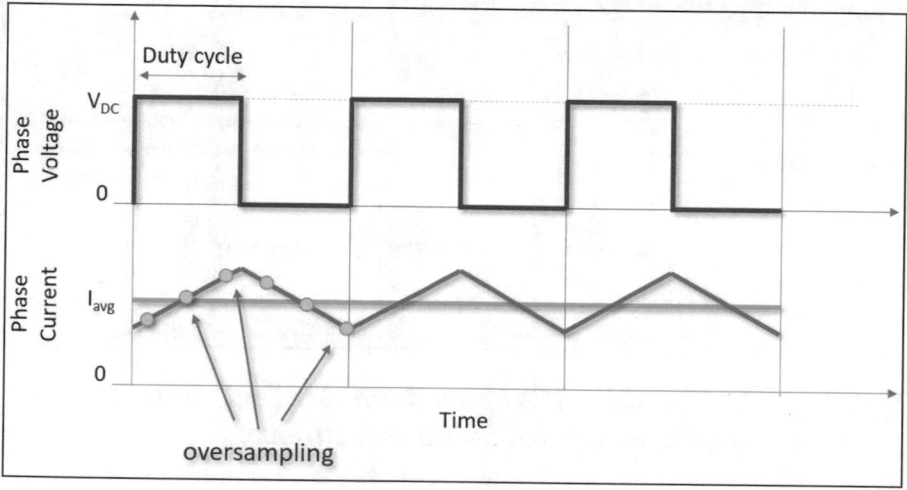

Figure 16-14. *Current in a motor phase (RL load) following the applied PWM voltage waveform*

The PWM voltage profile introduces yet another problem when sensing the phase currents: spikes (glitches) in the signal can occur at the transition edges both in the applied voltage and in the induced current and can be problematic. In critical cases, a spike in the sampled signal could trigger a false overcurrent alarm. In general, poor current sensing can lead to torque ripples in the motor and possibly position inaccuracy at the load. Sampling in the middle of the duty cycle, away from the edges, is the best solution. Filtering is also often implemented in practice. Some current sensors even feature integrated PWM rejection circuitry to reduce the output disturbance.

Sensing directly on the traces connected to the motor's windings is the best method to precisely measure the current flowing into each phase. The technique is called *in-line* sensing. Alternatively, in some simpler devices, the current is measured on the legs of the bridge, either on the low-side or the high-side, as shown in Figure 16-15.

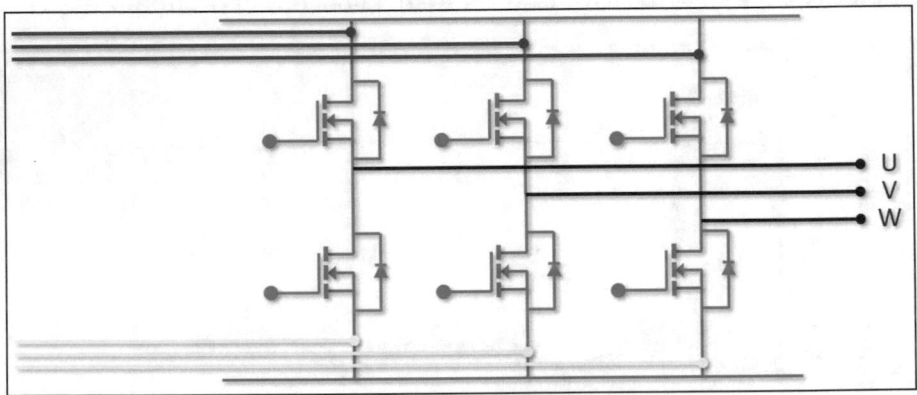

Figure 16-15. *Low-side (yellow) and high-side (red) current sensing are possible alternatives to real in-line measurements*

Low-side current sensing is the most common alternative to real in-line measurements. The main problem with that choice is that it can only be executed while the low-side switch of a leg is on; otherwise, no current flows through the sensing element. That poses additional requirements for the software, as current measurement must be synchronized with the PWM signal, which gets more challenging when a very short PWM duty is applied to the phase. Another shortcoming of low-side sensing is that it cannot detect short circuit faults on the load, between a motor winding and ground.

High-side sensing, on the other hand, is more similar to in-line sensing as it can detect short faults. When using a sensing resistor, the signal amplifier must be able to support the high common mode range (because the sensing happens at the DC bus voltage), as well as handle any voltage transients that may occur on the power rail. Those issues do not exist when using magnetic sensing because the amplifier is galvanically isolated from the load.

Sometimes we want to derive a value for the total motor current from the three individual measurements, e.g., to show it on the HMI. When doing high-side or low-side sensing, we simply add up the three components to each other. When measuring in-line currents, we need to take the signs into account and only add up the currents with the same sign (otherwise the total sum would be 0):

$$I_{MOT} = \max\left(0, i_U\right) + \max\left(0, i_V\right) + \max\left(0, i_W\right) \qquad (16\text{-}4)$$

Keep in mind that the motor current does not always correspond to the current supplied by the DC bus, especially during braking.

Also, not all the current flowing is always necessarily generating torque. In order to calculate the actual torque output by the motor, we need to consider only the quadrature component and then scale by the torque constant:

$$\tau = k_T\, I_q \qquad (16\text{-}5)$$

Summary

Building a motor driver is a challenging task but also an exciting and rewarding process. We started the journey by studying power switches, which are electronic devices used to deliver power from the DC bus to the motor according to specific commutation algorithms. Your task as a designer is to select the best switches based on their electrical characteristics and optimize their working behavior to maximize performance and minimize losses.

We then introduced gate drivers, whose task is to activate the switches according to the controller's commands and to protect them from abnormal situations. Each switch needs its right driver to work well.

Finally, we analyzed the current sensing stage, which is a critical and delicate section of the servo loop. Only high-speed and low-noise sensing allows for best quality control of the motor's motion.

CHAPTER 17

Power Management

Power management in any electric machine requires a great deal of attention: both from an operational point of view, to provide a steady supply of power to the electronics and motors, and also from a safety perspective, to make sure that the system does not turn into a hazardous fire under any circumstance.

The main power supply and its control are often integrated in the servo drives section of the robot, because that is where most power is needed. In other systems, it stands as an independent device. The primary function of the power supply controller unit is to provide a stable high-voltage DC bus to feed the motors. A secondary function is to spin off a low-power low-voltage bus to supply all electronics, both inside the servo drive and outside of it (main controller, encoder, IOs, etc.). Finally, a number of safety functions are also required to prevent incidents.

We will focus our analysis on the design of the DC power management section, regardless of whether the actual power is supplied by a rectified AC source or directly by a DC battery (Figure 17-1). Both cases are common in the industry, with batteries typically employed in low-power or mobile machines and robots.

© Fabrizio Frigeni 2023
F. Frigeni, *Industrial Robotics Control*, Maker Innovations Series,
https://doi.org/10.1007/978-1-4842-8989-1_17

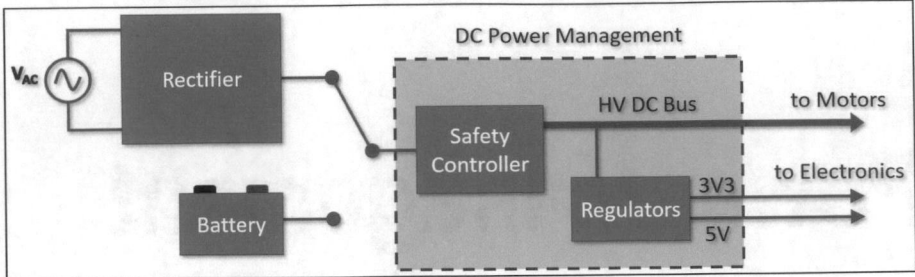

Figure 17-1. *Input power section*

DC Bus Voltage

The voltage of the DC bus depends on the application requirements and on the motors employed by the robot. A higher voltage level has the advantage of guaranteeing higher motor speeds and lower currents. However, it also poses more design challenges because of safety regulations. Large industrial robots typically hold their DC buses at several hundred volts. Smaller robots and the majority of hobbyist projects work at much lower levels (24V~60V) to simplify matters and avoid safety issues.

When working with high-voltage buses, a galvanic isolation between the low-voltage electronic circuits and the high-voltage section is important and often required by regulations. A typical example is an isolated gate driver that needs to keep the logic-level signals (including ground) separated from the switches side. The most common isolating solutions are optical (optocouplers), magnetic (integrated transformers), or capacitive. On the other hand, working at 60V DC or below is considered safe, and isolation is not needed.

The primary purpose of the DC bus voltage is to power motors, and therefore it needs to store enough energy to be able to respond quickly to any power request from the motors. The storage of energy is achieved by means of a large bulk capacitance. Typically, electrolytic capacitors are used because of their convenient capacitance-to-volume ratio. They do have quite a few drawbacks though: their actual capacitance value drifts over time; their shelf life is relatively short; their internal resistance is high, which means they dissipate significant power. However, nothing beats them when it comes to providing a large capacitance in a small amount of space and at a sensible price.

On a practical note, when designing the power stage for a motor driver, consider using many small capacitors in parallel instead of a single large one, because they are able to provide much higher instantaneous current levels, despite offering the same total capacitance.

One issue with having a large bulk capacitance is that a massive in-rush current is drawn when power is first applied at startup. The power supply is essentially shorted across a large empty charge reservoir to fill up: if no limitation is imposed, the initial current flow can be extremely high and damage the equipment. Also, it can cause the supply rails voltage to drop significantly, resulting in the entire system entering an undesired state and possibly affecting other subsystems powered by the same source. The solution is to introduce an in-rush current limiter to slowly pre-charge the capacitor bank, before the motor control stage is allowed to start up.

There are a few possible practical implementations for a current limiter. Figure 17-2 shows two common options. The first (on the left) uses a single N-channel MOSFET controlled by a gate driver with an internal charge pump. The switch is initially only partially turned on: it is operated in the high-resistive region to limit the in-rush current and slowly charge the bulk capacitance. The circuit works either in current control mode or in temperature control mode. In the first case, the current is kept constant, while the voltage across the capacitor bank increases over time:

$$I_{in-rush} = C_{bulk} \times \frac{dV_{bus}}{dt} \qquad (17\text{-}1)$$

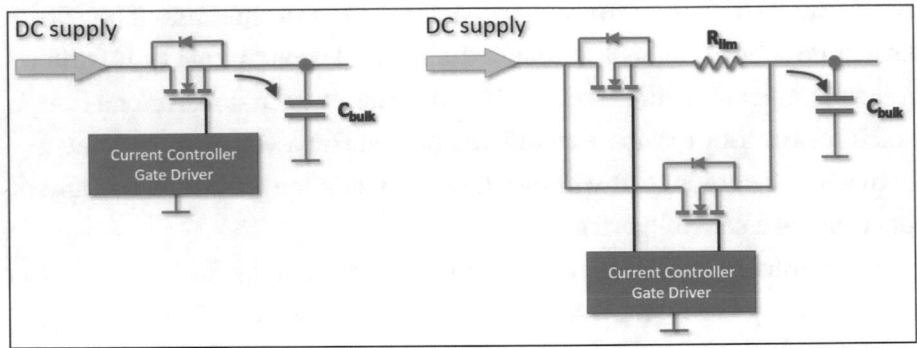

Figure 17-2. *Two possible implementations of in-rush current limiters*

In the second case, the junction temperature of the FET is monitored, and the device regulates the in-rush current to maintain the switch in its safe operation area. Regardless of the control mode, as more charge is stored in the bulk capacitors, the voltage difference between the power supply and the DC bus decreases. Eventually, when the C_{bulk} is fully charged, the switch is also fully open and opposes virtually no resistance to the flow of current needed to power the motors. This solution is relatively simple to implement, also considering that there are specialized chips on the market called *hot-swap* controllers, which integrate the gate driver and the current control together (e.g., LM5069 by TI).

The major drawback of this concept is that it requires a very large and expensive MOSFET, or more often a few of them in parallel, to allow the flow of enough energy to charge the bulk capacitors without burning out. In fact, the typical MOSFETs used in motor control bridges can sustain very high currents when fully conductive but perform very poorly when used in their high-resistive region (at low V_{GS}, where R_{DS} is high and a lot of power is dissipated into heat).

Figure 17-3 shows the channel current vs. voltage characteristic of a MOSFET (at 25°C). When the switch is fully open, the channel resistance is very low and so is V_{DS}: this condition can be held for virtually any amount of time (point A in the graphic, DC curve at $V_{DS} < 1V$, $I_D > 10A$). However, when the switch is only partially conductive, V_{DS}rises because of the large channel resistance: in this case, we can either allow only a very small amount of current to flow (point B in the graphic, DC curve at $V_{DS} = 60V$, $I_D \ll 1A$) or allow a large current but only for a very short amount of time (point C in the graphic, 100µs curve at $V_{DS} = 60V$, $I_D > 10A$), before overheating and permanent damage occur. Both conditions can only be used to charge small bulk capacitances.

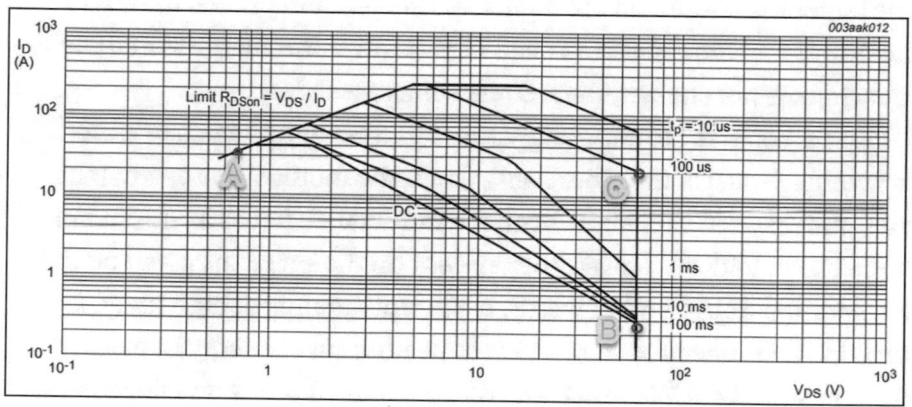

Figure 17-3. *MOSFETs can operate at high V_{DS} only for a very short amount of time*

The pre-charge of very large capacitance can be accomplished using a slightly different circuit, shown on the right side of Figure 17-2. In this case, two N-channel MOSFETs are deployed in parallel, and they are both operated in their fully conductive region. One switch is activated first to let the in-rush current flow through an external power resistor, which is able to sustain large power losses for a longer time. Once the C_{bulk} is charged up, the second switch is turned on to provide a low-impedance path for the motor current.

As soon as the DC bus voltage reaches above a safe threshold, we can switch on the secondary buses and activate the main control electronics. Once the capacitors are fully charged, we can start driving the motors. Interestingly though, the motors are not mere passive loads: their movement has a strong influence on the DC bus voltage itself.

A motor is partly an inductive load (see model shown in Figure 14-1), which means that it stores energy in form of an electromagnetic field while running. This energy is provided by the bus capacitors during acceleration. The larger the motor and the higher its acceleration profile, the larger the power that the capacitors need to provide. If the charge stored in the capacitors is not enough, the DC bus voltage will drop.

During operation, the energy stored in the motor increases with its actual speed. On the other hand, when the motor slows down, that energy is released in electric form: the inductance forces a current out of the motor with an intensity dependent on the speed drop and on the deceleration time. In other words, if a motor running at high speed is forced to an immediate stop, a very large amount of energy is released into a powerful current burst. Conversely, if the motor slows down with a low deceleration profile, the same energy is released over a longer time window, causing a lower output current. Note that this effect does not occur with pure resistive loads (e.g., an LED) because no energy is stored in there.

Critically, the output current must be allowed to flow somewhere. During the deceleration phase, the bridge is essentially operating in reverse, i.e., trying to generate an induced magnetic flux pulling against the direction of movement of the permanent magnets. The result is that the excess current flows backward to the DC bus: the motor regenerates energy into the supply.

The first immediate effect is that the bus capacitance gets recharged until full. Increasing the size of the bulk capacitance allows to soak in more charge and reduces bus voltage swings caused by regeneration.

For example, Figure 17-4 shows part of a small board I built to control three middle-power DC brushed motors with a total capacitor bank of 13mF. Large industrial robots have massive capacitor banks to keep output voltage as stable as possible.

Figure 17-4. *Capacitor bank on a DC bus shared between three motors*

However, increasing capacitance comes with a few drawbacks: capacitors are expensive and take large space on the board; also, they need to be safely pre-charged with a current limiter, especially when many motor drives are connected in parallel on the same DC bus.

A shared DC bus across several drives is always a good option to consider: energy is shared between the axes, and general consumption is optimized. One decelerating motor can power an accelerating one. After all, motors in robots often all move at different speeds because of the nonlinearities of inverse kinematics.

Occasionally, even a large oversized bulk capacitance can fill up, and the excess charge needs to be absorbed somewhere else. If the system is powered by a battery (e.g., mobile robots, AGVs, electric vehicles, etc.), then the motor conveniently recharges the battery while slowing down. Electric cars take full advantage of this feature, which also significantly reduces wear on their mechanical braking system. Industrial robots, however, are rarely powered by batteries and often rely on fixed power supply units. In that case, the energy can either be fed back to the main electric grid (if the power supply is designed with such feature) or must be burned off quickly in form of heat. Other less common options consist in storing the regenerated energy in a supercapacitor or in a flywheel.

Although burning electric power into heat is effectively a total waste, it is also the cheapest and most adopted solution in the field. The DC bus is connected to ground using a large power resistor. When the motor regenerates large amounts of energy and the capacitor bank cannot take in anymore, the bus voltage starts rising quickly. The microcontroller monitors this rise and triggers a switch to let the excess power flow through the braking resistor and dissipate as heat. Braking resistors must be sized accordingly to the expected working cycle of the motors and must be given enough safety margin to avoid overheating.

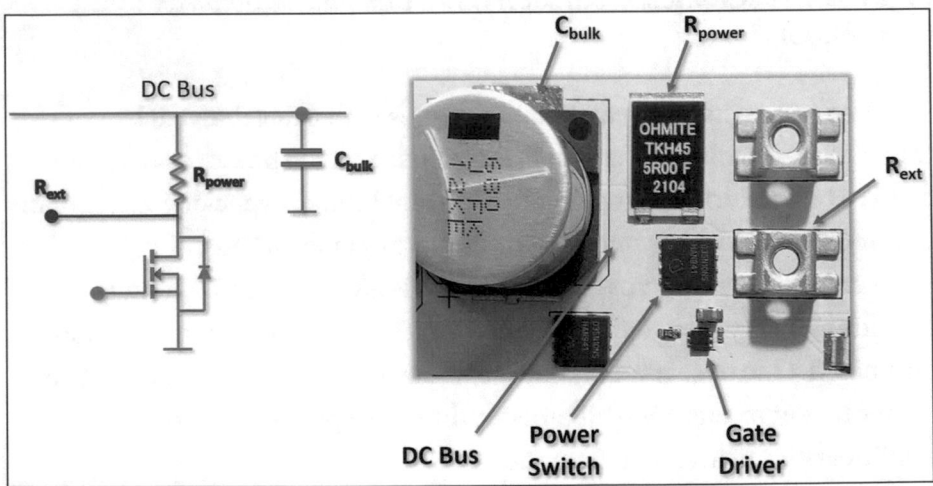

Figure 17-5. *Braking resistor schematic and actual board implementation*

Figure 17-5 shows the practical implementation of a braking resistor control circuit for low-power applications. The resistor is an SMD type and allows a large power dissipation conducting heat away through the metal substrate of the PCB. An output connection is provided to add an additional external resistor to dissipate even more power for applications that require quick motor stops. The switch is a power MOSFET, which needs its own gate driver. The microcontroller constantly monitors the DC bus voltage level and activates the braking resistor by PWM switching. Typically, a simple proportional control loop is good enough to determine the duty cycle according to the excess overvoltage.

On the other hand, the controller also needs to monitor for the opposite situation: an undervoltage on the DC bus. Even short undervoltage transients can be dangerous because they cause power to be temporarily removed from the control electronics and the results can be unpredictable. Possible cases for undervoltage behaviors can be as follows:

- **Pre-charge**: This is part of normal operation, as the voltage rises while the bulk capacitance is being charged. The main electronic controller should be in sleep state until the available voltage reaches above a safe threshold.

- **Lack of power**: This condition is typically caused by poor design. A sudden acceleration of all robot's motors linked on the same bus causes a peak current to be drawn. If the bulk capacitors cannot supply enough energy, then the DC bus voltage will start dropping significantly. Voltage ripples are normal, but they should never dip too low. In that case, it would mean that the drives (or at least the capacitor bank) are undersized for the job. The solution is either to limit the movement dynamics or to redesign the drives with larger bulk capacitance.

- **Short circuit**: An electric short somewhere on the wirings or internally in the motor can cause a voltage drop. The condition results in massive current being drawn from the power supply, and the safety functions should switch off the bridge immediately.

Protection Functions

When designing the power stage for any electric machine, we should always include a few safety features to make sure that nothing starts burning off. High voltage is significantly more critical to handle and requires its own set of regulations that we do not deal with here; low voltage can also result in hazardous situations for the devices, because the involved currents can still be quite high.

Some of the most common protection functions that should be included in the power stage of our robot are as follows:

- **Reverse polarity**: To prevent damage to the equipment when the user inadvertently applies the supply voltage at the wrong terminals

- **Circuit breaker**: Should trigger when a short circuit at the output causes a large current to flow

- **Current limiter**: Monitors the continuous average current to prevent overheating

- **Temperature monitor**: To prevent overheating due to any reason

- **Reverse current block**: (Optional) to prevent regenerating into a power supply

- **Over- and undervoltage**: To make sure that the DC bus voltage does not exceed its safety range (as already described in the previous section)

- **Bleeder resistor**: To provide a slow discharge path for the bulk capacitance after the system has been switched off

While there are integrated ICs that provide most of the listed functions into one single chip, they are usually limited to low-power applications. For high-power systems, as in the case of industrial robots, we need to implement a solution with several individual components.

Reverse polarity protection is probably the most basic requirement of any electronic system. Accidental wiring errors can always happen, and we should provide a way to avoid catastrophic and expensive consequences. Applying a negative voltage to the DC bus would cause the electrolytic capacitors to be reverse-biased, which results in damage and possible explosion. Technically, the reverse voltage across the bus would be initially clamped by the body diodes of the switches, which would conduct until they possibly burn down. In any case, a protection is needed.

While using a diode to block reverse voltage (see Figure 17-6) could be used for very basic and low-power applications, in practice that is not a reasonable solution for high-power applications, at least not until one-way superconducting diodes are available on large scale and high temperatures. A regular diode introduces a forward voltage drop V_f in its conduction area of operation, which causes a massive power loss of $P = V_f \times I$ when a large current I flows through. Additionally, a heatsink must be usually added to help with heat dissipation.

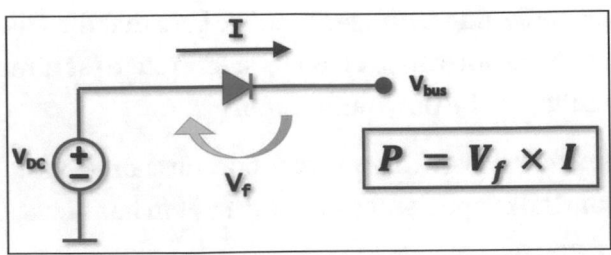

Figure 17-6. *Using a simple diode to block reverse polarity is not an efficient solution*

A much better approach is to use a MOSFET and allow it to conduce current in one direction only. Any reverse current due to accidental reverse polarity connection would be blocked. MOSFETs have a much lower impedance than diodes and therefore waste much less power in normal operation. A p-MOSFET would actually be the easiest choice (see Figure 17-7): the body diode initially conducts in forward direction, until the source reaches a higher voltage than the gate (which is initially held to ground); once V_{GS} is large enough, the channel opens up and the switch fully conducts. A Zener diode limits the maximum V_{GS} to protect the gate. In case of reverse polarity connection, the source voltage never goes higher than the gate, and the switch is never open to conduct in reverse. This solution is optimal for medium-power applications and is cheap to implement.

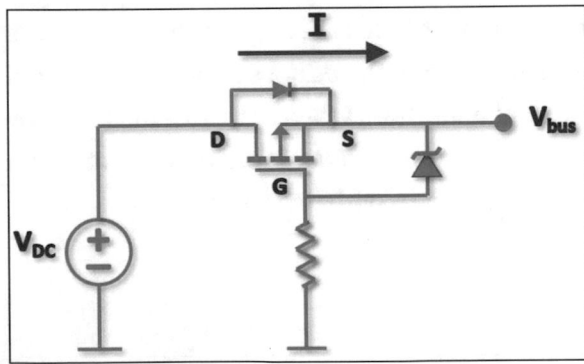

Figure 17-7. *A MOSFET is a more efficient choice for reverse polarity protection*

510

When the power requirements increase and the current flowing is very large (e.g., > 100A), then n-MOSFETs are the device of choice. Their channel resistance is much lower than p-MOSFETs (at parity of size and cost) and therefore offers a more efficient solution guaranteeing minimal power losses. The only issue is that an N-channel device requires a charge pump to drive its gate higher than the supply voltage, at which the source is tied. The resulting device is often referred to as *ideal diode* (see Figure 17-8) because it offers the same function as a diode but features a much lower voltage drop. An example of dedicated ideal diode controller is the LM74700 by TI.

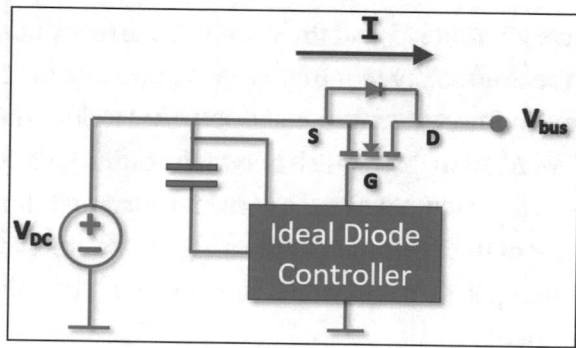

Figure 17-8. *Ideal diode: An N-channel device offers the lowest power loss*

Ideal diodes also act as **ORing** controllers: they allow multiple power sources to be connected in parallel so that uninterrupted supply is available to the load whenever any of the sources fail (see Figure 17-9). The ORings only allow current to flow from the supplies to the load and block any reverse current into the power sources.

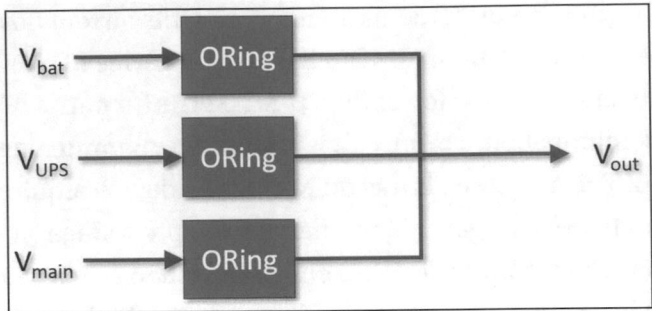

Figure 17-9. *ORing controllers deployed for power redundancy architecture*

Once we have guaranteed that the voltage is correctly applied to the DC bus, the second safety requirement we impose is to monitor the current flowing into the system and make sure that it does not exceed specific levels. We need to distinguish between continuous average current and peak current: a device can allow a nominal current to flow for an indefinite amount of time without overheating and ever needing to shut down; much higher peak current values are allowed, but only for very short bursts of time.

A simple fuse is always a sensible component to add in series to the power rail in order to prevent large current draws. However, besides being very cheap, it does not offer other advantages: it cannot perform very accurate current control, and it gets damaged when overcurrent occurs, requiring physical repair of the equipment.

An *e-fuse* is an electronic replacement (or complement) for traditional fuses and consists in a current sensing section and a switch (see Figure 17-10). The idea is to only allow a certain amount of current to flow through and turn the switch off when overcurrent occurs. The advantage of employing an electronic controller, instead of a discrete fuse, is that very precise current limits can be set, both for transient peak currents and also for continuous average values. In other words, the device works both as a fast circuit breaker against short circuits and as a current limiter to prevent overheating. Besides, a large overcurrent would simply cause the switch to turn off and disconnect the supply, without any damage to the e-fuse and without any need to replace components. The drawback of such a solution is cost, which increases with the size of the switch needed to control high-current systems.

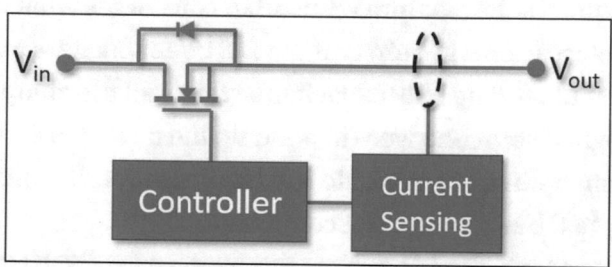

Figure 17-10. *An e-fuse can accurately monitor and limit current flow without being damaged*

The practical implementation of an e-fuse can either be realized with discrete components or even with integrated devices (e.g., TPS2595 by TI). Besides monitoring current, the device must also monitor the junction temperature of the FET, which should be typically kept below 150°C.

Actually, monitoring current flow should be done in both directions. We learned in the previous section that a decelerating motor can regenerate power into the DC bus and that the excess energy on the bus must be dissipated somewhere. Rechargeable batteries can naturally accept an input current (before they are fully charged), but standard AC-DC rectifiers cannot. We saw how braking resistors are the most common solution to this issue, but adding a protection to prevent unwanted current to flow back into the supply is also a good idea. Clearly, the reverse current protection is not needed if the power comes from a battery and regeneration can safely occur until the battery is fully charged (after that, the braking resistor will be forced to burn excess energy into heat).

Fortunately, the ideal diode solution implemented in Figure 17-8 also blocks reverse current. Note that the body diode of the switch is oriented from source to drain, so to block any regenerated reverse current even when the system is powered off, which can happen when the motors are forced to move by an external load.

One additional safety feature we need to consider is what happens to that massive electric energy stored in the DC bus capacitors as soon as the system is shut off. The bulk capacitance holds all the charge and keeps the voltage high. If someone were to open up the device to service it, a potentially dangerous voltage would still be present on the bus even if the power supply had been entirely disconnected.

For safety reasons, the DC bus capacitors must be discharged through a high-impedance path, called *bleeder resistor*. The resistor must be high enough to prevent quick discharge during normal operation but must be low enough to allow the DC bus voltage to drop below safe levels in a relative short time (i.e., 5 minutes) after the electric power has been shut off. That prevents danger to an operator accessing the equipment for service or repair, as long as the safety discharge time has elapsed.

In practical applications, we can take advantage of the DC bus voltage monitor to also act as a bleeder resistor. Figure 17-11 shows a voltage divider using resistors R_1 and R_2 that generates a low-voltage access point for the ADC port of a microcontroller to measure the actual voltage on the bus. The C_1 capacitor is used to smooth out ripples and noise.

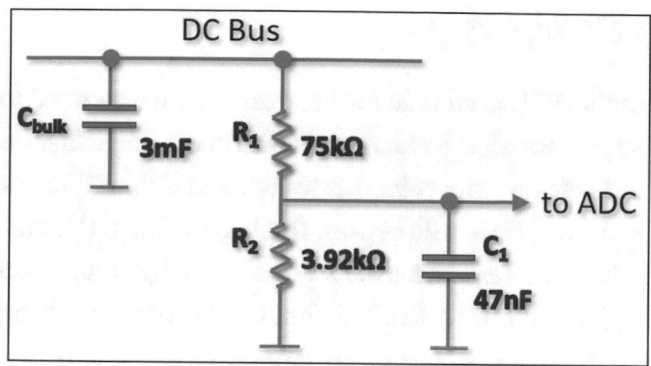

Figure 17-11. *The DC bus voltage monitor also serves as bleeder resistor*

When power is switched off, the $R_1 + R_2$ series allows the charge stored in C_{bulk} to drain away in a reasonable time. The discharge rate of the initial capacitor voltage V_0 through the resistive path R is given by the following:

$$V = V_0 e^{-\frac{t}{RC}} \qquad (17\text{-}2)$$

For a numerical example, let us assume $V_0 = 60V$, $R = 78.92k\Omega$, $C = 3mF$. It is easy to verify that the DC bus voltage has dropped to negligible levels (below 10V) after about 5 minutes of discharge (see Table 17-1).

Table 17-1. *Example of discharge rate of bulk capacitor through bleeder resistor*

Time [s]	0	60	120	180	300
DC bus voltage [V]	60	41	28	19	9

Usually, a bright warning sticker is placed on the outside case to act as a reminder: "Wait at least 3 to 5 minutes after disconnecting power before opening the device for maintenance."

Voltage Converter

Besides the main DC bus voltage for the motors, we also need to provide a low-voltage supply for all the electronic controllers and other components in the system. Typically, most electronics work at either 5V or 3.3V, so we need a way to derive those values from the high-voltage DC bus.

There are two different kinds of DC-to-DC voltage regulators: linear and switching (see Figure 17-12). The first kind is very simple and cheap and requires only a couple of external capacitors to work. However, its idea of operation is basically to waste the excess voltage into heat, which is not the most efficient solution. Linear regulators are best for low-voltage drops and low-power applications, where a very stable output voltage is required. The typical example is powering a microcontroller.

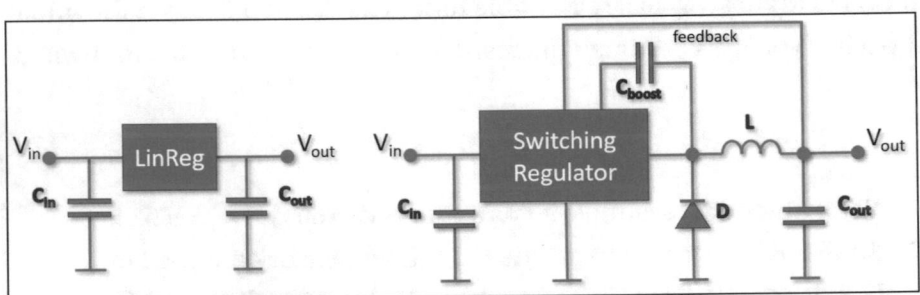

Figure 17-12. *Simplified schematics for linear regulator (left) vs. switching regulator (right)*

Switching regulators, on the other hand, are much more complex devices that use an integrated PWM-controlled power switch to modulate output voltage in a closed-loop scheme. They are expensive, they require a large number of external components to be arranged with careful PCB layout, and they provide a noisy output voltage with non-negligible ripples, which can only be partially smoothed out with a large capacitor placed at their output. However, they are incredibly efficient and generate little power loss even at large voltage drops.

The typical solution is to take the best of the two worlds and combine them together (see Figure 17-13). We first use a switching regulator to convert the main DC bus voltage down to lower levels (e.g., bring a DC bus of 60V down to a 12V bus) and then use a couple of linear regulators to power the small electronics with clean stable voltages (e.g., a 3.3V bus for the MCU and a 5V bus for the encoder).

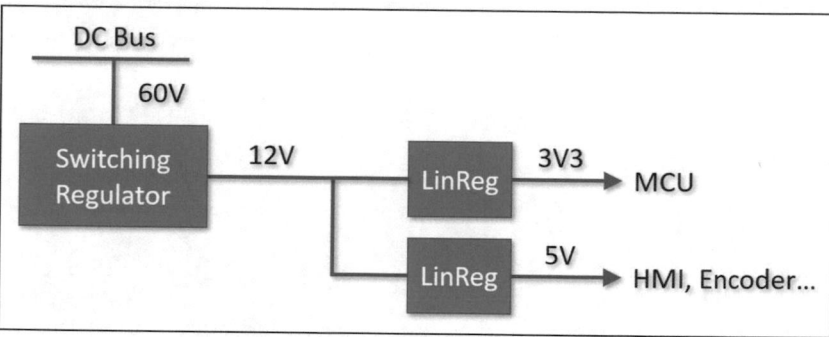

Figure 17-13. *Combining switching and linear regulators for best efficiency*

When selecting the voltage regulators, make sure that they can provide enough output current for the required application. Also, place generous copper area to allow for proper cooling.

Summary

Power management is the process of providing a stable and safe voltage source to all the system components, from the control electronics to the motors. Regardless of what energy source we use at the front end (either battery or AC supply), we need to generate a main DC bus, from which everything will be powered: the control system via DC/DC converters and the motors via the power switches in the inverters.

The power section is also the most critical in terms of safety functions, and a number of monitoring features must be built-in to guarantee safe operations in all circumstances. We described a few circuits to limit current and voltage to protect the system from hazards.

CHAPTER 18

Main Controller

The main controller is the robot's central brain that performs all the motion calculations and sends the set values to the drives. Such a centralized architecture is common in multi-axes systems where interpolated movements are required. The main controller receives input commands from the HMI and IOs, plans a path for the robot, generates a motion profile for each axis using the kinematic model, and monitors the servo drives to make sure that the robot is performing the correct movements.

From the hardware point of view, the main controller includes a microprocessor running a real-time operating system, a memory bank to store configuration parameters and user programs, and a few interfaces to the HMI, the IOs, and the drives. We will analyze each of these parts in the next sections so that you can build your own electronic board to control the robot.

Microcontroller

When selecting the core processor, the choice for hobbyists typically comes down to a microcontroller. You could use a CPU or an FPGA for more performance, but the truth is that nothing beats a modern microcontroller in terms of simplicity of use, low price, wealth of integrated peripherals, low-power consumption, and support from the community.

© Fabrizio Frigeni 2023
F. Frigeni, *Industrial Robotics Control*, Maker Innovations Series,
https://doi.org/10.1007/978-1-4842-8989-1_18

A large variety of 32-bit microcontrollers are available from different vendors (ESP32, PIC32, STM32, AT32, and others), and all share very similar features. Figure 18-1 shows a couple of examples I built using different kinds of controllers.

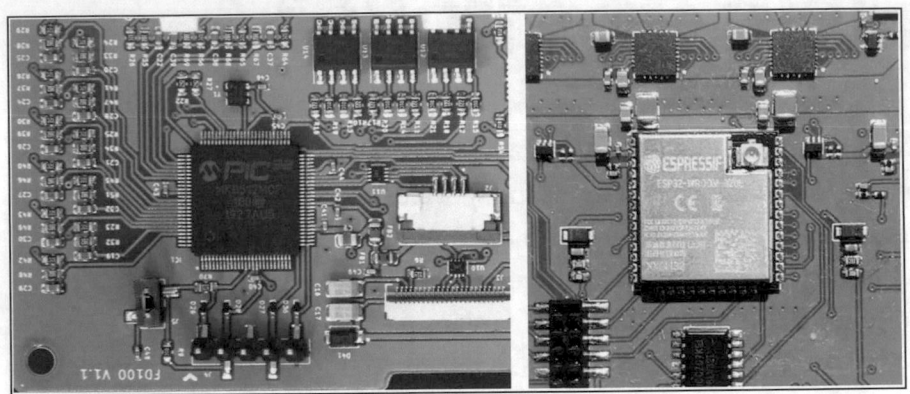

Figure 18-1. *Examples of boards using the PIC32 (left) and ESP32 (right)*

I personally tend to use the ESP32 for most of my projects, which really is an SoC device (system on a chip). It is ideal for IoT products because of its integrated connectivity (both Wi-Fi and Bluetooth); it offers a very rich range of peripherals (serial, SPI, I2C, PWM for motor control, pulse counter for quadrature encoders, CAN, ADCs, and many others); it has a large internal flash memory (currently up to 32MB) with security encryption features.

The controller firmware runs on a real-time operative system (RTOS) which supports multitasking architecture. In other words, the control software can be structured in different tasks, each responsible for a specific functionality and running with its own priority. Additionally, the chip is a dual-core processor, so that you can let one core focus entirely on the critical motion operations and leave all the other tasks (e.g., HMI and connectivity) to the second core.

Besides the pure technical reasons, the ESP32 is a great choice also because of its low price, large availability, trustable reliability, user-friendly software, and large worldwide user community. The one drawback of these chips is the low number of IO pins, which limits their use to smaller products. Sometimes I parallelize two or three MCUs in the same application to multiply computing resources and IO lines. Alternatively, you can use a simple IO expander with an SPI or I2C link to the MCU.

In any case, the choice of hardware often comes down to personal preferences, or specific customer requirements, and should not really affect the final product.

Microcontrollers require a sufficient and stable power supply to run safely. Any glitches in the power line could temporarily affect the processor's operation and in worse case cause a system reset. Best practice to avoid issues is to add decoupling capacitors (typically 100nF ceramic components), which should be placed as close to the pins as possible on the same side of the board, in order to filter out noise and stabilize the power source. When significant high-frequency noise is present on the power lines, ferrite beads can also be inserted as additional filtering components.

Another sensitive part is the oscillator used to generate the clock signal for the microcontroller. Interferences to the oscillator should be prevented; otherwise, all time-sensitive functions of the microcontroller could be affected (high-frequency communication interfaces in particular). Ideally, the area around the oscillator on both sides of the board should be isolated from surrounding circuits using a large grounded copper pour and avoiding routing any high-speed signals nearby. As the crystal oscillator is particularly sensitive, placing radiation-emitting devices (e.g., switching voltage converters or unshielded inductors) nearby should also be avoided to prevent interference.

Finally, the microcontroller is the place where you download the firmware you program and all the configuration data for the robot, so you want to provide an easy and safe access to it. Firmware upgrades are often required, especially during the initial development phase. The most common procedure to do that is an *update on the fly*: we first transmit the new version to the microcontroller and let it store it somewhere in its flash memory; then we let the existing firmware know that a new version is available and that it should be used upon the next reboot. This procedure is safe, because if something goes wrong during the download or boot of the new version, the controller can always switch back to the old version that is still stored in memory. If using the ESP32, a firmware upgrade can be conveniently done over Wi-Fi through the robot's HMI application.

When building commercial products, we also like to protect the software we write into the microcontroller because it contains our own know-how and is probably the most valuable part of the entire project. Most controllers on the market offer flash encryption protection, which means that no one else except for yourself can upload meaningful information from the flash content: both your program and your data are safe for being copied.

IOs

Robots typically carry tools to perform certain tasks, and those tools normally require to be controlled and monitored synchronously to the movement executed by the robot: e.g., switch on/off a welding gun, open/close a gripper, activate a painting brush and read its actual fluid pressure, and so on.

The simplest way for the robot's controller to exchange signals with its tools (and also with other machines in the same working cells) is to use IO modules, which stands for input/output. Input channels can read a signal, either *digital* (an on/off state) or *analog* (a numeric value). Output channels are normally digital, and they can trigger a simple command (on/off switch). Most microcontrollers offer the option to fully customize via software the function of each port (input/output; digital/analog) and use the name **GPIO** (general purpose input/output) to describe that feature.

One characteristic of IO channels is that they represent a direct interface between the sensitive internal microcontroller and the rough outside environment. The link can be dangerous for the inside electronics and must be designed with care.

Figure 18-2 shows the most basic circuit to read a digital signal of known fixed voltage. There are quite a few external components added to the interface. The resistors are used to build a voltage divider to shift the original signal level (e.g., 24V) to a safe level for the microcontroller, typically 3.3V. The capacitor serves as a low-pass filter to smooth down the signal and remove noise glitches according to the cutoff frequency $f_c = (2\pi RC)^{-1}$: higher capacitance values increase the filtering effect but reduce the speed response of the input port. Additionally, a software filter is typically used to remove the bouncing effect of a mechanical switch.

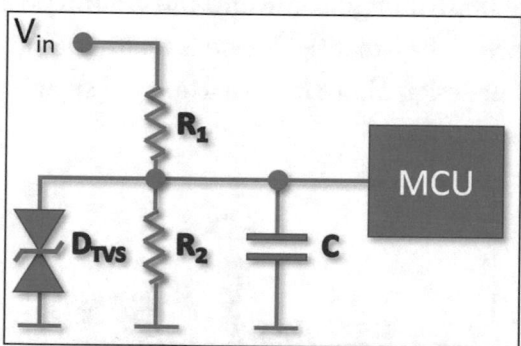

Figure 18-2. *Digital input interface*

The **TVS** (transient voltage suppression) diode is an important protection element used to clamp possible voltage surges caused by electrostatic discharges (EDS). Most microcontroller pins are already internally protected against relatively low-power discharges that might happen during the manufacturing and assembly phases of your electronic board. However, if you want your product to be certified to work in harsh industrial environments, you need additional protections for some serious high-power discharges: fast rise time, long duration, high peak current, and a repeated series of pulses with both polarities (see the IEC 61000-4-2 standard for more details).

Note that by placing the TVS across the MCU pin instead of directly at the input voltage, we take advantage of the R1 resistor as a useful current limiter for the diode. Also, keep in mind that while TVS have very quick response times (in the order of few nanoseconds), they are not designed to block continuous overvoltage conditions; those are better handled by Zener diodes.

For further isolation between the internal electronics and the outside environment, a better practice adopted in the industry is to galvanically isolate the two sides from each other, e.g., by means of optical devices called optocouplers (see Figure 18-3) or bidirectional capacitive isolating barriers. Note that both the signal line and the ground plate are isolated. The speed response of the isolating device is an important parameter to consider during the design phase in case the input signals are expected to switch at high frequencies.

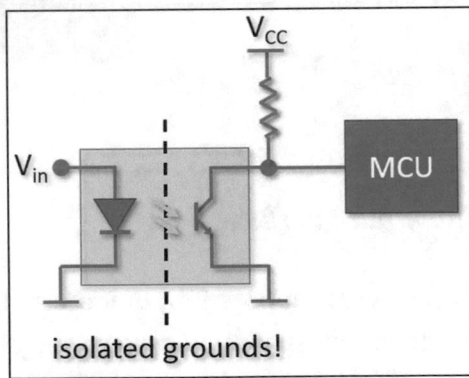

Figure 18-3. *Optocouplers can be used to electrically isolate external signals from the internal electronics*

Reading analog signals can usually be done in a similar way as for digital signals, using a resistor divider to step down the voltage and the RC combination to filter out noise. That is the typical combination used to read signals out of a low output impedance IC, e.g., a Hall effect current sensor. However, care must be taken when dealing with high-impedance sources (e.g., a simple joystick potentiometer for drive-by-wire applications), whose output signal could be significantly affected by the impedance of the analog input interface. In that case, adding a high input impedance voltage follower between the sensor and the input is a good idea (see Figure 18-4). Make sure that the speed of the follower matches the expected bandwidth of the incoming signal; otherwise, you will lose useful information.

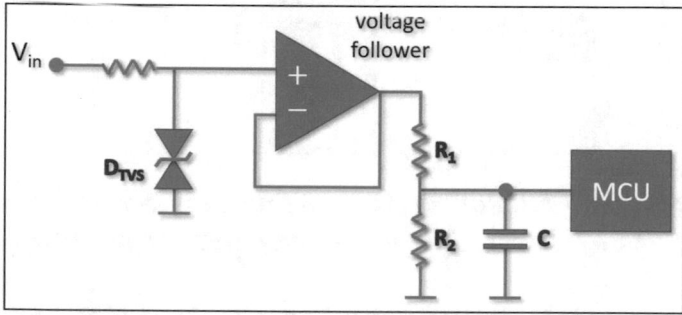

Figure 18-4. *Adding a voltage-follower increases the port's impedance, preventing the input signal from being affected*

We now look at the circuit to set digital outputs and switch on/off external devices. All microcontrollers have digital output pins, which can typically source a few milliamps of current at 3.3V. That level of power is not enough to drive any serious load and needs to be amplified accordingly. Depending on the power and speed requirements, there are two solutions we can adopt: if the load requires low-speed switching and draws little current (a few Amps), then we can use a logic-level MOSFET directly driven by the microcontroller (see Figure 18-5); if the load needs high-power and high-frequency switching, then a gate driver is the best solution, exactly as we described for the servo drives in Chapter 16. The reason why a microcontroller can only drive at low speed is that its output current is too little to quickly charge the gate of a large MOSFET device. Conversely, a gate driver capable of sourcing and sinking a couple of Amps of gate current can quickly charge powerful switches for high-power applications.

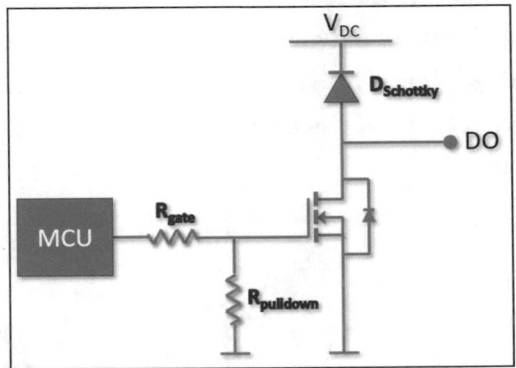

Figure 18-5. *Low-side switch circuit using a logic-level MOSFET directly driven by the MCU for low-power and low-speed applications*

Using a low-side N-channel switch connects the load to ground when conducting and leaves it floating when turned off. A gate resistance R_{gate} must be included to limit the gate current during transitions, and a $R_{pulldown}$ must be included to force the switch in off state when the system is in undefined state (e.g., at startup). Multiple switches can also be tied together in parallel to support higher current operation.

When driving inductive loads (i.e., solenoids, relays, and motors), a flywheel diode to the supply rail must be included in the design (see $D_{Schottky}$ in Figure 18-5). The reason is that the inductance of the load stores electrical energy, which holds the flowing current constant during normal operation. When the applied power is switched off, that energy cannot suddenly disappear: the current must keep flowing somewhere to draw away the stored charge. The flywheel diode provides a path to dissipate the current in a circular loop across the load. If we did not provide that path, the inductance would generate a large voltage spike and likely damage the switch.

If the digital output circuit is meant to drive loads connected to a high-voltage DC bus, then it is a better idea to completely separate that part of the circuit from the microcontroller by means of a galvanically isolated gate driver. On the other hand, when interfacing with low-power devices and sensors embedded on the same board, it is possible to link them directly to the microcontroller without any additional external switches and protections.

Finally, when in need of a very large number of connections, IO expanders chips come into rescue. They essentially replicate the functions of the microcontroller's IO ports and use a serial interface (typically SPI or I2C) to let the main controller read and write values (see Figure 18-6). Some of these chips (e.g., TLA2518 by TI) even allow each port to be fully configurable either as digital input, or analog input, or digital output for a total flexible solution.

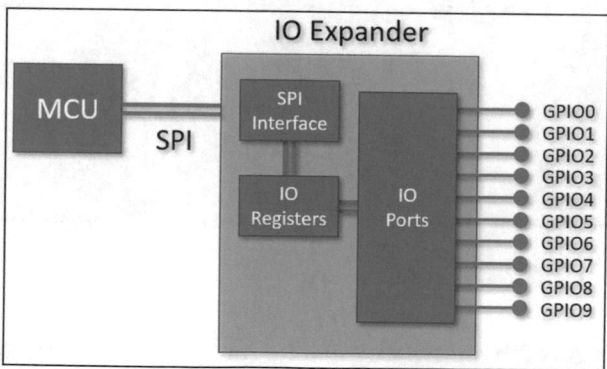

Figure 18-6. *Configurable IO expanders add flexibility to small MCUs*

Fieldbus

Besides offering simple IO channels, a microcontroller also needs to be able to exchange data at high speed with other devices, most importantly the servo drives. The common adopted solution is a **fieldbus**, which is essentially a local area network within all the connected devices, where each node is able to read and write information from the others in form of data frames. Fieldbuses typically feature a high immunity from noise and have built-in error detection, which makes them robust even for the most demanding environments.

One of the simplest fieldbus protocols, which is already integrated in most microcontrollers, is **CAN** (Controller Area Network). It is a very old and technically obsolete protocol but still widely used for its simplicity. The CAN is based on two channels (see Figure 18-7): a CAN_HIGH and a CAN_LOW line. All devices on the network are connected to the two channels and exchange data via messages.

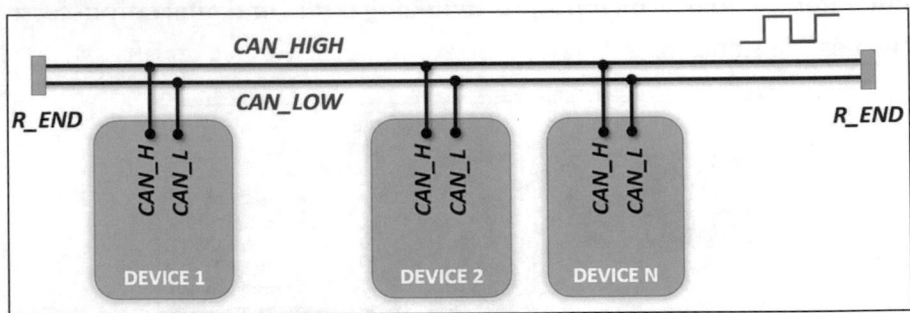

Figure 18-7. *CAN bus architecture*

The two ends of the network must be terminated by a resistor (typically $R_{END} = 120\Omega$) to dampen the signal and prevent it from bouncing back, which would otherwise create interferences and degrade the transmission.

Although most microcontrollers offer a CAN software interface, they cannot directly hook up to the CAN differential bus. An intermediate transceiver is needed in between, to translate the logic signals TXD and RXD (transmit data and receive data) into the differential lines CAN_H and CAN_L on the bus (see Figure 18-8).

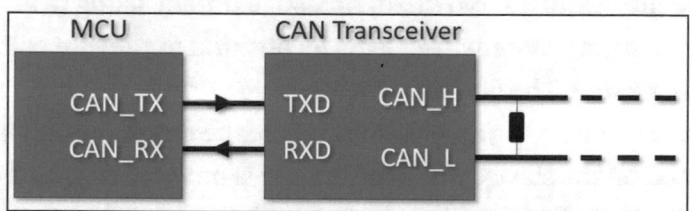

Figure 18-8. *A CAN transceiver is needed to translate logic signal into the differential bus*

A large variety of other protocols more advanced than CAN exist in the industrial world, each with different transmission speeds, frame sizes, and other features. Two of the most common are PROFIBUS and EtherCAT. However, these protocols are specifically tied to one product manufacturer and rely on proprietary technologies. As a result, a common effort by many industrial automation players led to the development of an open cross-platform standard, not related to any specific system: **OCP UA** (Open Platform Communications Unified Architecture). OPC UA is the default communication layer adopted by the Industry 4.0 standard and is also gaining ground within IoT devices.

If you build a robot controller and want your system to be recognized as Industry 4.0 compliant, then you need to embed an OPC UA server in the main controller, so that any OPC client device out there can connect to the robot and exchange data with it. There are several options you can follow; some are expensive, while others are free (e.g., there are even open source C stacks that you can download and install on a microcontroller).

Finally, if you use the ESP32 and want to take full advantage of its integrated wireless connectivity, I suggest trying out their ESP-NOW protocol: it is a fast and simple wireless communication protocol that allows multiple controllers to exchange data without the need of an external network. You can use it to send peer-to-peer messages (e.g., from the main controller to a servo drive) or also in broadcasting (e.g., each cart on a linear transport system disclosing its position to prevent collisions with all other carts). The communication is not strictly deterministic but is totally fine for monitoring applications where the entire motion is planned and executed on the slaves, while the master is only sending commands and checking that all runs well every few milliseconds or so. In those cases, perfect time determinism is not needed.

Integrated Solution

The most common hardware architecture to control robots in the industrial world is to have a separate main controller serving as a master and several slave drives to control the motors, as was previously shown in Figure 13-1.

However, when designing the control architecture for very small systems (i.e., machines with up to three axes and low-power requirements), it is common to merge all the electronics into a single integrated component. The main controller, the IO modules, and the servo drives can all be embedded into a single board, resulting in considerable savings of space and cost. The outcome is a product often labeled as **smart motion controller**, where the adjective "smart" highlights the ability to perform functionalities that are usually decentralized to an external controller (e.g., interpolated path planning of the robot's axes).

Integrated motion controllers also belong to the class of *edge devices*, in the sense that all the calculations are decentralized at the very edge of the hardware control chain without the need of a central powerful

CPU. Edge devices are widely used in the concept of Industry 4.0: they are deployed to collect and analyze data from the machine and provide real-time information to the manufacturer, who can then make appropriate decisions to optimize the production process. Also, they offer *condition monitoring* functionalities, showing if a machine needs maintenance because it is wearing out, allowing the user to react to potentially dangerous and costly situations before they occur (*predictive maintenance*).

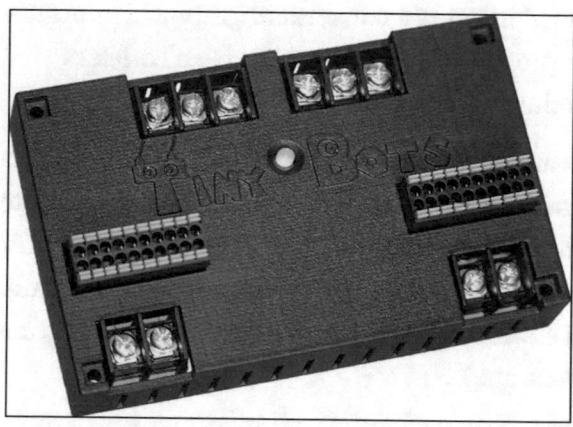

Figure 18-9. *An integrated smart motion control built by the author*

Figure 18-9 shows an example product I recently built to control robotic axes up to 1KW each, with a DC bus of 60V maximum. Some key features of this product include the following:

- **Flexibility**: The board can be quickly reconfigured via a remote application to activate different kinds of motor control algorithms for each individual axis (brushed, stepper, brushless, linear), various encoder interfaces, and customizable IO channels to match the needs of different sensors and actuators.

- **Modularity**: Several devices can be linked together via a fieldbus connection (either wired CAN or wireless ESP-NOW) to control multi-axes machines. For instance, an anthropomorphic six-axes manipulator needs three of these boards, one of which acts as a master and the other two as slaves.

- **Codeless programming**: All configuration and user programming operations are performed via a remote application using a convenient graphical interface: robot programming is simplified to a codeless procedure.

- **Software core**: The board is a powerful edge device, able to independently run all sorts of motion control algorithms (several robot's kinematic models, AGVs control with omni wheels drive, auxiliary gear and CAM slave axes, collision avoidance for linear transport systems, etc.).

- **IoT**: The device has an open Wi-Fi interface, which allows for remote configuration and monitoring. It offers convenient access to all its internal parameters and data for collection and analysis purposes.

The hardware simplification and the software flexibility make integrated solutions very attractive for low-power machines. Single axes controllers can be mounted directly onto the motors further reducing space requirements. As a practical example, a small SCARA robot (schematically shown in Figure 18-10) only needs a few of these boards to function entirely (drives D1...D3 control the motors M1...M3), removing the need for an external main controller and an electric cabinet. By taking advantage of the wireless fieldbus, the only wiring needed in the robot's body is the DC bus supply (typically clamped at 48V for safety reasons).

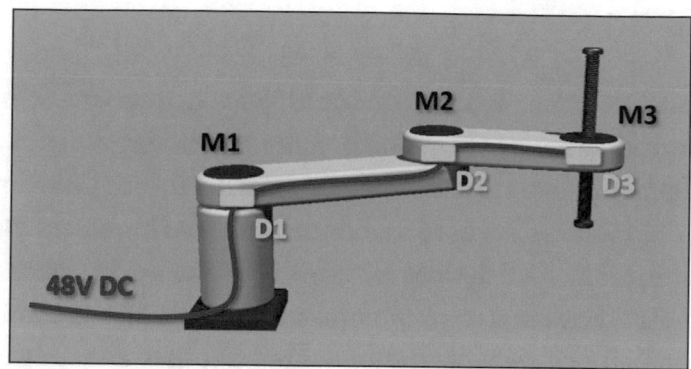

Figure 18-10. *Integrated solutions simplify wirings and save cost*

Besides robotics, another application where this kind of integrated controllers is widely used is linear motion transporters. Each cart is fitted with an independent control board and runs its own control program (motion, IOs, collision detection) and only needs a power connection to the supply rails (see Figure 18-11). Configuration and monitoring are performed over a wireless network. A traditional hardwired solution with a centralized controller would be impossible for rails longer than just a few meters.

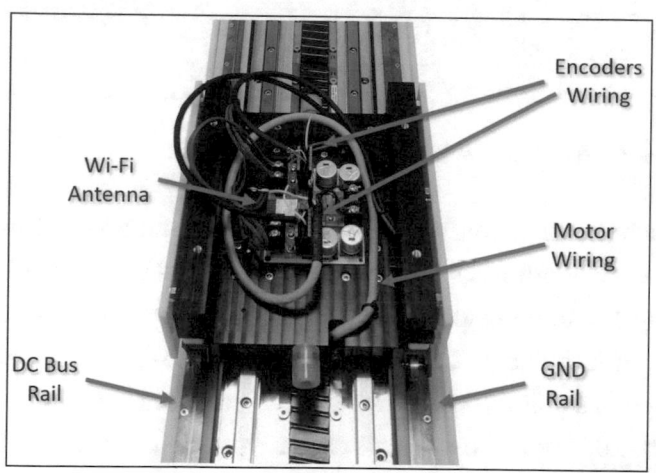

Figure 18-11. *Linear transport system with wireless integrated controller for each cart*

Display

The last function of the robot's main controller is to interface with the operator and allow for configuration parameters' changes, commands input, and general status monitoring (see HMI Section in Chapter 9).

Robots' manufacturers have traditionally offered handheld touch panels or simpler keypad devices to access the software interface. These devices typically have their own internal controller, which is configured as slave of the robot's main controller. Sometimes they even rely on more powerful processors running a full-fledged operative system (e.g., Windows IoT). The communication to the HMI is not time-critical and can be solved with standard TCP/IP protocol over an Ethernet cable (see Figure 18-12).

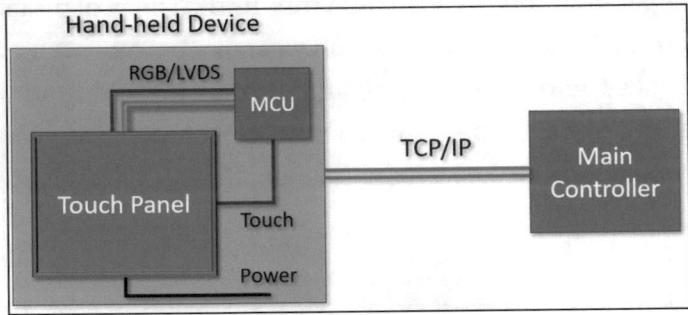

Figure 18-12.   *Architecture of an HMI display*

The touch panel needs its own power supply and an interface to receive the visualization data from the controller (typically RGB or LVDS). Alternatively, for small displays, you can even rely on a much simpler SPI connection with the help of an intermediate SPI-to-RGB converter. The raw cost of such a solution is higher, but the control firmware is definitely simpler to handle, and the wiring is certainly more convenient. Finally, a digital feedback channel for the capacitive touch screen is required to read the position of the operator's fingers.

Touch panels can be bought for very affordable prices from several distributors and are available in a large variety of physical sizes, shapes, resolutions, and interfaces (see an example in Figure 18-13).

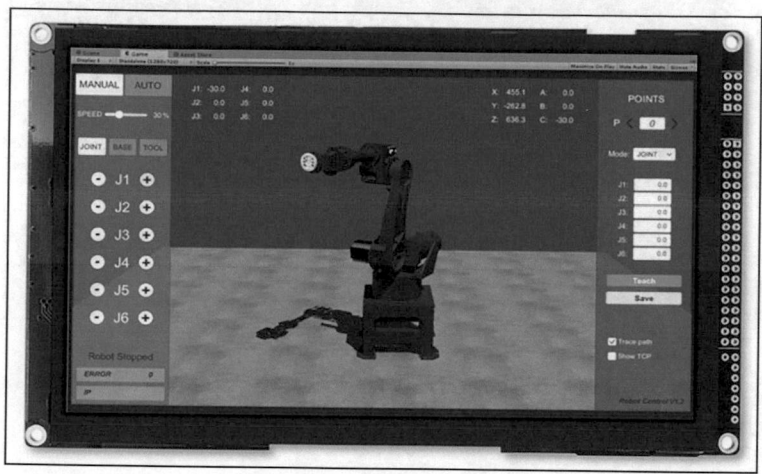

Figure 18-13. *A 7-inch TFT display with capacitive touch and SPI interface*

More modern robots have ditched the wired panel altogether and simply rely on widely available consumer tablets to handle the HMI (see Figure 18-14), connecting to the main controller via Wi-Fi or Bluetooth. The manufacturer only needs to provide an interfacing application that can run on any user device. This solution largely reduces costs and increases flexibility.

Figure 18-14. *A low-cost Android tablet running the HMI for a robot and using a direct Wi-Fi link to the controller*

Summary

The computing core of any robot is its electronic brain. We use microcontrollers for hobbyist projects because of their low cost, simplicity of operation, and large helpful communities of users. Their computing power is ever increasing and their connectivity always getting faster. The trend is to take advantage of that performance to decentralize data collection and analysis to each edge device.

In this chapter, we showed how to connect an MCU to physical sensors and actuators via I/O channels and how to exchange data back to the main network using fieldbuses. Wireless technology is also gaining ground, especially in mobile robots and large linear transport systems.

Other popular trends are the integration of logic control and servo drives into smart motion controller devices and the remote monitoring and configuration via consumer tablets for a user-friendly and modern looking HMIs.

CHAPTER 19

Fabrication

There is nothing more exciting than holding in your hands a piece of
hardware that you personally designed, whether an electronic circuit
board or a robotic mechanical joint. In this final chapter, we quickly go
through the main steps required to produce electronics and mechanical
components for a robot (or any other machine, for that matter) using
widely available CAD software.

PCB Design

At this point, you have probably completed the schematic of an electronic
circuit based on the designs described in the previous chapters. It is now
time to turn it into a real *printed circuit board* (PCB) and make a final
product out of it.

The process involves the following steps:

- **Component selection**: This is a lengthy step where
  you need to choose each electronic component for the
  board and compile the *bill of materials* (BOM).

- **Layout design**: You use a dedicated software to place
  all the electronic components of your circuit on a
  canvas and connect them with each other to generate
  the final design of the board.

© Fabrizio Frigeni 2023
F. Frigeni, *Industrial Robotics Control*, Maker Innovations Series,
https://doi.org/10.1007/978-1-4842-8989-1_19

- **Board manufacturing**: You send your layout files to a professional manufacturer, who will build the board for you in a couple of days.

- **Components assembly**: You send your BOM to the manufacturer, who typically purchases the components for you and solders them on the board.

Some hobbyists like to manufacture and/or assemble the board by themselves; however, outsourcing those tasks to a professional manufacturer is much easier, faster, and cheaper. If you plan to produce high-quality boards in high volumes, you practically have no other choice.

The first step consists in selecting the right components for the board. You might have six N-channel MOSFETs in that motor driving bridge you just designed, but once you actually go purchase those switches, you will literally find hundreds of different options from several manufacturers. Besides looking at the obvious electrical characteristics (e.g., rated voltage, current, resistance, etc.), you also need to filter the available options according to more practical factors: *price* (because your product needs to be affordable), *inventory* (because basing a production on rare components is risky), *size* (in case you have specific space limitations), *mounting type* (surface mount devices are much cheaper to solder than through hole components but are less resistant to mechanical stress), *tolerance* (tight tolerances against temperature variations are expensive but required in some cases: e.g., current sensing resistors), and *operating temperature* (e.g., some power boards are expected to work at much higher temperatures than simple logic boards).

In general, it is always a good idea to select two or three compatible options, so that if your main choice becomes unavailable at a later stage, you can quickly switch to another component without the need of redesigning the entire board. Components can be bought either directly from the original manufacturers (best when in large batches) or from large distributors (e.g., Digikey, LSCS, Mouser, etc.). The most convenient option is to have the PCB manufacturer purchase all the components for you and assemble them directly after fabrication.

Once you have chosen the components, the next step you need to accomplish is to produce a good layout design of your board. There are a number of software packages that you can use to turn a schematic circuit into a PCB layout. Some are free, while others are very expensive but probably offer more features and support. Some are web-based, while others need to be installed on your computer and come with a large number of libraries. A few common examples to choose from are Altium, Eagle, EasyEDA, KiCad, etc., the last one being my personal favorite.

Regardless of which software you use, the concept is similar for all of them. Each electronic component in the circuit schematic is transformed into its equivalent *footprint*, while each wire connecting two components is turned into a copper *trace*. Your job is to place the footprints on an initially empty canvas in a sensible way and then connect them using traces (see Figure 19-1 for a simple example).

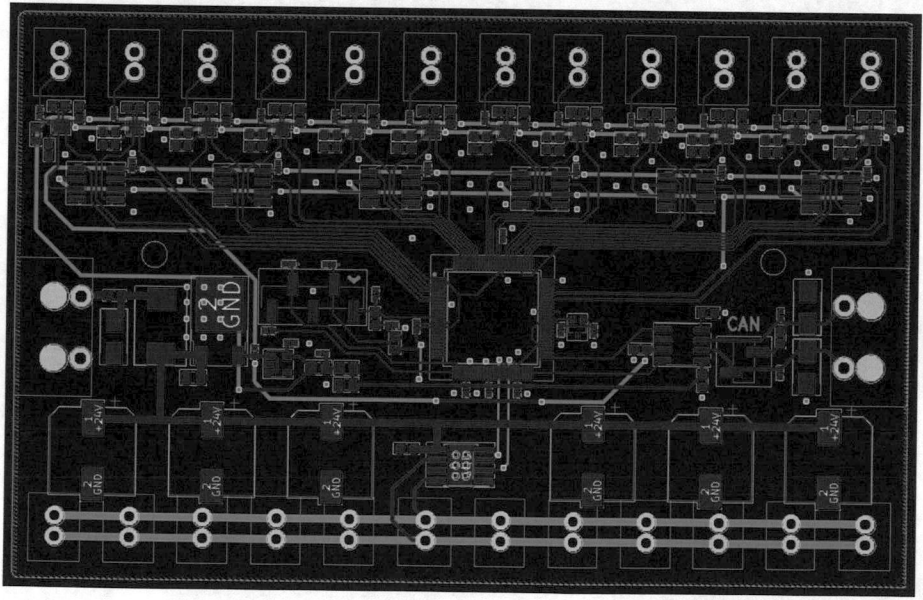

Figure 19-1. *A PCB layout consists in placing components footprints and connecting them with copper traces*

Depending on the size and complexity of a board, traces can run on
different layers across the board. The simplest boards only have two layers:
one in the front and one in the back. The example shown in Figure 19-1,
being a very small board with only a few components, only requires two
copper layers: the top shown in red and the bottom in green.

However, when wiring larger boards with several components, you
quickly run out of free paths, as the traces start clogging up the space
on each layer. The solution is to add more internal copper layers in
the structure of the PCB and use them to build more traces. Boards for
common industrial products typically have four to ten layers. Figure 19-2
shows a zoomed-up detailed view of a four-layer board (top layer red,
middle layers yellow and purple, bottom layer green). Wiring such boards
with only two layers would have been impossible.

Figure 19-2. Details of four-layer PCB layout

Interconnecting traces on different layers is done using *vias* (vertical
interconnect access), which are basically holes drilled in the PCB and
plated with copper to allow for electrical (and thermal) connectivity
between layers. Vias can be drilled from the top layer all the way through to
the bottom layer (*through vias*), but they can also connect only one outer

layer with one or more inner layers without exiting on the other side of the board (*blind vias*), or they can even be entirely invisible from the outside and only connect inner layers (*buried vias*). All three kinds of vias are shown in Figure 19-3. Buried vias allow for higher density of connectivity but are also more expensive to fabricate. Vias are normally empty inside but can also be filled with copper (at an extra cost) to increase thermal and electrical conductivity, as is often the requirements for power applications.

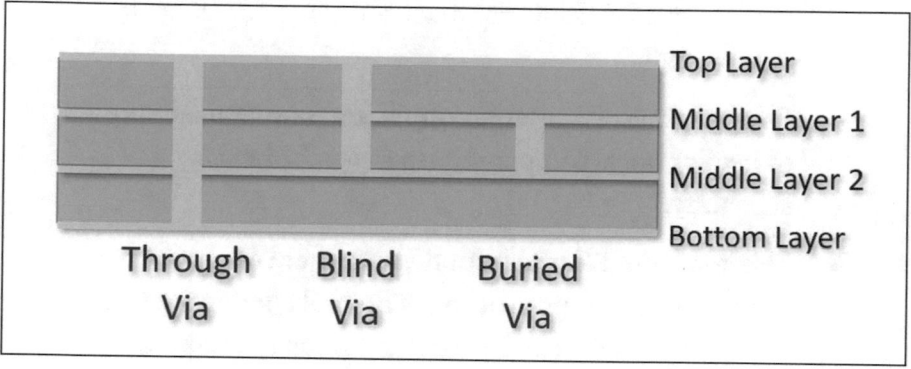

Figure 19-3. *Vias are used to connect traces running on different layers*

The thickness of each copper layer can also be adjusted depending on the application. The standard size is 35µm, which works fine for low-power signals. However, if the traces are carrying large currents (e.g., to the motor windings), then a thicker copper layer is desirable to prevent overheating and damage to the board. Thicknesses of 70µm, 105µm, and more are common, although increasingly expensive. Given the chosen thickness, a simple trace calculator program can be used to determine the minimum required trace width to allow a certain maximum current to flow.

As often happens with power electronics, the ability to dissipate heat is critical. A PCB can quickly heat up if its components and traces need to carry high currents. The generated heat must be dissipated to prevent damage to the components and to the board itself. There are a few rules to keep in mind during the PCB design in order to optimize heat dissipation:

- **Smart layout**: Avoid placing too many components that generate heat next to each other. A higher density makes the board more compact but also more difficult to keep cool.

- **Copper areas**: Place large copper areas around and/ or below critical components to allow the heat to sink away.

- **Thermal vias**: Electrical connectivity between layers is not the only purpose of vias. Thermal conduction is also a strategic function, especially when the vias are placed directly under the component's body (*in-pad vias*, shown in Figure 19-4). In that case, the vias are typically filled to avoid the solder paste flowing away during assembly. During operation, the heat generated by the component is quickly drawn away by the vias and dissipated through the other layers.

Figure 19-4. *Thermal vias placed under a component help carry the heat away*

- **Metal substrate**: Power electronic boards are commonly built on a metal substrate, as shown in Figure 19-5, typically aluminum, although copper alloys are also possible. The resulting thermal conductivity is greatly enhanced if compared to standard boards built on fiberglass substrate. The metal board is also mechanically much more robust and can withstand stronger vibrations without bending or braking. The metal substrate can be mounted directly on the machine's body for further heat sinking capability.

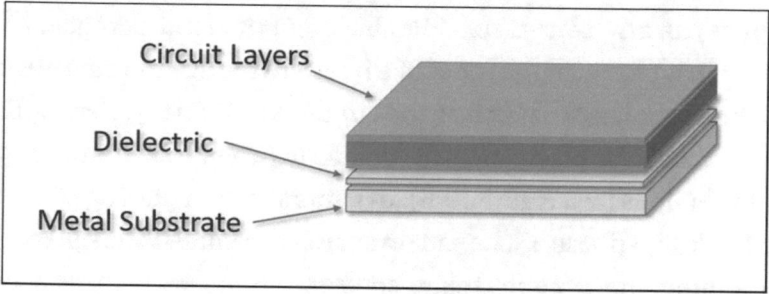

Figure 19-5. *A metal substrate can quickly conduct heat away from the PCB*

Besides heat management, another important aspect of designing a PCB is electromagnetic compatibility (EMC). This is a complex topic with strict regulations for products to be released on the market. The goal is to prevent electromagnetic interference between our electronic circuit board and other devices operating nearby.

A PCB can emit radiation intentionally (e.g., from Wi-Fi or Bluetooth transmitters) but also unintentionally (e.g., from the high-frequency switching transistors of the servo drive). Also, high-speed signals (e.g., SPI communication lines) propagating through long traces with sharp bends generate electromagnetic waves. A general suggestion to minimize unwanted radiation is to keep the conductive traces as short as possible and their bending as smooth as possible. Long traces typically act as stray inductances and have the tendency to emit radiations and also pick up electromagnetic noise from external sources, which can cause interference. High-speed logic signals traveling through traces with sharp angles can easily reflect off the edges also generating significant interferences.

Adding large areas of copper to the PCB design strongly helps shielding against electromagnetic fields. Sometimes entire layers of the board are dedicated to that purpose (*ground layers*); incidentally, they help with heat dissipation as well.

Electromagnetic interference (EMI) does not only travel through air: power lines can also carry a considerable amount of interference. EMI filters are often placed at the input of a board to reduce the amount of external noise making its way into the circuit. Additionally, decoupling capacitors should be placed next to each IC to protect them from glitches on the power lines (see Figure 19-6). In general, when placing components during the design phase, make sure to search the datasheets for specific layout requirements from the manufacturer.

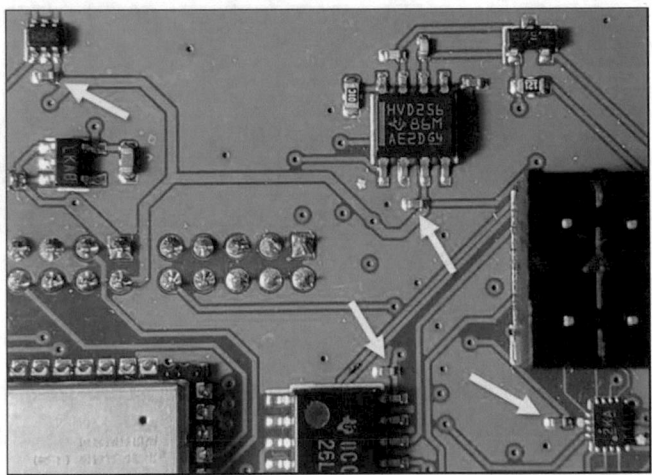

Figure 19-6. *Decoupling capacitors help protecting ICs from glitches on the power lines*

The footprints for each component must carefully match the physical size and pin pattern specified on the datasheet. Some manufacturers provide footprints; most do not. You can quickly draw them out yourself and build your own library of favorite components. Alternatively, you can search on third-party libraries (e.g., Ultra Librarian, SnapEDA, etc.), though you should carefully double-check the footprints you download because some of them are incorrect. A mistake found during the design phase is much cheaper and quicker to fix than when it is found after the board is manufactured.

Once you have completed the design, the last thing to do is to run a rule check. That is an automated software check that will tell you if the design contains fabrication mistakes: traces too thin or too close together, components overlapping each other, etc. If all works out fine, you can now enjoy a final 3D preview of the board: for example, Figure 19-7 shows the preview of the same board whose layout design was previously shown in Figure 19-1.

Figure 19-7. *Computer-generated preview of the board after completing the layout*

At this point, the board is ready for fabrication. You need to export the fabrication files (also known as *Gerber* files) and send them to your favorite manufacturer. There are plenty of options around for quick-turn prototyping as well as high-volume production: a couple I have used very often and found reliable is PCBWay and JLCPCB. Before submitting your project to them, you should always check their manufacturing capabilities and make sure that they can deliver what you need (e.g., minimum trace width and spacing, copper thickness, drill sizes, surface finishing, vias filling, etc.). More demanding requirements are usually available at extra cost. The manufacturing of a small size board in low volumes normally takes only a couple of days to complete (see Figure 19-8).

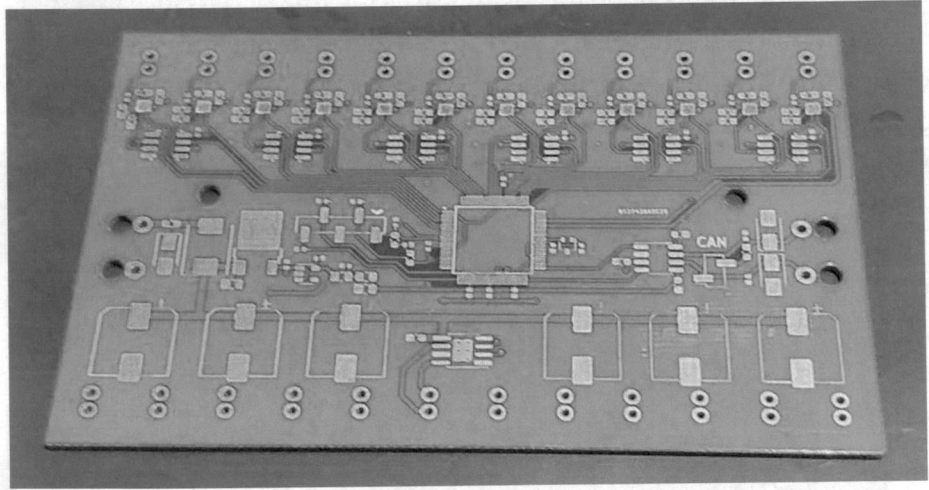

Figure 19-8. *Manufactured board before assembling. The exposed pads where the components will be soldered are clearly visible*

The last step is assembly, i.e., soldering the components on the board. This is something many hobbyists like to do on their own, but again, nothing beats the convenience, speed, and accuracy of having the board manufacturer take care of the assembly phase for you. When dealing with large production batches and tiny BGA (ball grid array) components, then automatic assembly is a must.

Electronic components are placed on the board by a high-speed CNC machine and soldered together using a process called *reflow soldering*. First, solder paste (a mixture of powdered solder and flux) is applied to the board using a mask so that only the pads are exposed; then the components are placed on the board where they held in place by the paste; finally, the board is heated up to melt the solder and ensure permanent connection between the components and the pads.

The board is then optically inspected for soldering mistakes; optionally, an automated X-ray inspection and a customized functional test are also offered. Finally, the board is washed to remove all flux and dirt from the manufacturing process, then dried, packaged, and shipped to your address.

Figure 19-9. *Final board after components assembly*

When you receive the board (see Figure 19-9), you should carefully check the most important connections with a multimeter, and when you are ready to apply power, do so with a current-limited supply to prevent nasty surprises. It goes without saying that all the hardware and software functions must be thoroughly tested before using the product in a real application setting.

Mechanics

We have focused throughout this book on the control of robots with the perspective of an automation engineer, who is typically more involved in the mathematical modeling, programming, and electronics design. The mechanics of a robot are usually designed and built by machine constructors and provided to the automation department as finished parts.

However, it is always interesting to be able to work interdisciplinary and at least have a sense of what is involved in the design and fabrication of mechanical parts. Whether you want to play with a hobby robotic arm, or test a simple prototype for research purposes, or even design an entirely new product for your startup, you will need to get your hands dirty and build something by yourself.

The first step is choosing appropriate materials, depending on the complexity of the design, the requirements of the final product, and the skills you have available. Over time, I have built several hobby robots using a number of different materials: in my beginner years, I started with simple wood or even LEGO blocks, since they are easy to work with and do not require any complex or costly manufacturing processes. Their flexibility, however, is quite limited. Later, I evolved to more advanced materials, plastic, aluminum, and steel, using a combination of CNC machining and 3D printing manufacturing processes.

Aluminum is cheap and easy to machine, arguably offers the best strength to weight ratio, and looks very attractive indeed when anodized. Steel is more expensive but much stronger. The most common stainless steel is type 304. For some critical parts, where mechanical stress is expected to be particularly high, you can consider using steel type 45, which is reinforced with additional carbon and offers increased strength. In general, a combination of different materials can be used for different parts, according to their needs, always keeping in mind the optimization of weight and cost. As for plastics, nylon is a black, good-looking material, which I typically use for the outside cover of machines and for electronic cases.

The fabrication process is similar to what we described for PCBs: you first use a CAD software to design the part and then submit the files to your favorite manufacturer who will produce the component and ship it to you within a couple of days. There is a large variety of software packages, within different price and functionality ranges, that can be used for to design 3D objects. I personally use Fusion360 (see Figure 19-10), but any other option is equally valid. Besides exporting the fabrication files for the manufacturer, you can also export object meshes and use them in Unity for simulation purposes (see Section on Importing CAD Models in Chapter 12).

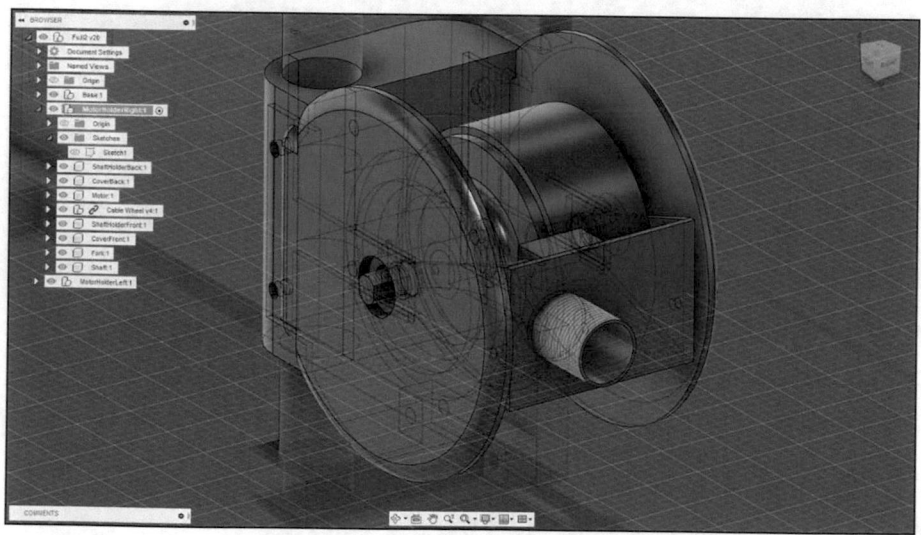

Figure 19-10. *Motorized machine joint designed by the author*

Most manufacturers provide a convenient online interface to select detailed parameters about the process: material, tolerances, surface finishing, and color, plus many other customized requirements (often at extra cost).

Plastic processing is usually either 3D additive manufacturing (i.e., 3D printing) or injection molding. The first works best for prototyping and small batches, while the second becomes convenient with very large production volumes due to the cost of the mold.

Metal processing, on the other hand, is mostly a subtractive process: it starts with a raw piece of metal, which gets machined (e.g., milling, turning, cutting, bending) to remove the unwanted volume and generate the final part. There are, however, complex components that cannot be realized with standard machining processes. In those cases, additive manufacturing is also used for metal parts: 3D metal printing allows the creation of small complex components in short times (see Figure 19-11).

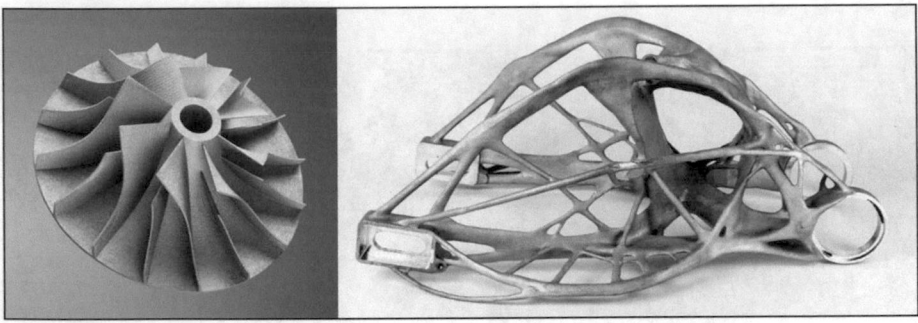

Figure 19-11. *3D metal printing can quickly generate complex parts that would be difficult or impossible to produce with standard subtractive manufacturing*

When designing mechanical components, special attentions is required to optimize their physical characteristics (least material use, lowest weight and cost) within the strength and torsion constraints imposed by the application. The optimization algorithm often takes a reinforced learning approach (see Machine Learning Section in Chapter 12) to generate parts with complex shapes, curved surfaces, and several holes to achieve massive weight reductions. For example, the component shown in Figure 19-11 on the right side was produced using this optimization technique (known as *generative design*) and then 3D printed.

Robotic arms can take great advantage from generative design, because weight reduction on the links allows for better movement speed and accuracy and possibly for smaller and cheaper motors.

Summary

The final step of a product generating process is the transformation of an abstract design into a tangible object. In robotics, we mainly deal with the fabrication of electronic boards and mechanical parts. In this chapter, we described the steps required to design a PCB, from its electronic layout to its actual manufacturing process, with attention to practical tips to optimize stability and performance of the board. Also, we briefly introduced some possible materials you can use to build the physical robot, from plastic to metal.

With this final chapter we conclude the book. I hope you have learned a few basic concepts and some helpful tricks on how to design, program, and build your own robotics controller. Good luck and have fun!

Correction to: Industrial Robotics Control

Correction to:

Fabrizio Frigeni, *Industrial Robotics Control*
https://doi.org/10.1007/978-1-4842-8989-1

This book was published without Series ID, Print ISSN number & Electronic ISSN Number. This has now been updated in the book with the Series ID - 17311, Print ISSN: 2948-2542 & Electronic ISSN: 2948-2550.

The updated version of this book can be found at
https://doi.org/10.1007/978-1-4842-8989-1

© Fabrizio Frigeni 2023
F. Frigeni, *Industrial Robotics Control*, Maker Innovations Series,
https://doi.org/10.1007/978-1-4842-8989-1_20

APPENDIX

Kinematic Models

We solved the kinematic model for the standard six-axes anthropomorphic robot in detail in Chapter 3 and Chapter 4. In this appendix, we briefly show the solutions for the other kinds of industrial robots presented in Mechanical Configurations Section in Chapter 1.

Please note that, in order to understand the techniques and notations used in this appendix, you need to be familiar with the content introduced when solving the standard six-axes robot. I suggest you get a good grasp of that solution's details before tackling these other robots.

We solve all the kinematic models using the base and TCP frames as reference. If you need to introduce an additional tool or zero frame, you can follow the same procedures described for the six-axes robot in the Sections on Zero Frame and Tool Frame in Chapter 3 (FK) and Chapter 4 (IK). When working with four-axes robots, the A and B angles do not exist, and we only consider the C angle for rotations around the vertical Z axis.

© Fabrizio Frigeni 2023
F. Frigeni, *Industrial Robotics Control*, Maker Innovations Series,
https://doi.org/10.1007/978-1-4842-8989-1

COBOT

Cobots (collaborative robots) are very popular these days because they are very simple to set up and deploy and especially because they are safe to deploy right beside human operators without being confined behind protection cages.

Their mechanical structure is a bit different than the standard six-axes manipulator that we analyzed in-depth in this book, and the kinematic model needs to be adjusted accordingly.

Figure A-1 shows a Cobot model in its home position and the names assigned to the mechanical parameters for each link. You can also choose a different home position and give different names to the parameters, but the concept does not change.

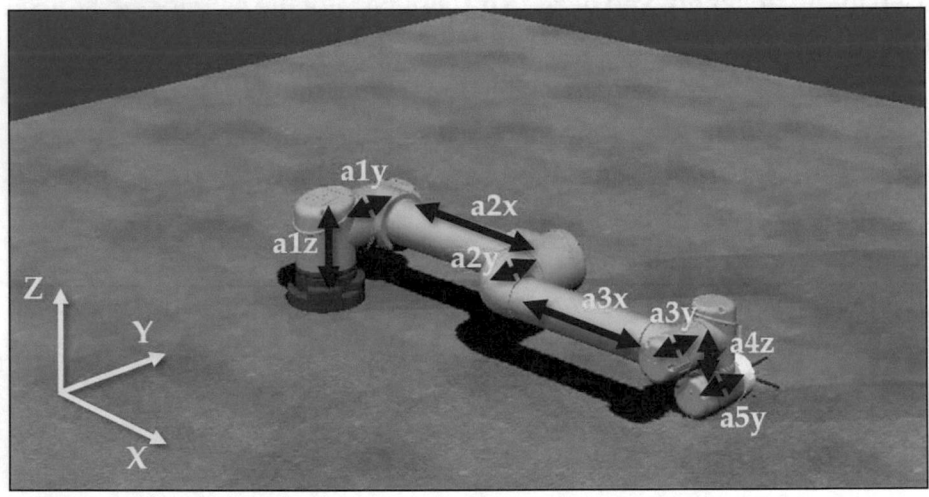

Figure A-1. *Six-axes Cobot in its home position*

The first joint J_1 rotates around the vertical Z axis (Figure A-2).

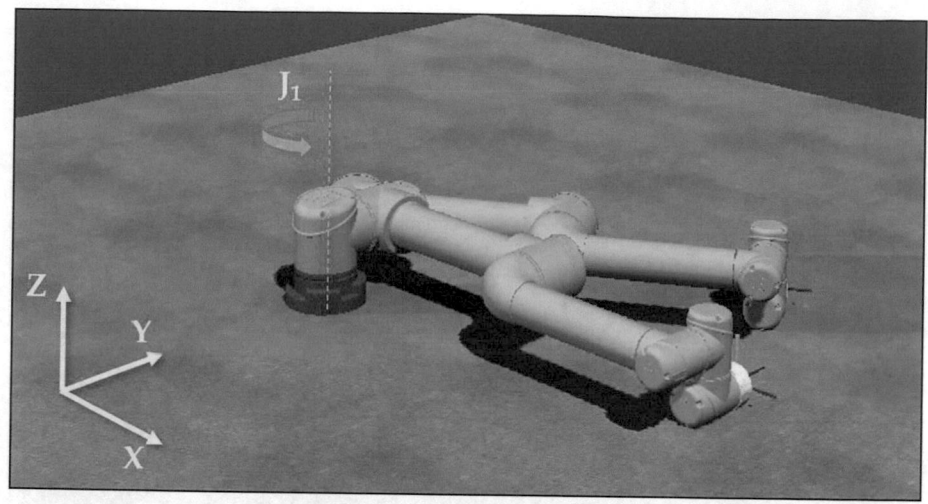

Figure A-2. *The first joint rotates around the Z axis*

Then we move along the Z and Y axes by $a1z$ and $a1y$ to reach the second joint J_2, which rotates around the Y axis (Figure A-3).

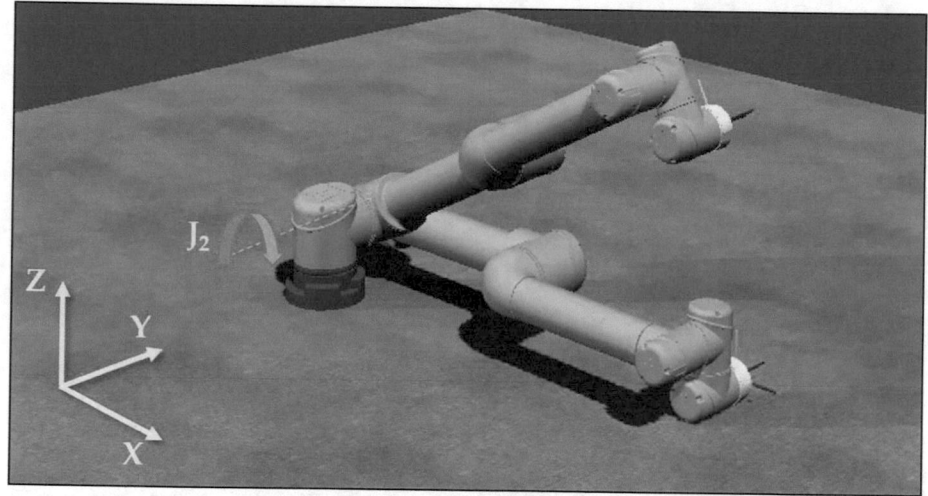

Figure A-3. *The second joint rotates around the Y axis*

We further move along the X and Y axes by *a2x* and *a2y* to find the third joint J_3, which also rotates around the Y axis (Figure A-4).

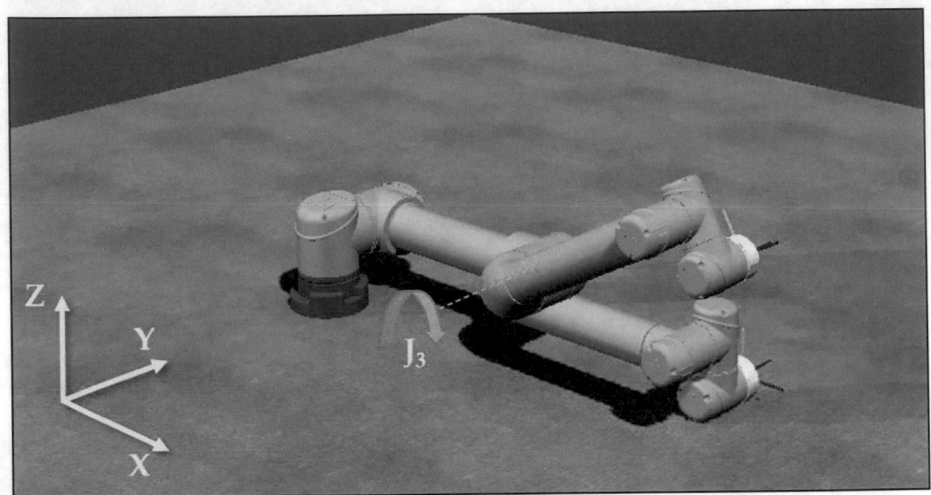

Figure A-4. *The third joint rotates around the Y axis*

Then again add *a3x* and *a3y* to reach the fourth joint J_4, also rotating around Y (Figure A-5).

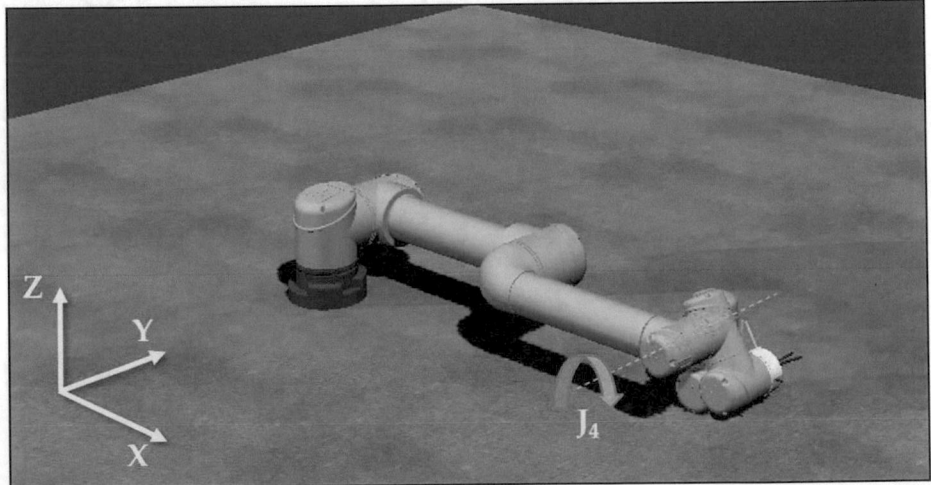

Figure A-5. *The fourth joint rotates around the Y axis*

The fifth joint J_5 rotates around Z and is offset by $a4z$ from J_4 (Figure A-6).

Figure A-6. *The fifth joint rotates around the Z axis*

Finally, we add $a5y$ to reach the last joint J_6, which rotates around Y (Figure A-7).

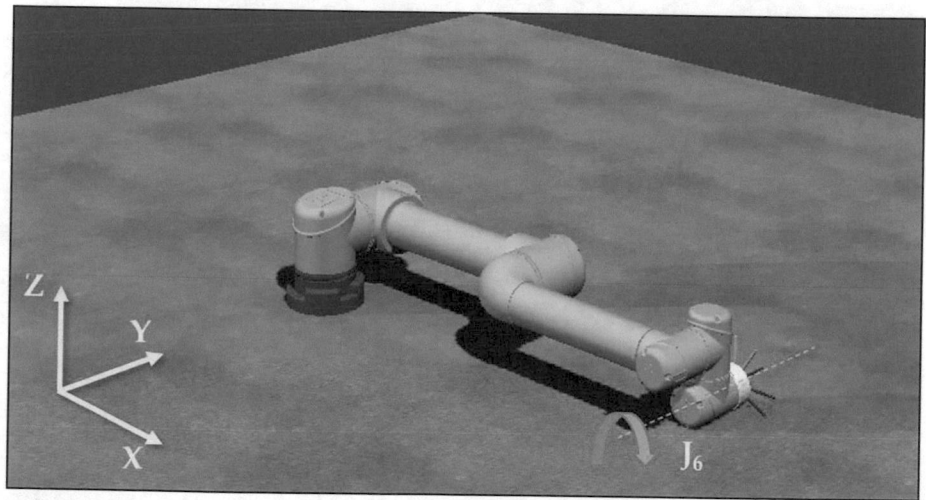

Figure A-7. *The sixth joint rotates around the Y axis*

No fewer than *four out of six joints rotate around the Y axis*, while the other two rotate around the Z axis. There is no rotation around the X axis. Such a mechanical structure is quite different from the standard six-axes manipulator we are used to.

To solve the direct kinematics, we follow the usual procedure: we build the homogeneous matrix for each joint taking into account the rotation and translation caused by the corresponding mechanical link. See Chapter 2 for details.

For J_1, we simply need a rotation around Z.

$$T_1 = \begin{bmatrix} c1 & -s1 & 0 & 0 \\ s1 & c1 & 0 & 0 \\ 0 & 0 & 1 & 0 \\ 0 & 0 & 0 & 1 \end{bmatrix} \tag{A-1}$$

For joints J_2, J_3, and J_4, we have rotations around Y and additional position offsets. Careful with the third joint: the value $a2y$ is actually negative, because the link points backward, in opposite direction of the Y axis.

$$T_2 = \begin{bmatrix} c2 & 0 & s2 & 0 \\ 0 & 1 & 0 & a1y \\ -s2 & 0 & c2 & a1z \\ 0 & 0 & 0 & 1 \end{bmatrix} \tag{A-2}$$

$$T_3 = \begin{bmatrix} c3 & 0 & s3 & a2x \\ 0 & 1 & 0 & a2y \\ -s3 & 0 & c3 & 0 \\ 0 & 0 & 0 & 1 \end{bmatrix} \tag{A-3}$$

$$T_4 = \begin{bmatrix} c4 & 0 & s4 & a3x \\ 0 & 1 & 0 & a3y \\ -s4 & 0 & c4 & 0 \\ 0 & 0 & 0 & 1 \end{bmatrix} \tag{A-4}$$

For J_5 we add a rotation around Z. Here also, make sure that $a4z$ is negative because it points downward.

$$T_5 = \begin{bmatrix} c5 & -s5 & 0 & 0 \\ s5 & c5 & 0 & 0 \\ 0 & 0 & 1 & a4z \\ 0 & 0 & 0 & 1 \end{bmatrix} \tag{A-5}$$

Finally, J_6 introduces one more rotation around Y.

$$T_6 = \begin{bmatrix} c6 & 0 & s6 & 0 \\ 0 & 1 & 0 & a5y \\ -s6 & 0 & c6 & 0 \\ 0 & 0 & 0 & 1 \end{bmatrix} \tag{A-6}$$

To find out the TCP position as seen from the base frame, we need to chain-multiply the six homogeneous matrices together:

$$TCP(X,Y,Z) = T_1 T_2 T_3 T_4 T_5 T_6 O = T_{16} \begin{bmatrix} 0 \\ 0 \\ 0 \\ 1 \end{bmatrix} \tag{A-7}$$

If you expand all the calculations, you can verify that the [X, Y, Z] position coordinates of the TCP are as follows:

$$\begin{aligned} TCP_X = & -a1y\, s_1 + a2x\, c_1 c_2 - a2y\, s_1 + a3x\, c_1 c_{23} - a3y\, s_1 \\ & + a4z\, c_1 S_{234} - a5y \left(s_1 c_5 + c_1 s_5 c_{234} \right) \end{aligned} \tag{A-8}$$

$$\begin{aligned} TCP_Y = & \; a1y\, c_1 + a2x\, s_1 c_2 + a2y\, c_1 + a3x\, s_1 c_{23} + a3y\, c_1 \\ & + a4z\, s_1 S_{234} + a5y \left(c_1 c_5 - s_1 s_5 c_{234} \right) \end{aligned} \tag{A-9}$$

$$TCP_Z = a1z - a2x\, s_2 - a3x\, s_{23} + a4z\, c_{234} + a5y\, s_5 S_{234} \tag{A-10}$$

Notice the compact notation $s_{234} = \sin(J_2 + J_3 + J_4)$.

The total rotation of the TCP with respect to the base frame is the product of all the joint rotation matrices:

$$TCP(A,B,C) = R_1 R_2 R_3 R_4 R_5 R_6 \tag{A-11}$$

By expanding the product, you should find the following rotation matrix:

$$R_{TCP} = \begin{bmatrix} -(s_1 s_5 - c_1 c_5 c_{234})c_6 - s_6 s_{234} c_1 & -s_1 c_5 - s_5 c_1 c_{234} & -(s_1 s_5 - c_1 c_5 c_{234})s_6 + s_{234} c_1 c_6 \\ (s_1 c_5 c_{234} + s_5 c_1)c_6 - s_1 s_6 s_{234} & -s_1 s_5 c_{234} + c_1 c_5 & (s_1 c_5 c_{234} + s_5 c_1)s_6 + s_1 s_{234} c_6 \\ -s_6 c_{234} - s_{234} c_5 c_6 & s_5 s_{234} & -s_6 s_{234} c_5 + c_6 c_{234} \end{bmatrix} \tag{A-12}$$

The TCP rotation matrix can be decomposed in either Euler angles or quaternions according to your needs, as explained in Chapter 2 and Chapter 5, respectively.

Notice how the mounting flange of the last joint is oriented along the Y axis, as opposed to facing the X axis for the standard manipulator. Keep that in mind when calculating offsets for the tool.

Solving the inverse kinematics can be done in different ways: some approaches are more geometric; others are more analytical. We present here one of the possible solutions, definitely a valid one, but not the only one.

This step is much harder than solving the direct transformations: it is not even guaranteed to have a solution, and when a solution exists, it is not necessarily unique. Actually, given a specific TCP pose, there are eight possible joints solutions (and many more if you consider all the $\pm 2\pi$ offsets for each joint).

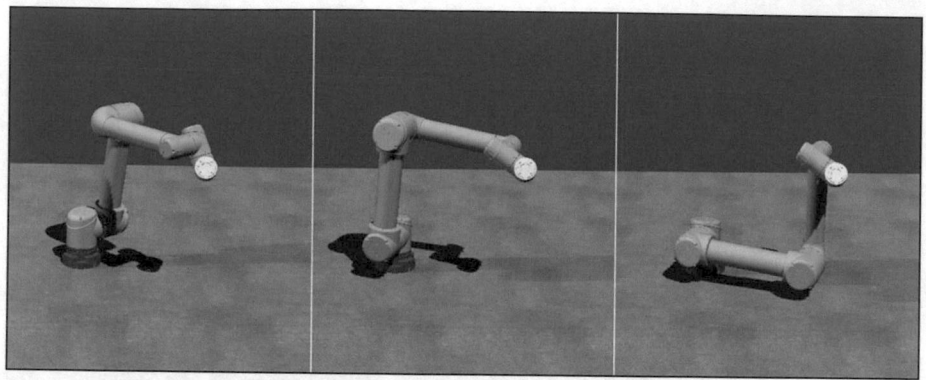

Figure A-8. *Different joint configurations reach the same TCP pose*

Figure A-8 shows a Cobot in different joint configuration reaching the same TCP pose. In the first two cases, the difference depends on the position of the first joint: we call them LEFT and RIGHT. In the third case, the configuration of the second and third joint is different: we call it DOWN, as opposed to UP for the first two examples.

Let's start by finding the value of the first joint. The orientation frame of the TCP is given by the rotation matrix $R_{TCP} = [n, o, a]$, which can be quickly composed from the given Euler angles [A, B, C]. Recall that $[n, o, a]$ are the column vectors of the rotation matrix and represent the orientation of the frame axes at the TCP:

$$R_{TCP} = \begin{bmatrix} n & o & a \end{bmatrix} = \begin{bmatrix} n_x & o_x & a_x \\ n_y & o_y & a_y \\ n_z & o_z & a_z \end{bmatrix} \tag{A-13}$$

The column vector o is oriented along the local Y axis and is facing out of the mounting flange (see Figure A-9). If we move from the TCP backward along that direction by a distance of $a5y$, we find the wrist point:

$$P_5 = TCP - o \, a5y \tag{A-14}$$

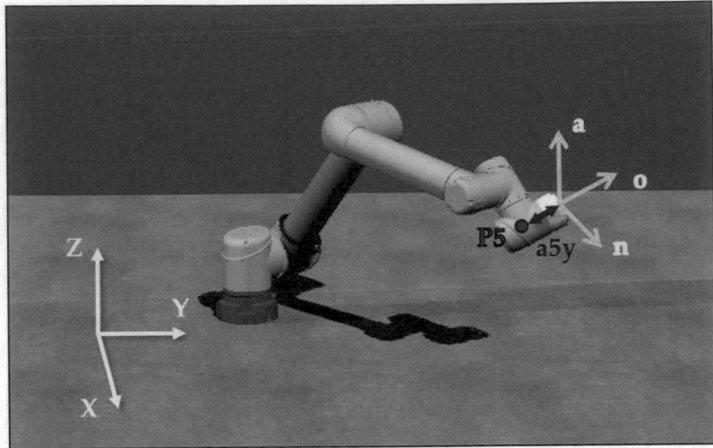

Figure A-9. *Finding the wrist point from the TCP*

Observe that P_5 is mechanically restricted to lie on a plane that never intersects the origin of the robot. It is easier to see that from the top, as shown in Figure A-10.

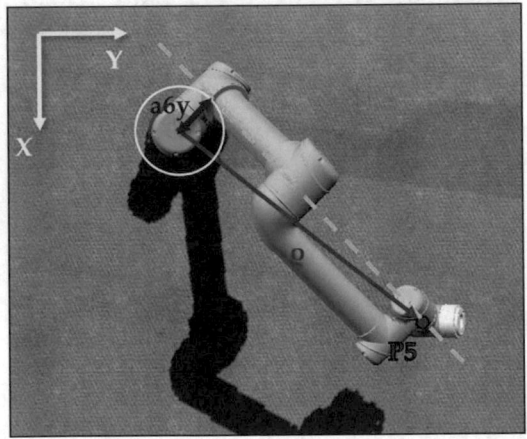

Figure A-10. *The wrist point moves on a plane that does not contain the base origin*

We call ρ the distance between P_5 and the base origin of the robot on the two-dimensional XY plane. We can state that no inverse solution exists for values of ρ smaller than $a6y$, which is the sum of all the links along the Y axis $(a1y + a2y + a3y)$, all taken with their correct signs (remember that $a2y$ is usually negative).

$$\rho = \sqrt{P_{5_X}^{\,2} + P_{5_Y}^{\,2}} > a6y = a1y + a2y + a3y \qquad (A\text{-}15)$$

In practice, that means that the robot cannot physically reach any TCP position which would force the wrist center point to be too close to the base origin. On the other hand, given a valid target position for P_5, we can quickly find the value of J_1.

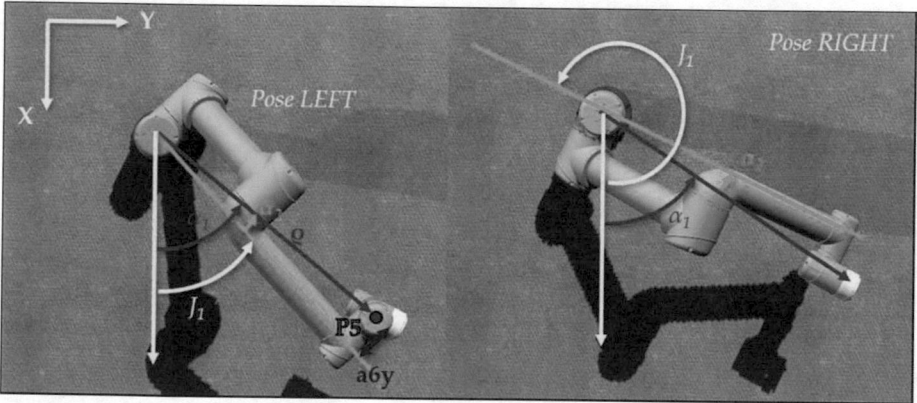

Figure A-11. *Solving for the first joint*

There are actually two configurations that we can choose for the first joint, either with the arm LEFT or RIGHT, which are rotated at about 180 degrees from each other. Not exactly 180 degrees, because the plane of P_5 does not pass through the origin, as opposed to the case of the standard six-axes manipulator.

Let's consider the standard LEFT configuration first (Figure A-11 left). The value of J_1 (in yellow) is the difference between α_1 (in purple) and α_2 (in green).

The angle α_1 is determined by the position of P_5:

$$\alpha_1 = atan2\left(P_{5_y}, P_{5_x}\right) \tag{A-16}$$

The angle α_2 is constrained by $a6y$ and ρ:

$$\sin(\alpha_2) = a6y / \rho \tag{A-17}$$

As usual, while one could solve via the arcsine function directly, I suggest implementing your code with the arctangent function for better computational stability.

The final solution for J_1 in the left configuration is as follows:

$$J_1 = \alpha_1 - \alpha_2 \tag{A-18}$$

In case the selected solution must be the RIGHT configuration, J_1 has to turn all the way around (see Figure A-11 right).

$$J_1 = \alpha_1 + \alpha_2 + \pi \tag{A-19}$$

If no forcing constraint is given by the operator, we can choose either one of the two solutions, typically the one closer to the current joint value at the start of the movement.

Given J_1 we can now find J_5. There are two different possible approaches that led to the same result: a geometric and an algebraic one. We choose the latter, because it is computationally more efficient. The idea is to extract the value of the joint from the robot's rotation matrices.

Recall that the total TCP rotation matrix is the product of the six rotation matrices of each joint:

$$R_{TCP} = R_1 R_2 R_3 R_4 R_5 R_6 \tag{A-20}$$

Multiply both sides by the inverse rotation matrix of the first joint:

$$R_1^{-1} R_{TCP} = R_2 R_3 R_4 R_5 R_6 \tag{A-21}$$

We already derived all the individual rotation matrices while solving the direct transformations. Using the properties that the inverse of a rotation matrix is simply its transpose ($R_i^{-1} = R_i^T$) and recalling Equation (A-13), we can quickly expand Equation (A-21) into the following expression:

$$\begin{bmatrix} c_1 & s_1 & 0 \\ -s_1 & c_1 & 0 \\ 0 & 0 & 1 \end{bmatrix} \begin{bmatrix} n_x & o_x & a_x \\ n_y & o_y & a_y \\ n_z & o_z & a_z \end{bmatrix} = \begin{bmatrix} \cdots & \cdots & \cdots \\ s_5 c_6 & c_5 & s_5 s_6 \\ \cdots & \cdots & \cdots \end{bmatrix} \tag{A-22}$$

$$\begin{bmatrix} \cdots & \cdots & \cdots \\ -s_1 n_x + c_1 n_y & -s_1 o_x + c_1 o_y & -s_1 a_x + c_1 a_y \\ \cdots & \cdots & \cdots \end{bmatrix} = \begin{bmatrix} \cdots & \cdots & \cdots \\ s_5 c_6 & c_5 & s_5 s_6 \\ \cdots & \cdots & \cdots \end{bmatrix} \tag{A-23}$$

We only focused on the second row ignoring the rest of the matrix, because we spot a lonely c_5, which will lead us to finding J_5. The elements on both sides of the identity must be equal, including the central element on the second column:

$$\cos(J_5) = -o_x \sin(J_1) + o_y \cos(J_1) \tag{A-24}$$

Since we already know J_1, we can immediately calculate J_5.

The sign of J_5 is not uniquely defined: there are two possible robot's configurations, one with J_5 positive and another one with J_5 negative. Both solutions are technically valid, but we will choose only one later on, based on the values we find for the remaining joint angles.

We now move on to finding J_6. Let's consider again the same matrix equivalence in Equation (A-23): since now both J_1 and J_5 are known, we can extract J_6 from the elements on the first and third columns:

$$s_5 c_6 = -s_1 n_x + c_1 n_y \qquad \text{(A-25)}$$

$$s_5 s_6 = -s_1 a_x + c_1 a_y \qquad \text{(A-26)}$$

Given the sine and cosine of J_6, we immediately can find its value with the arctangent function:

$$J_6 = atan2 \left(\frac{-s_1 a_x + c_1 a_y}{s_5}, \frac{-s_1 n_x + c_1 n_y}{s_5} \right) \qquad \text{(A-27)}$$

There is always one unique solution for J_6, as long as J_5 is not zero. Otherwise, the denominator is 0 and the robot is in a singularity. Mathematically, it means that there are infinite valid solutions for J_6. We can simply force it equal to any value we desire, typically the current value of the physical joint at the beginning of the movement.

The next step is to find J_2, J_3, and J_4, which requires quite a lengthy procedure. The trick is looking at the robot from the side and working on the projected plane along the first joint, as shown in Figure A-12. We observe that J_2, J_3, and J_4 form a kinematic chain with three rotational joints: we need to find the points P_2 and P_4 and from there solve the triangle.

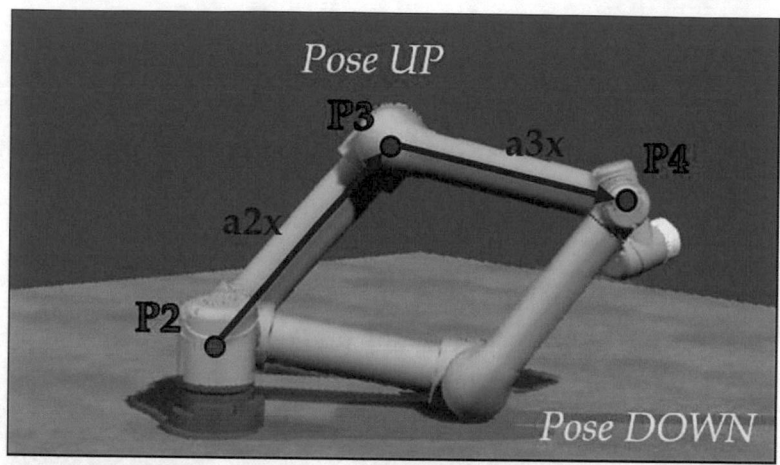

Figure A-12. *The projection plane to find J_2 and J_3*

Just keep in mind that there are two possible options, either UP or DOWN. We will solve both configurations.

Let's start from P_2. Observe in Figure A-13 that the first joint of the robot automatically fixes the position of point P_2, which lies on the rotation axis of J_2.

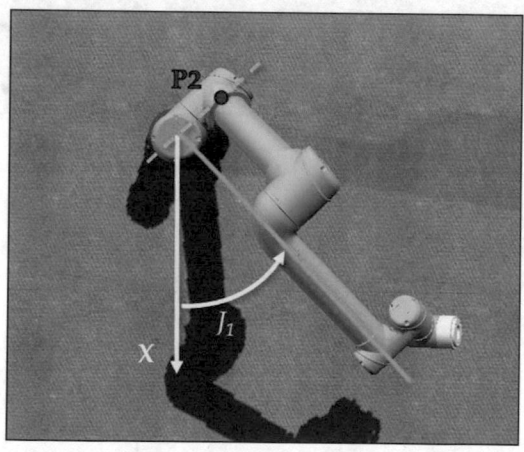

Figure A-13. *Calculating P_2 from J_1*

567

We use the forward kinematics to solve for P_2:

$$P_2 = T_1 \begin{bmatrix} 0 \\ aly \\ alz \\ 1 \end{bmatrix} = \begin{bmatrix} -s_1 aly \\ c_1 aly \\ alz \\ 1 \end{bmatrix} \tag{A-28}$$

To find P_4, we move on the other end of the robot (Figure A-14). P_4 lies on the rotation axis of J_4, and its position and orientation are fixed by J_5 and J_6, both of which we already know.

Figure A-14. *Calculating P_4 from the wrist point P_5*

Starting from the orientation of the TCP and removing the rotations for J_5 and J_6, we find the orientation of the frame $\begin{bmatrix} x_4 & y_4 & z_4 \end{bmatrix}$ at point P_4:

$$R_{P4} = R_{TCP} \begin{bmatrix} R_5 R_6 \end{bmatrix}^T = \begin{bmatrix} n_x & o_x & a_x \\ n_y & o_y & a_y \\ n_z & o_z & a_z \end{bmatrix} \begin{bmatrix} c_5 c_6 & s_5 c_6 & -s_6 \\ -s_5 & c_5 & 0 \\ c_5 s_6 & s_5 s_6 & c_6 \end{bmatrix} = \begin{bmatrix} x_4 & y_4 & z_4 \end{bmatrix} \tag{A-29}$$

Then we can also calculate the position of P_4 starting from P_5 and move back along the Z axis of the local frame of P_4:

$$P_4 = P_5 - a4z\ z_4$$

(A-30)

At this point, we have both P_2 and P_4, and we can start solving the kinematic chain between the two of them. We call their geometrical distance P_{42}. The projection of P_{42} on the side plane along the first joint is as follows:

$$\lambda = \sqrt{P_{42X}^2 + P_{42Y}^2 + P_{42Z}^2 - a7y^2}$$

(A-31)

$$a7y = a2y + a3y$$

(A-32)

Figure A-15 shows a top view to clarify.

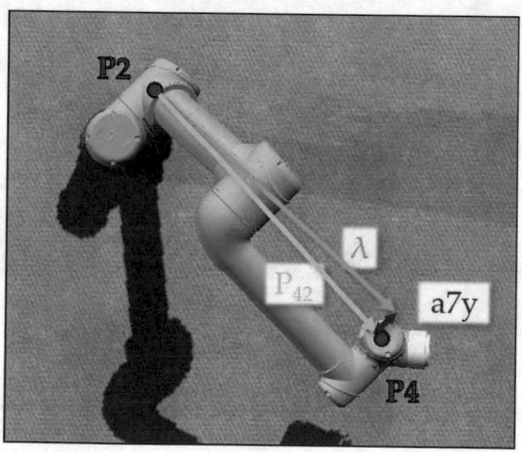

Figure A-15. *Top view to find the projected distance between* P_2 *and* P_4

The small difference along the Y axis ($a7y$) is caused by a mechanical offset, the sum of $a2y$ and $a3y$. Remember that $a2y$ is usually negative.

In order to have a physically valid solution, we need to make sure that λ can be reached using the two links $a2x$ and $a3x$. Two conditions must be verified; otherwise, the inverse transformation function will fail:

$$\lambda < a2x + a3x \tag{A-33}$$

$$\lambda > |a2x - a3x| \tag{A-34}$$

Now that the three sides of the triangle $[a2x, a3x, \lambda]$ are all known, solving for J_2 and J_3 is a matter of basic trigonometry (see Figure A-16).

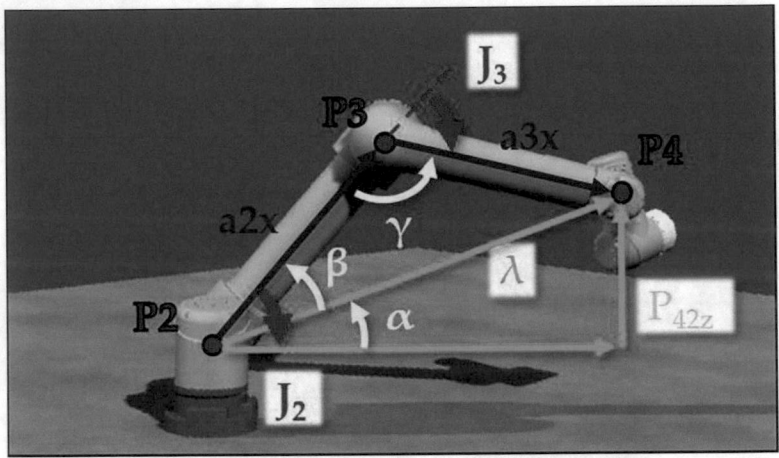

Figure A-16. *Geometrical solution for J_2 and J_3*

We first calculate α and then derive β and γ using the law of cosines:

$$\alpha = atan2\left(P_{42_z}, \sqrt{P_{42_x}^2 + P_{42_y}^2 - a7y^2}\right) \tag{A-35}$$

$$cos\beta = \frac{\lambda^2 + a2x^2 - a3x^2}{2\,\lambda\,a2x} \tag{A-36}$$

$$cos\gamma = \frac{a2x^2 + a3x^2 - \lambda^2}{2\,a2x\,a3x} \tag{A-37}$$

Following the right-hand rule for the angles, we obtain the following values for the joints in the default UP configuration:

$$J_2 = -\alpha - \beta \tag{A-38}$$

$$J_3 = -(\pi - \gamma) \tag{A-39}$$

For the DOWN configuration,

$$J_2 = -\alpha + \beta \tag{A-40}$$

$$J_3 = \pi - \gamma \tag{A-41}$$

The only joint we are missing at this point is J_4. The solution is shown in Figure A-17 and requires finding the difference in orientation between the X axes of the frames in P_2 and P_4. We call that difference α_{14}, and also note that

$$\alpha_{14} = J_2 + J_3 + J_4 \tag{A-42}$$

In other words, once we know α_{14}, we can complete the solution of the inverse kinematics by applying the following:

$$J_4 = \alpha_{14} - J_2 - J_3 \tag{A-43}$$

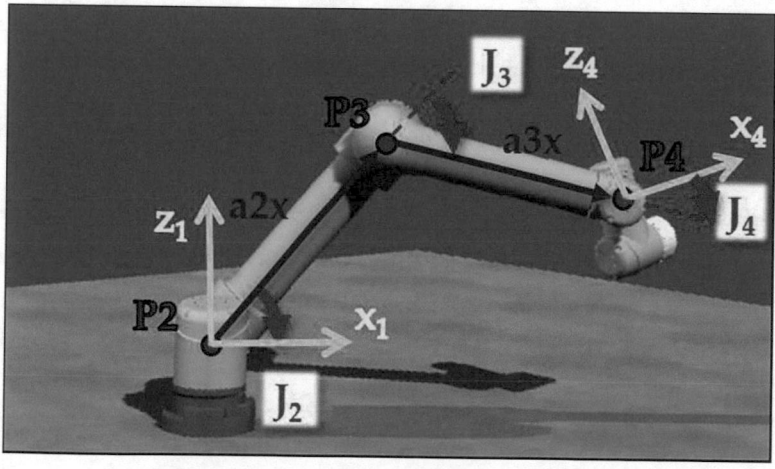

Figure A-17. *Finding the missing angle J_4*

The vector x_1 is simply the direction of the local X axis at point P_2, which is parallel to ground and only depends on the first joint:

$$x_1 = \begin{bmatrix} c_1 \\ s_1 \\ 0 \end{bmatrix} \tag{A-44}$$

The vector x_4 is the direction of the local X axis at point P_4, which we calculated earlier on in Equation (A-29), when we derived the orientation of point P_4:

$$x_4 = \begin{bmatrix} c_5 c_6 n_x - s_5 o_x + c_5 s_6 a_x \\ c_5 c_6 n_y - s_5 o_y + c_5 s_6 a_y \\ c_5 c_6 n_z - s_5 o_z + c_5 s_6 a_z \end{bmatrix} \tag{A-45}$$

We also need the direction normal to the plane containing these two vectors. We call it **n**, and it is simply given by the local Y axis at point P_2:

$$n = \begin{bmatrix} s_1 \\ -c_1 \\ 0 \end{bmatrix} \tag{A-46}$$

The angle between two vectors in a three-dimensional space, lying on the same plane, is given by the following equivalence:

$$\alpha_{14} = atan2\big(det[\boldsymbol{x1},\boldsymbol{x4},\boldsymbol{n}],dot[\boldsymbol{x1},\boldsymbol{x4}]\big) \tag{A-47}$$

The determinant of the $[\boldsymbol{x1}, \boldsymbol{x4}, \boldsymbol{n}]$ matrix gives the sine of the angle between x_1 and x_4, while the dot product $(\boldsymbol{x1} \bullet \boldsymbol{x4})$ gives the cosine. We use the arctangent function to extract the correct value of the angle.

It is critical to *normalize* all the three vectors before carrying on the calculations.

Using Equation (A-43), we can finally find J_4. Now we have the values for all the six joint angles, although remember that we end up with two different sets of solutions because of the two possible signs for J_5 after Equation (A-24). Select the one forced by the operator or the one closest to the current configuration of the robot before starting the movement.

SCARA

The SCARA robot can be built in different mechanical configurations according to its application requirements. The most typical option comes with four joints (three rotational and one translational) and offers four degrees of freedom: the TCP can be positioned along the [X, Y, Z] axes and rotated around the C coordinate.

Figure A-18 shows a SCARA model with four joint axes in its home position. It is conventional to place the base frame of the robot at the beginning of the first link avoiding any initial vertical offset from the base column.

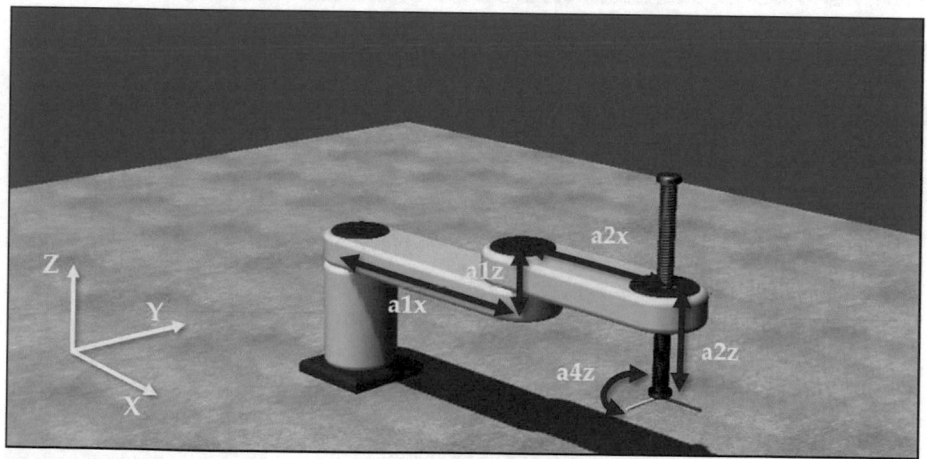

Figure A-18. *SCARA robot in its home position*

The last parameter $a4z$ is actually a mechanical coupling between the last two joints. Since most industrial SCARA robots are built that way, we include the parameter directly in the default transformations. If your robot does not have such a coupling, you can safely set the $a4z$ coefficient equal to 0.

The first two joints J_1 and J_2 rotate around the vertical Z axis and are entirely responsible for positioning the [X, Y] coordinates of the TCP (see Figure A-19 and Figure A-20).

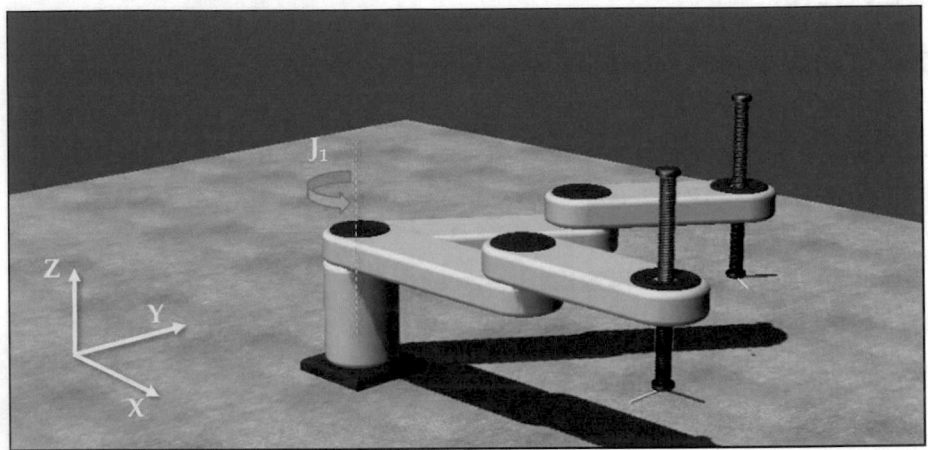

Figure A-19.   *The first joint rotates around the Z axis*

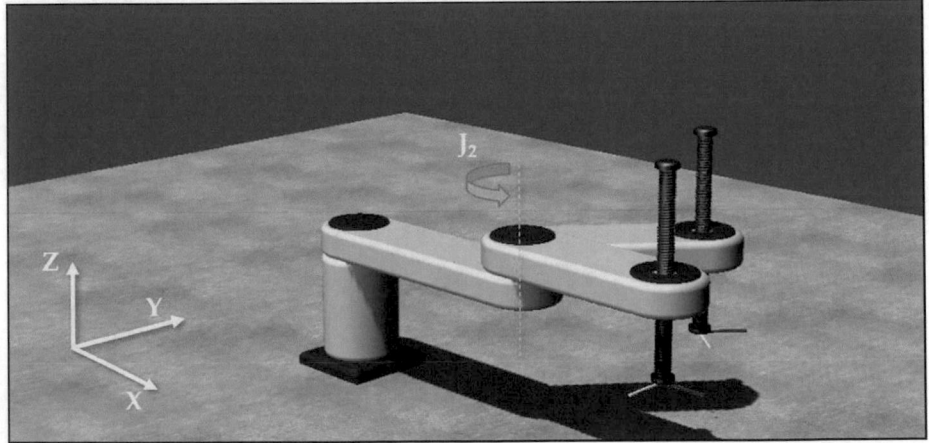

Figure A-20.   *The second joint rotates around the Z axis*

The third joint J_3 translates the TCP along the vertical Z axis (see Figure A-21).

Figure A-21. *The third joint slides along the Z axis*

Finally, the fourth joint J_4 introduces an additional rotation around the Z axis (see Figure A-22).

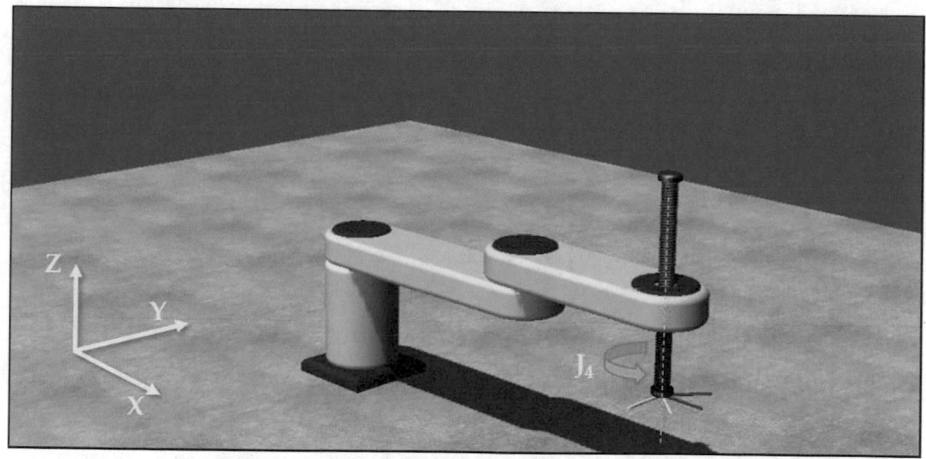

Figure A-22. *The fourth joint rotates around the Z axis*

Although the SCARA's mechanical structure is simple enough to be quickly solved geometrically, we still adopt the common approach of composing all the homogeneous matrices for each joint to find the solution to the direct kinematics. See Chapter 2 for details.

The first joint J_1 introduces a rotation around Z.

$$T_1 = \begin{bmatrix} c_1 & -s_1 & 0 & 0 \\ s_1 & c_1 & 0 & 0 \\ 0 & 0 & 1 & 0 \\ 0 & 0 & 0 & 1 \end{bmatrix} \tag{A-48}$$

The second joint J_2 also rotates around Z, after an offset along the X and Z axes.

$$T_2 = \begin{bmatrix} c_2 & -s_2 & 0 & a1x \\ s_2 & c_2 & 0 & 0 \\ 0 & 0 & 1 & a1z \\ 0 & 0 & 0 & 1 \end{bmatrix} \tag{A-A9}$$

The third joint J_3 is of translational type and acts along the vertical Z axis.

$$T_3 = \begin{bmatrix} 1 & 0 & 0 & a2x \\ 0 & 1 & 0 & 0 \\ 0 & 0 & 1 & a2z + J_3 \\ 0 & 0 & 0 & 1 \end{bmatrix} \tag{A-50}$$

Finally, the last joint J_4 introduces an additional rotation around Z. In most robots, this rotation is achieved by means of a screw-type mechanical shaft, which has the side effect of lifting and lowering the TCP while rotating. The additional vertical movement must be accounted for in the offset part of the homogeneous matrix. The coupling coefficient $a4z$

specifies how much vertical displacement is caused by a complete rotation of J_4. Use 360 for degrees or 2π if your input values are in radians.

$$T_4 = \begin{bmatrix} c_4 & -s_4 & 0 & 0 \\ s_4 & c_4 & 0 & 0 \\ 0 & 0 & 1 & \dfrac{a4z\,J_4}{360} \\ 0 & 0 & 0 & 1 \end{bmatrix} \tag{A-51}$$

To find out the TCP position as seen from the base frame, we need to chain-multiply the four homogeneous matrices together:

$$TCP(X,Y,Z) = T_1 T_2 T_3 T_4 O = T_{14} \begin{bmatrix} 0 \\ 0 \\ 0 \\ 1 \end{bmatrix} \tag{A-52}$$

Expanding the calculations leads to the following:

$$TCP_X = a1x\,c_1 + a2x\,c_{12} \tag{A-53}$$

$$TCP_Y = a1x\,s_1 + a2x\,s_{12} \tag{A-54}$$

$$TCP_Z = a1z + a2z + J_3 + a4z\,J_4 \tag{A-55}$$

As usual, we adopted the compact notation $s_{12} = \sin(J_1 + J_2)$.

To find the rotation of the TCP, we could multiply together the rotation matrices and extract the only relevant Euler angle C. However, it is much faster to observe the geometry of the problem on the horizontal [X, Y] plane (see Figure A-23), and notice that the TCP rotation is built up by the series chain of the three rotational joints:

$$TCP_C = J_1 + J_2 + J_4 \tag{A-56}$$

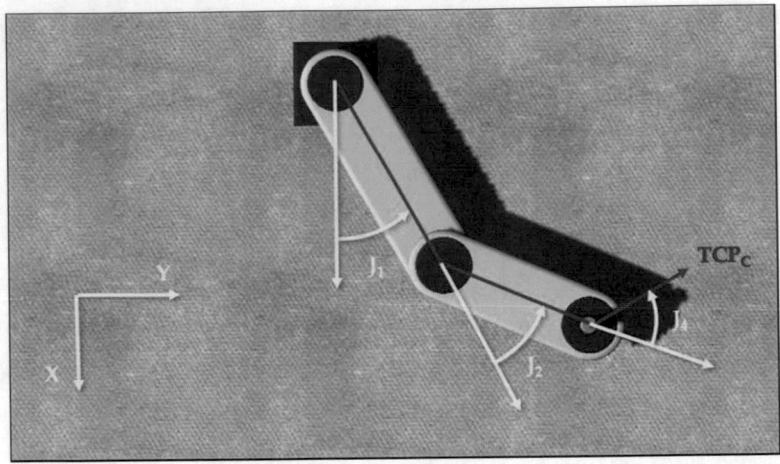

Figure A-23. *Top view of the SCARA robot*

The **inverse** transformations are best solved geometrically. Given the position and orientation [X, Y, Z, C] of the TCP, we need to find the four joint angles.

We start with the first two joints J_1 and J_2 and observe that they are responsible for determining the position of the TCP on the [X, Y] plane (see Figure A-24). We also notice that the solution is not unique: there are two possible options for the arm configuration, either LEFT or RIGHT, to reach the same TCP position.

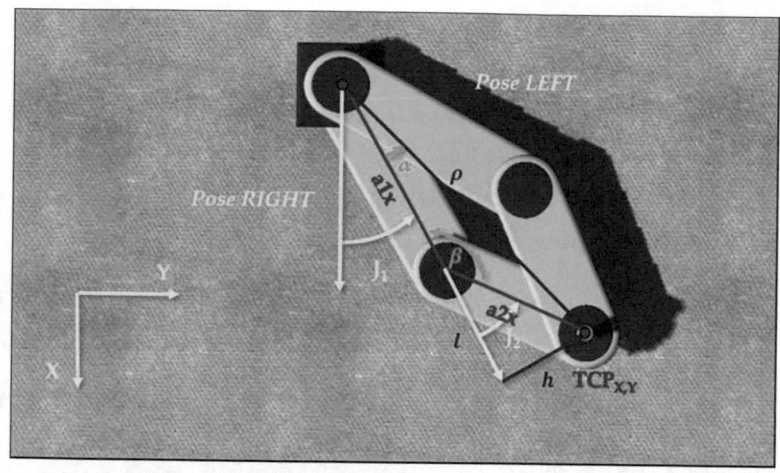

Figure A-24. *Solving for J$_1$ and J$_2$*

We call ρ the two-dimensional distance between the TCP and the base frame on the [X, Y] plane.

$$\rho = \sqrt{X^2 + Y^2}$$

(A-57)

It is possible to find a solution to the inverse transformations only if the two links $a1x$ and $a2x$ can reach the distance ρ.

$$\rho \leq a1x + a2x$$

(A-58)

$$\rho \geq |a1x - a2x|$$

(A-59)

The angle between the base X axis and the vector ρ is given by the following:

$$\theta = atan2(Y, X)$$

(A-60)

579

The two possible configurations of the arm are symmetric with respect to the vector ρ, forming two equal triangles with sides $a1x$, $a2x$, ρ. We call α the angle between $a1x$ and ρ and β the angle between $a1x$ and $a2x$ (both angles are in green in Figure A-24).

Now we can express the values of the first two joints as functions of α and β. For the RIGHT pose, we have the following:

$$J_1 = \theta - \alpha \tag{A-61}$$

$$J_2 = \pi - \beta \tag{A-62}$$

For the LEFT pose,

$$J_1 = \theta + \alpha \tag{A-63}$$

$$J_2 = \beta - \pi \tag{A-64}$$

All we are missing at this point are the values for α and β. The angle β can be quickly found using the law of cosines in the triangle $a1x$, $a2x$, ρ.

$$cos\beta = \frac{a1x^2 + a2x^2 - \rho^2}{2\,a1x\,a2x} \tag{A-65}$$

As usual, I suggest using the arctangent for better numerical stability.

Once J_2 is known, we can derive the projections of the second link (l and h):

$$l = a2x\,c_2 \tag{A-66}$$

$$h = a2x\,s_2 \tag{A-67}$$

Finally, we find α:

$$\alpha = atan2(h, a1x + l) \tag{A-68}$$

Using either Equation (A-61) or (A-63), we can calculate J_1.

Solving the last two joints is much more straightforward. From Equation (A-56), we immediately find J_4:

$$J_4 = C - J_1 - J_2 \tag{A-69}$$

Finally, J_3 is the only joint that affects the vertical displacement of the TCP from the base frame, with the addition of the fixed mechanical offsets ($a1z$, $a2z$) and the (optional) coupling effect from J_4.

$$J_3 = Z - a1z - a2z - a4z\, J_4 \tag{A-70}$$

PALLETIZER

The palletizer is a four-axes robotic arm typically deployed in loading/unloading applications, where large and heavy loads need to be transferred from conveyors to pallets. With only four joints, this robot can control four degrees of freedom of its TCP: three position coordinates [X, Y, Z] and one rotation angle C around the vertical Z axis. Figure A-25 shows a palletizer model in its home position.

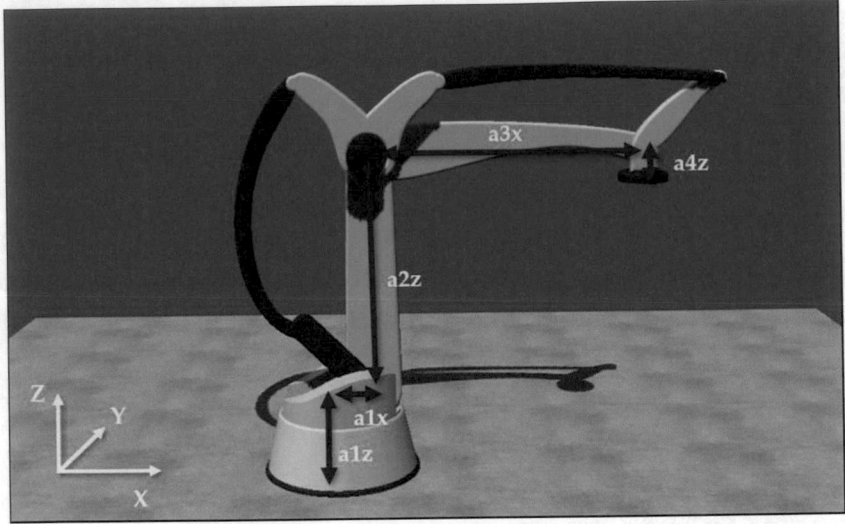

Figure A-25. *Palletizer robot in its home position*

The striking characteristic of its mechanical structure is that it is able to keep the load always parallel to ground regardless of the value of its joint axes. This feature is very helpful when handling packages, so that they can be picked or placed in any position in space without any need to electronically control their tilt. The TCP always stays horizontal to ground, thanks to a double parallelogram structure in the links of the robot. Figure A-26 and Figure A-27 show this feature when joints J_2 and J_3 are rotated.

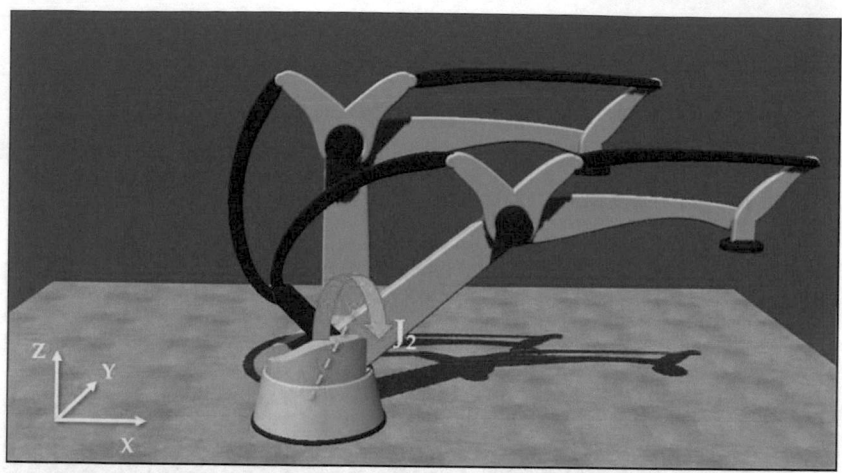

Figure A-26. *The second joint rotates around the Y axis*

The first parallelogram also introduces a mechanical coupling between joints J_2 and J_3. Specifically, a rotation of J_2 causes a rotation of the J_3 physical axis by the same angle in reverse direction, even when the motors in J_3 do not move. When solving the direct and inverse transformations, we will need to introduce a compensation for the effect of this mechanical coupling, as we learned in the related Sections in Chapters 3 and 4.

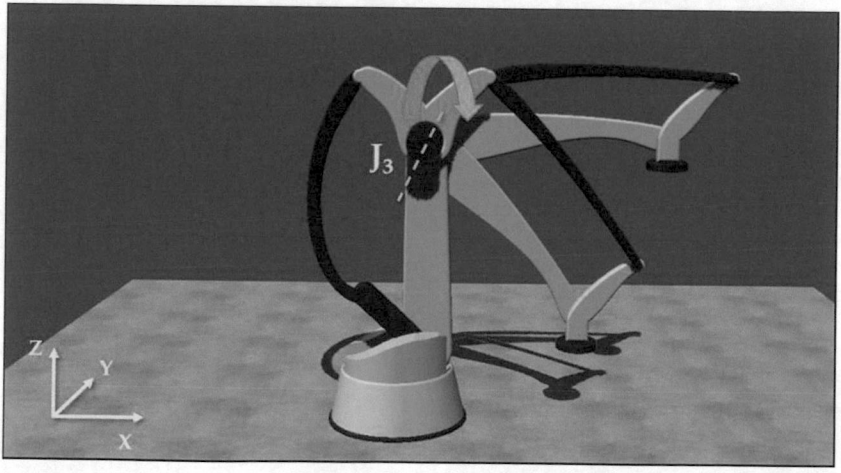

Figure A-27. *The third joint rotates around the Y axis*

The second effect of the mechanical parallelograms is to keep the TCP frame parallel to the base frame. The movement acts on the passive joint at the wrist point WP by introducing a rotation of $-(J_2 + J_3)$, as shown in Figure A-28. There is no motor acting on this joint, but we still need to keep track of its passive rotation in the transformations.

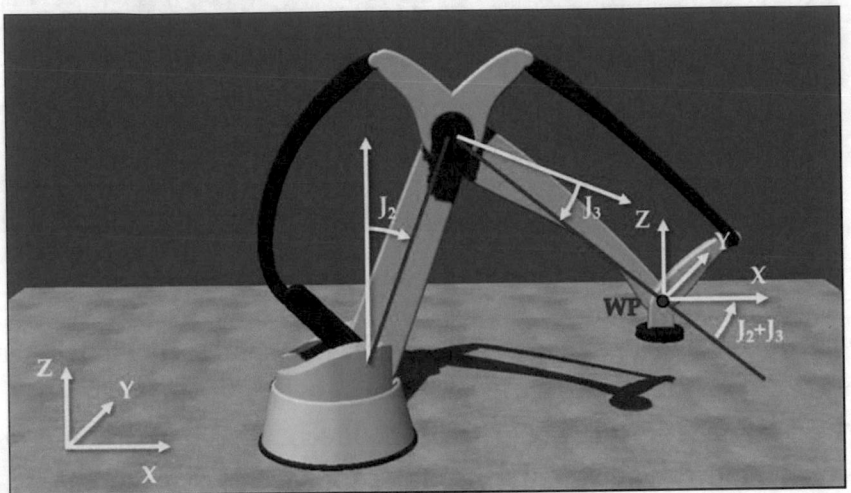

Figure A-28. *The mechanical links add a rotation opposite to $(J_2 + J_3)$ at the WP*

Finally, the first and last joints (J_1 and J_4) of the palletizer robot both rotate around the vertical Z axis (see Figure A-29).

Figure A-29. *The first joint rotates around the Z axis*

In particular, J_4 does not have any effect on the position of the TCP. It only affects its rotation angle C.

Despite being an open-chain kinematic, the extra mechanical links make this robot a hybrid between serial and parallel models. Nevertheless, the transformations are typically solved using a standard serial kinematic model and then adding coupling coefficients between the joints.

The direct kinematics could be solved geometrically because of the simple structure of the palletizer. However, we keep following the same notation as for the other robots, and we present here all the homogeneous transformations for each axis.

The first joint J_1 introduces a rotation around Z.

$$T_1 = \begin{bmatrix} c_1 & -s_1 & 0 & 0 \\ s_1 & c_1 & 0 & 0 \\ 0 & 0 & 1 & 0 \\ 0 & 0 & 0 & 1 \end{bmatrix} \tag{A-71}$$

The second joint J_2 rotates around the Y axis, after an offset along the X and Z axes.

$$T_2 = \begin{bmatrix} c_2 & 0 & s_2 & a1x \\ 0 & 1 & 0 & 0 \\ -s_2 & 0 & c_2 & a1z \\ 0 & 0 & 0 & 1 \end{bmatrix} \tag{A-72}$$

The third joint J_3 also rotates around the Y axis after a vertical offset along Z.

$$T_3 = \begin{bmatrix} c_3 & 0 & s_3 & 0 \\ 0 & 1 & 0 & 0 \\ -s_3 & 0 & c_3 & a2z \\ 0 & 0 & 0 & 1 \end{bmatrix} \tag{A-73}$$

The mechanical links introduce a negative rotation around the Y axis centered at the wrist point. The magnitude of the passive rotation is the opposite of the sum of J_2 and J_3.

$$T_4 = \begin{bmatrix} c_{23} & 0 & -s_{23} & a3x \\ 0 & 1 & 0 & 0 \\ s_{23} & 0 & c_{23} & 0 \\ 0 & 0 & 0 & 1 \end{bmatrix} \tag{A-74}$$

Finally, the last joint J_4 introduces an additional rotation around Z. The translational offset between WP and TCP can be both along the X and Z directions, depending on the mechanical built. Keep in mind that $a4z$ is usually negative, translating toward ground.

$$T_5 = \begin{bmatrix} c_4 & -s_4 & 0 & a4x \\ s_4 & c_4 & 0 & 0 \\ 0 & 0 & 1 & a4z \\ 0 & 0 & 0 & 1 \end{bmatrix} \tag{A-75}$$

To find out the TCP position as seen from the base frame, we need to chain-multiply the five homogeneous matrices together:

$$TCP(X,Y,Z) = T_1 T_2 T_3 T_4 T_5 O = T_{15} \begin{bmatrix} 0 \\ 0 \\ 0 \\ 1 \end{bmatrix} \tag{A-76}$$

Expanding the calculations leads to the following:

$$TCP_X = c_1 \left(a1x + a2zs_2 + a3xc_{23} + a4x \right) \tag{A-77}$$

$$TCP_Y = s_1 \left(a1x + a2zs_2 + a3xc_{23} + a4x \right) \tag{A-78}$$

$$TCP_Z = a1z + a2z\, c_2 - a3x\, s_{23} + a4z \tag{A-79}$$

When feeding the joint angle values to the direct transformation functions, remember to add the effect of the coupling between J_2 and J_3:

$$J_3 \leftarrow J_3 - J_2 \tag{A-80}$$

The rotation angle C can be extracted from the rotation matrix, or simply geometrically derived, by observing that only the first and last joint have any effect on it:

$$TCP_C = J_1 + J_4 \tag{A-81}$$

The inverse transformations can be derived geometrically observing the robot from the side, projected on a plane along the first joint (see Figure A-30). We start by deriving the position of the wrist point from the TCP coordinates [X, Y, Z]:

$$WP_{XY} = TCP_{XY} - a4x = \sqrt{X^2 + Y^2} - a4x \tag{A-82}$$

$$WP_Z = Z - a4z \tag{A-83}$$

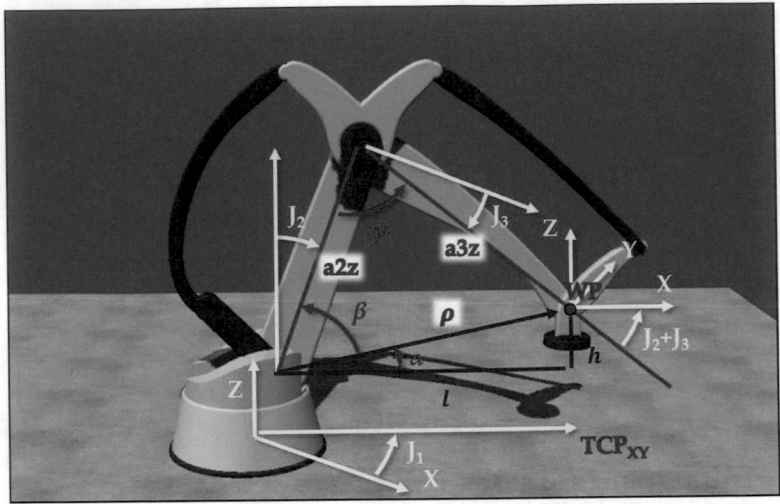

Figure A-30. *Solving the inverse kinematics for the palletizer robot*

Then, we find all the geometrical parameters we need in order to solve the triangle $(a2z, a3z, \rho)$.

$$l = WP_{XY} - a1x = \sqrt{X^2 + Y^2} - a4x - a1x \qquad \text{(A-84)}$$

$$h = WP_Z - a1z = Z - a4z - a1z \qquad \text{(A-85)}$$

$$\rho = \sqrt{l^2 + h^2} \qquad \text{(A-86)}$$

The condition we need to impose, for the transformations to return a valid solution, is that the arm links $a2z$ and $a3x$ can physically stretch enough to reach the wrist point.

$$\rho \le a2z + a3x \qquad \text{(A-87)}$$

$$\rho \ge |a2z - a3x| \qquad \text{(A-88)}$$

We can now find all the unknown angles α, β, γ. The last two, using the law of cosines,

$$\alpha = \text{atan2}(h,l) \tag{A-89}$$

$$\cos\beta = \frac{\rho^2 + a2z^2 - a3x^2}{2\,a2z\,\rho} \tag{A-90}$$

$$\cos\gamma = \frac{a2z^2 + a3x^2 - \rho^2}{2\,a2z\,a3x} \tag{A-91}$$

As usual, I suggest using the arctangent for better numerical stability. At this point, we can quickly derive the values for J_2 and J_3:

$$J_2 = 90 - \alpha - \beta \tag{A-92}$$

$$J_3 = 90 - \gamma \tag{A-93}$$

Unlike most other robots, the palletizer does not allow different configurations of its joint to reach the same TCP pose. The values found in Equations (A-92) and (A-93) are unique.

The result of the inverse transformations for J_2 and J_3 must be adjusted to compensate for the mechanical coupling:

$$J_3 \leftarrow J_3 + J_2 \tag{A-94}$$

Solving for J_1 is straightforward, as it is the angle formed by the two coordinates [X, Y] of the TCP:

$$J_1 = \text{atan2}(Y,X) \tag{A-95}$$

Finally, from Equation (A-81), we calculate the angle of the last joint:

$$J_4 = C - J_1 \tag{A-96}$$

DELTA

A Delta (or Tripod) robot is a parallel kinematic chain typically employed in applications with small workspace and high-speed requirements.

Figure A-31 shows a Delta robot with three joint axes in its home position. The three rotational joints J_i ($i = 1..3$) are located on the upper platform and directly actuate the upper arms h_i, which link points A_i to B_i. The lower arms k_i link points B_i to C_i and are built using parallelogram structures, which automatically keep the lower platform parallel to the upper one. Forcing the TCP to have a constant orientation makes the kinematic model much easier to solve.

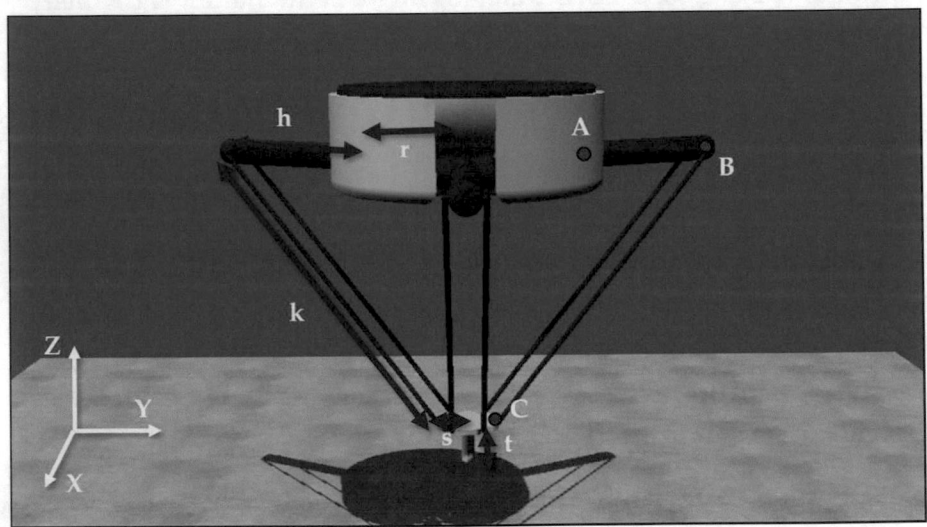

Figure A-31. *Delta robot in its home position*

Most of the delta robots found in the industry feature an additional fourth axis to add a rotation capability to the TCP around the vertical Z axis. This extra joint acts directly on the TCP and does not affect the rest of the kinematic model, so we did not show it in the figures. The motor rotating the TCP is rarely mounted directly on the lower platform because that would increase its weight and reduce the dynamic performance of the robot. The motor is usually placed on the upper platform and connected to the TCP by means of a central flexible link.

The position of the TCP is often translated from the lower platform by an offset t. We call the center of the lower platform the wrist point (WP) and derive its position from the TCP as follows:

$$WP = TCP + t\, Z \tag{A-97}$$

The center of the upper platform is the origin of the robot's base frame. The first joint J_1 is oriented along the X axis, while the two other joints J_2 and J_3 are oriented, respectively, at 120 and 240 degrees. The three axes thus form an equilateral triangle (see Figure A-32). The distance between the origin and the joints centers (A_i) is r. Similarly, on the lower platform, the distance between the WP and the (passive) joints centers (C_i) is s.

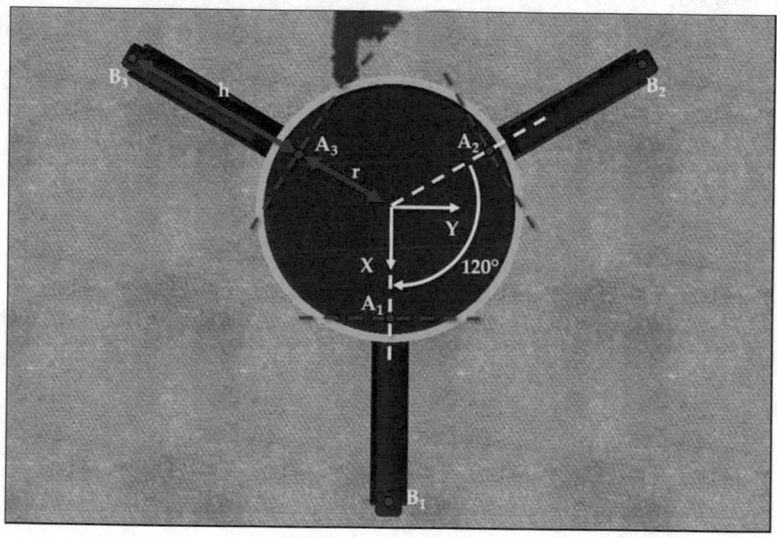

Figure A-32. *Top view of the upper platform*

A movement of the actuated joints causes a translational displacement of the lower platform along the positional [X, Y, Z] coordinates. See Figure A-33 for some examples.

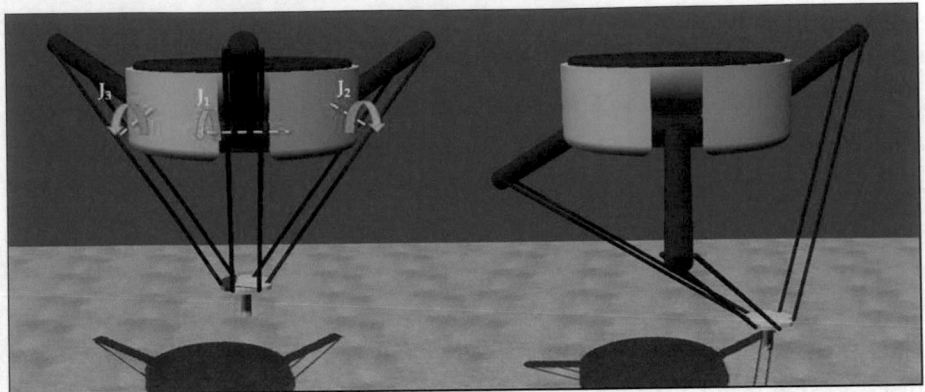

Figure A-33. *The lower platform always remains parallel to the upper one*

Solving the direct kinematics consists in finding the position of the WP given the angles of the joint axes (see Figure A-34). The procedure requires the following steps:

- Calculate the position of the B_i points.

- Calculate the position of the P_i points (offset from B_i by s along the joint planes).

- Find WP as the intersection of the three spheres centered in P_i of radius k.

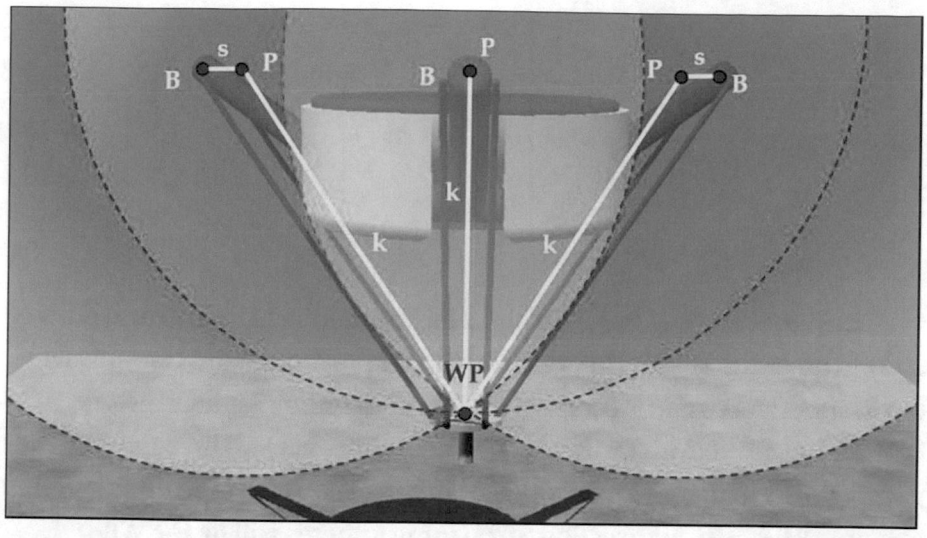

Figure A-34. *One intersection of the three spheres is the wrist point*

Adopting the usual compact notation $c_1 = \cos J_1$, the B_1 point along the X axis has coordinates:

$$B_1 = \begin{bmatrix} r + hc_1 \\ 0 \\ -hs_1 \end{bmatrix} \tag{A-98}$$

Similarly, we find the B_2 and B_3 points, which lie on the joint planes rotated at ±120 degrees around the Z axis for the J_1 plane:

$$B_2 = \begin{bmatrix} -\dfrac{1}{2}(r + hc_2) \\ \dfrac{\sqrt{3}}{2}(r + hc_2) \\ -hs_2 \end{bmatrix} \tag{A-99}$$

$$B_3 = \begin{bmatrix} -\dfrac{1}{2}(r+hc_3) \\[2ex] -\dfrac{\sqrt{3}}{2}(r+hc_3) \\[2ex] -hs_3 \end{bmatrix} \tag{A-100}$$

The passive universal joints in B_i allow rotations of the lower k_i links around an entire sphere. In other words, the position of the C_i points will be on spherical surfaces centered on B_i and of radius k. Since the WP is offset by s from C_i along the three joint axes planes, we first offset the points B_i by s to find the points P_i, and then we construct the three spheres centered in P_i. The intersection of the three spheres will be the WP.

$$P_1 = \begin{bmatrix} r+hc_1-s \\[1ex] 0 \\[1ex] -hs_1 \end{bmatrix} \tag{A-101}$$

$$P_2 = \begin{bmatrix} -\dfrac{1}{2}(r+hc_2-s) \\[2ex] \dfrac{\sqrt{3}}{2}(r+hc_2-s) \\[2ex] -hs_2 \end{bmatrix} \tag{A-102}$$

$$P_3 = \begin{bmatrix} -\dfrac{1}{2}(r+hc_3-s) \\[2ex] -\dfrac{\sqrt{3}}{2}(r+hc_3-s) \\[2ex] -hs_3 \end{bmatrix} \tag{A-103}$$

There exist various possible ways of solving the problem of intersecting three spheres, both algebraic and geometric. One option involves solving a system of three quadratic equations in [X, Y, Z]. Another approach uses a technique called *trilateration*, which we show here in detail.

Given three spheres of radius k centered in P_i, let us find their intersection point [X, Y, Z]. It can be shown that there are actually two intersection points, both lying on a line perpendicular to the plane containing the three center points. Mathematically, there are also cases in which the spheres only intersect in one single point or do not intersect at all, in case the centers are spaced too far apart from each other. However, the arms of the delta robot are designed specifically to avoid these complications. The lower arms k_i are longer than the upper arms h_i, so the spheres always intersect. The fact that they intersect in two points is also irrelevant, because we always choose the location for WP lower than the upper base platform.

Figure A-35 shows the plane containing the three center points. We define the two vectors $\boldsymbol{U} = P_2 - P_1$ and $\boldsymbol{V} = P_3 - P_1$. From simple vector properties, we can derive the $\left[\hat{\boldsymbol{x}}, \hat{\boldsymbol{y}}, \hat{\boldsymbol{z}} \right]$ frame:

$$\hat{\boldsymbol{x}} = \frac{\boldsymbol{U}}{\|\boldsymbol{U}\|} \tag{A-104}$$

$$\hat{\boldsymbol{y}} = \frac{\boldsymbol{V} - l\,\hat{\boldsymbol{x}}}{\|\boldsymbol{V} - l\,\hat{\boldsymbol{x}}\|} \tag{A-105}$$

$$\hat{\boldsymbol{z}} = \hat{\boldsymbol{x}} \times \hat{\boldsymbol{y}} \tag{A-106}$$

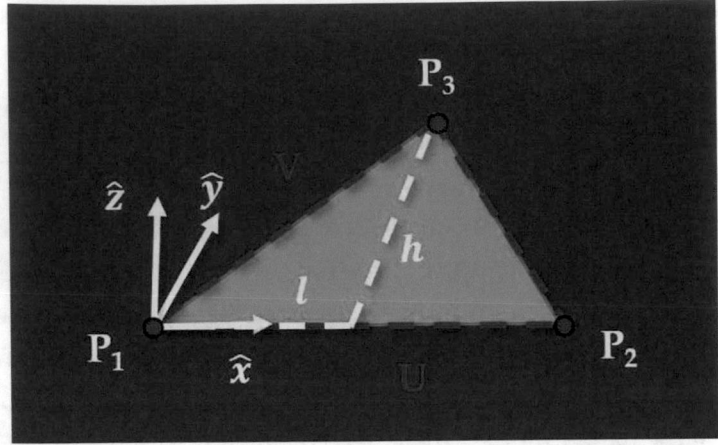

Figure A-35. *Plane containing the three center points*

The projections of **V** over the $\hat{x}$ and $\hat{y}$ axes are as follows:

$$l = \hat{x} \cdot V \qquad \text{(A-107)}$$

$$h = \hat{y} \cdot V \qquad \text{(A-108)}$$

It can be proven that the coordinates of the intersection point in the $\left[\hat{x}, \hat{y}, \hat{z}\right]$ frame are as follows:

$$x = \frac{\|U\|^2}{2\|U\|} \qquad \text{(A-109)}$$

$$y = \frac{l^2 + h^2 - 2lx}{2h} \qquad \text{(A-110)}$$

$$z = \pm\sqrt{k^2 - x^2 - y^2} \qquad \text{(A-111)}$$

Notice that there are two solutions for the z coordinate. We already mentioned before that the only physically relevant value for the WP is the one below the upper platform of the robot. Also, the expression under the square root must be positive in order to return two distinct real values.

That is always the case with standard industrial robots, because they are built to avoid mechanical violations.

Finally, we need to transform the WP coordinates back into the robot's base frame:

$$WP = P_1 + x\,\hat{x} + y\,\hat{y} + z\,\hat{z}$$

(A-112)

Adding the last vertical offset will return the actual TCP position:

$$TCP = WP - t\,Z$$

(A-113)

Solving the inverse transformations is somewhat simpler. The position of the wrist point is given, which fixes the lower platform in space. The upper platform is also fixed by default, so all we need to do is find the values of the joint angles J_i for which the upper arm and the lower arm connect with each other. Given the symmetry of the robot, the problem can be solved individually for each axis.

The upper arms can only rotate around their pivots A, forcing the points B to describe circles in space. The lower arms, on the other hand, can more freely swing around the universal joints in C, allowing the point B to describe spheres in space. The solution to the inverse kinematics for each joint is the intersection between the upper circle and the lower sphere, as shown in Figure A-36.

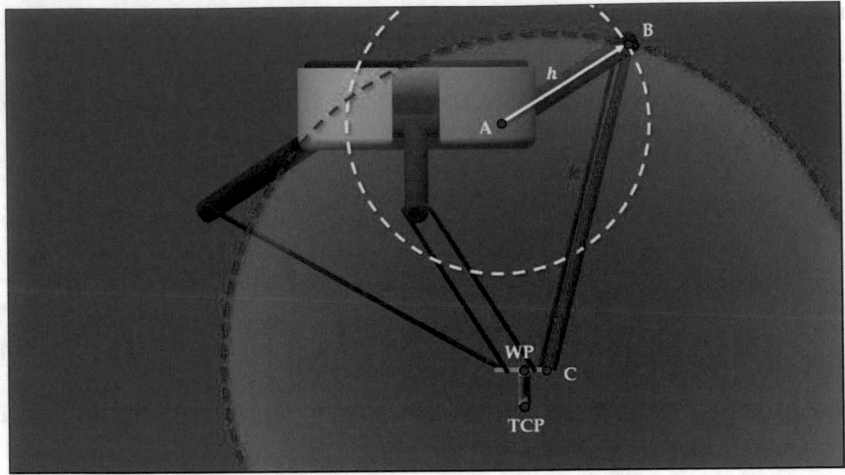

Figure A-36. *Intersection between upper circle (yellow) and lower sphere (red)*

Actually, if we project the geometry of the problem onto a plane containing the joint axis (e.g., the XZ plane for J_1), the lower sphere degenerates into a circle, and the solution becomes the intersection between two circles, which is straightforward to find. Notice that out of the two possible intersections, we pick the one with the robot's arm pointing outward, not inward (see Figure A-37).

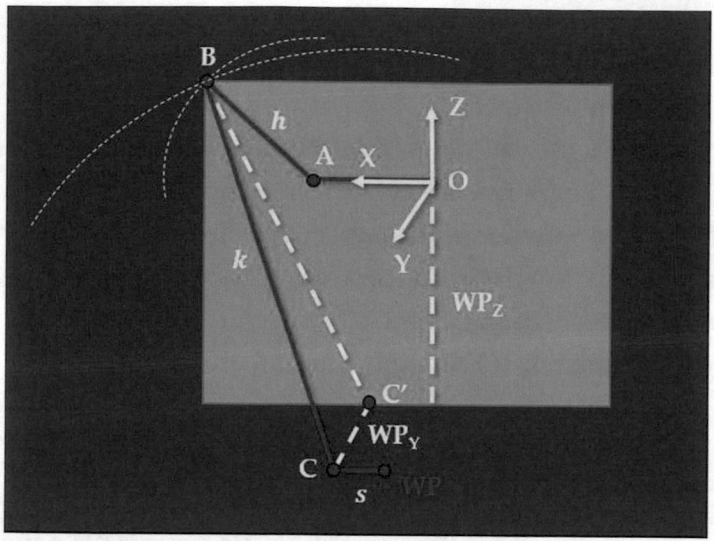

Figure A-37. *Projecting the problem on the XZ plane*

Let's consider the XZ plane and solve for J_1:

- The upper circle is centered in $A = [r, 0]$ and has radius h.

- The lower circle is centered in $C' = [WP_X + s, WP_Z]$ and has radius $\sqrt{k^2 - WP_Y^2}$.

The resulting equations of the two circles are as follows:

$$\begin{cases} (x-r)^2 + z^2 = h^2 \\ (x - WP_X - s)^2 + (z - WP_Z)^2 = k^2 - WP_Y^2 \end{cases} \tag{A-114}$$

The system can be quickly solved by extracting z from the first equation and plugging it into the second to find x. If two distinct real solutions exist, we choose the one with the largest x coordinate to keep the robot's arm on the outside of the platform. The unique result is point $B = [B_X, B_Z]$, which forces the joint angle to be as follows:

$$J_1 = \text{atan2}(B_Z, B_X) \tag{A-115}$$

599

If no real solution to the system is found, then the given TCP position is outside of the reachable robot's workspace, and the inverse transformation function must return an error.

To find J_2 and J_3, we can simply follow the same procedure used for J_1, but each time the XZ plane needs to be rotated by 120 degrees around the vertical Z axis. The equation of the upper circle does not change. However, the coordinates of the wrist point used in the lower circle need to be adjusted to the new frames.

For J_2, use a rotation of 120 degrees from J_1 around Z:

$$WP_2 = R_{120}WP = \begin{bmatrix} -\dfrac{1}{2} & -\dfrac{\sqrt{3}}{2} & 0 \\ \dfrac{\sqrt{3}}{2} & -\dfrac{1}{2} & 0 \\ 0 & 0 & 1 \end{bmatrix} \begin{bmatrix} WP_X \\ WP_Y \\ WP_Z \end{bmatrix} = \begin{bmatrix} -\dfrac{1}{2}WP_X - \dfrac{\sqrt{3}}{2}WP_Y \\ \dfrac{\sqrt{3}}{2}WP_X - \dfrac{1}{2}WP_Y \\ WP_Z \end{bmatrix} \tag{A-116}$$

For J_3 use a rotation of -120 degrees from J_1 around Z:

$$WP_3 = R_{-120}WP = \begin{bmatrix} -\dfrac{1}{2} & \dfrac{\sqrt{3}}{2} & 0 \\ -\dfrac{\sqrt{3}}{2} & -\dfrac{1}{2} & 0 \\ 0 & 0 & 1 \end{bmatrix} \begin{bmatrix} WP_X \\ WP_Y \\ WP_Z \end{bmatrix} = \begin{bmatrix} -\dfrac{1}{2}WP_X + \dfrac{\sqrt{3}}{2}WP_Y \\ -\dfrac{\sqrt{3}}{2}WP_X - \dfrac{1}{2}WP_Y \\ WP_Z \end{bmatrix} \tag{A-117}$$

We can plug these values in Equation (A-114) and find the joint angles from Equation (A-115).

Index

A

AB quadrature signal, 457

Absolute encoders, 350, 453–455, 461, 464

Acceleration, 253
    limited profile, 184, 185
    profile, 195

Actual mechanical configuration, 56

Aggressive tuning, 353

Algorithm, 2, 12, 13, 16, 77, 136–138, 148, 200, 214, 217, 245, 246, 249, 250, 254, 300, 305, 316, 318, 336, 363, 382, 383, 386–389, 391, 393, 394, 398–400, 402, 404, 416, 419, 440, 449, 451, 466, 467, 469, 470, 482, 489, 497, 531, 532, 550

Aluminum, 548

Angular speed, 214

Anthropomorphic robot, 8, 9

Anti-windup, 302

Asynchronous/passive decay, 464, 482

Autoencoders, 407

Automatic movement programming, 269

B

Back electromotive force (BEMF), 426, 427, 436, 444

Back-propagation, 411

Ball grid array (BGA), 546

Bang-bang approach, 207

Bezier curves, 130, 134, 139, 140, 199–201

Bezier profile
    acceleration limit, 201
    computational complexity, 203
    constant time distance, 199
    cubic spline development, 199
    De Casteljau's algorithm, 200
    dynamic values, 202
    equations, 201, 202
    higher order spline, 201
    mechanical limitations, 202
    movement duration, 202
    path distance, 201
    peak acceleration, 201
    peak jerk, 201
    polynomial approach, 203
    seven-zones S-curve, 203
    smooth position profiles, 199
    total acceleration time, 203

Bezier splines, 130

Bilateral filter, 400

© Fabrizio Frigeni 2023
F. Frigeni, *Industrial Robotics Control*, Maker Innovations Series,
https://doi.org/10.1007/978-1-4842-8989-1

S

T

Printed in the United States
by Baker & Taylor Publisher Services